Optimisation Theory
Applications in OR and Economics

MJ Fryer and JV Greenman
Department of Mathematics
University of Essex

Edward Arnold

© MJ Fryer and JV Greenman 1987

First published in Great Britain 1987 by
Edward Arnold (Publishers) Ltd, 41 Bedford Square, London WC1B 3DQ

Edward Arnold (Australia) Pty Ltd, 80 Waverley Road, Caulfield East,
 Victoria 3145, Australia

Edward Arnold, 3 East Read Street, Baltimore, Maryland 21202, U.S.A.

British Library Cataloguing in Publication Data

Fryer, M.J.
 Optimization theory.
 1. Mathematical optimization
 I. Title II. Greenman, J.V.
 515   QA402.5

ISBN 0-7131-3527-1

All rights reserved. No part of this publication may be reproduced,
stored in a retrieval system, or transmitted in any form or by any means,
electronic, photocopying, recording, or otherwise, without the prior
permission of Edward Arnold (Publishers) Ltd.

Text set in 10/11pt Times Monophoto
by Macmillan India Ltd., Bangalore 560 025
Printed and bound in Great Britain
by Whitstable Litho Ltd, Kent

Preface

Optimisation theory has become increasingly important in the solution of complex problems arising in Economics, Operational Research and Engineering. Its wide-ranging applicability has led to a rapid growth in its development, particularly since 1947 when Dantzig introduced his Simplex method for solving a Linear Program. This growth has been accompanied by a greater understanding of the underlying structure of the theory through the use of the Lagrange formalism, which enabled the different methods and topics to be viewed as aspects of the same unified theory.

Our intention in this text is to present this Lagrangean theory at a level which will be accessible to a wide readership amongst students of Economics, Operational Research and Management Science. We therefore do not attempt to provide rigorous arguments to justify the different steps in the development of the theory, but support that development by geometric reasoning and many worked examples. We assume our readers to have a working knowledge of the differential and integral Calculus of one variable and an understanding of the basics of Algebra, including techniques for solving linear simultaneous equations. It would also be of advantage for the reader to have some acquaintance with partial differentiation, elementary matrix theory and simple planar curve sketching. However, we have included enough of an introduction to these topics for them not to be an essential prerequisite. Sections of the text that involve a more intense use of mathematics have been starred and can be omitted on first reading.

The basic Lagrangean theory is developed in the first three chapters of the text. We start with a revision of the mathematics of unconstrained optimisation, for simplicity restricting attention to functions of one and two variables. In the second chapter we introduce the technique of Lagrange multipliers enabling us to reduce an optimisation problem subject to equality constraints to an unconstrained problem. The extension of the method to include inequality constraints is undertaken in the third chapter. The theory is developed further in Chapter 4 to derive the primal–dual structure that is of importance, for example, in linear and geometric programming.

In the remaining chapters we apply the Lagrangean method to solve particular types of optimisation problem. In Chapter 5, we derive the powerful Simplex method for solving linear programs, while in Chapters 6 and 7 we derive the Hamiltonian method for solving problems in discrete and continuous optimal control theory; subjects that are becoming increasingly important in applications and in the undergraduate syllabus. In the final 'postscript' chapter we tie together some of the loose ends and suggest topics for further reading.

The development of the theory is punctuated by exercise sets, consisting of straightforward problems similar to the worked examples in the text, with solutions provided for the starred questions—about half the total. To help in these exercises relevant solution methods are summarised in a set of 'Procedures', whose purpose is strictly limited to the task at hand. At the end of each chapter there is a Problem Set whose purpose is to give additional practice in solving standard problems and also to test the reader's understanding of the material with less straightforward questions. We cannot stress enough how important it is for students to attempt as many exercises as possible before continuing with the text.

This text has evolved from an integrated set of lecture courses given for a number of years in the Mathematics Department at Essex University to students studying Economics and Operational Research. It was the challenge of presenting this material to this audience that led us to the conclusion that a textbook of this kind was necessary.

We would like to acknowledge the assistance of some of our colleagues who read parts of the manuscript and made many helpful comments, especially in the applications to economic theory. We do, of course, bear full responsibility for any errors that remain in the text. Finally, we would like to thank our Publishers for encouraging us to write this book.

MJF
JVG
1985

Contents

Preface   iii

1 The Basics: Unconstrained Optimisation   1

1.1 Introduction   1
1.2 Stationary Points for Functions of One Variable   3
1.3 Applications   10
1.4 Stationary Points for Functions of Two Variables   12
1.5 Applications   17
1.6 Contour Maps: An Introduction   20
1.7 Exotic Contours   23
Problem Set 1   27
References   28

2 Optimisation with Equality Constraints   30

2.1 Introduction   30
2.2 Constrained Stationary Points   31
2.3 The Lagrange Multiplier Method   33
2.4 Singularities   37
2.5 Other Aspects of the Lagrange Method   38
2.6 Generalisation of the Lagrange Method   41
2.7 Applications   47
Problem Set 2   50
References   52

3 Inequality Constraints   54

3.1 Introduction   54
3.2 The One-Variable Case   55
3.3 The Two-Variable Case: A Single Inequality Constraint   59
3.4 The Two-Variable Case: Several Inequality Constraints   62
3.5 The Sign of the Lagrange Multiplier   66
3.6 The Complementary Slackness Conditions   69
3.7 An Application: The Coal Transportation Problem   72
3.8 The Kuhn–Tucker Conditions   74
Problem Set 3   76
References   77

4 Duality   79

4.1 Introduction   79
4.2 Some Simple Examples   81
4.3 Higher Dimensional Examples   87
4.4 The Dual Linear Program   90

vi Contents

4.5	Convexity and Concavity	94
4.6	The Primal–Dual Structure	100
4.7	Geometric Programming: An Application of Duality Theory	103
4.8	The Geometric Program with Constraints	108
Problem Set 4		110
References		112

5 Linear Programming 114

5.1	Introduction	114
5.2	Slack Variables	116
5.3	The Kuhn–Tucker Conditions	118
5.4	Development of the Algorithm	120
5.5	Streamlining the Algorithm	121
5.6	The Simplex Tabular Algorithm	123
5.7	Applying the Simplex Tabular Algorithm	127
5.8	Sensitivity Analysis	133
5.9	Applications	135
Problem Set 5		140
References		141

6 Optimal Control Theory: The Basics 142

6.1	Introduction	142
6.2	Optimal Control Theory: The Discrete Case	144
6.3	The Pioneer Problem	150
6.4	Continuous Optimal Control Theory	153
6.5	An Exhaustible Resource Problem	160
6.6	Phase Space	162
6.7	Ramsey's Growth Model	171
Problem Set 6		173
References		176

7 Optimal Control: Generalisations 177

7.1	Introduction	177
7.2	Bounds on the Control Variable	177
7.3	An Application in Renewable Resource Theory	182
7.4	Terminal Conditions	186
7.5	Further Generalisations	192
7.6	Further Applications	199
7.7	The Calculus of Variations	202
Problem Set 7		208
References		211

8 Postscript 213

8.1	Introduction	213
8.2	Finite State Systems	214
8.3	Dynamic Programming	220
8.4	Stochastic Programming	224
8.5	The Theory in Perspective	225
References		227

Appendix   **229**

A   Linear Simultaneous Equations   229
B   Summation Notation   231
C   Matrices   232
D   Limits of Functions   234
E   Geometric Vectors   237
F   Conic Sections   240
G   Differential Equations   242
H   Hyperbolic Functions   244
I   Some Classical Problems   246
J   The Chain Rule and the Variational Principle   249

Solutions   **256**

Chapter 1   256
Chapter 2   261
Chapter 3   264
Chapter 4   270
Chapter 5   274
Chapter 6   281
Chapter 7   285

Index   **291**

1
The Basics: Unconstrained Optimisation

1.1 Introduction

Intuitively obvious solutions to practical problems are not always as efficient as we might first suppose. This is certainly true of the problem of how a labourer should dig a ditch which was analysed by the eminent 19th century American, Frederick Taylor[1]. Previously it had been argued that a labourer should earn his wages by removing each time a heaped shovelful of earth, but Taylor realised that this might not be the best strategy. If the labourer dug out a smaller amount each time he would get less tired, could continue digging for a longer period, and finish the ditch earlier. He backed up his argument by analysing the problem scientifically and by experiment and calculation he was able to determine the optimal amount to be removed with each shovel load.

This is probably one of the earliest examples of the use of rigorous analysis in the solution of a problem involving the organisation of people and resources. The demand for this type of thorough analysis did not, however, become significant until organisations had grown to such a size that their problems could no longer be satisfactorily solved by 'rule of thumb'. This stage was reached by the Military during the Second World War when the logistic problems of handling men and materials in preparation for and in actual combat became particularly severe. After the war the mathematicians and scientists involved in solving these problems went back into civilian life and found that large Government Departments and newly formed Public Corporations were eager to use their skills. This was especially true of the State Electricity Companies which had to plan future investment in new plant and supply networks and, on a day to day basis, allocate demand between existing plant.

There are three basic steps in applying mathematics to the solution of these types of problems. The first is the 'modelling' stage in which the problem is translated into precise mathematical terms. The second is the identification and use of the relevant algorithm to solve that mathematical problem and the third the interpretation of the mathematical results and their implications for the original problem. This book is primarily concerned with the second of these stages, with optimisation algorithms and their derivation from the underlying mathematical theory. However, to illustrate the very important modelling step and in the process construct a model which will be of use to us in the subsequent discussion we will consider a simplified version of the Electricity Allocation Problem mentioned above.

Imagine we are on a Caribbean island which has just two power stations, A and B, and that at a particular time of the day the electricity demanded by its customers amounts to 2 megawatts (much less than peak demand). If we denote the amount of electricity generated by power station A by x (megawatts) then the other power station, B, must generate the remaining $2 - x$ (megawatts). There are a variety of ways in which the demand can be met depending on the value of x. The Electricity Company could:

(a) choose $x = 2$, so that A is producing all the power,
(b) choose $x = 0$, so that B produces all the power, or

2   *The basics: unconstrained optimisation*

(c) choose x to have some intermediate value, so that both power stations are contributing.

Before it can pick the best allocation the company has to decide precisely what it means by 'best'. It has no direct control over the actual demand for electricity (which it is legally bound to satisfy), but it does have some control over the cost of meeting that demand. It might, therefore, be in the interests of the company to meet that demand at the minimum possible cost. To find the allocation of power between the power stations that will achieve this, we must first specify the cost function for each power station.

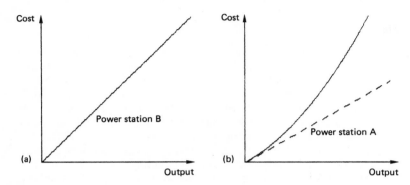

Fig. 1.1

Suppose the cost, C_B, of running power station B (neglecting fixed overheads) is proportional to its output (Fig. 1.1a), where the constant of proportionality depends on the monetary unit used. For simplicity, let us take this constant to be one, so that:

$$C_B = 1(2-x).$$

In contrast, suppose that power station A produces electricity at only half the cost at low output, but that when this output increases further the cost rises disproportionately (as in Fig. 1.1b), according to the formula:

$$C_A = (\text{output})/2 + (\text{output})^2/2$$

i.e. $C_A = x/2 + x^2/2.$

Adding these two costs together we find the total cost of generating 2 megawatts is given by:

$$C_A + C_B = x/2 + x^2/2 + (2-x)$$
$$= x^2/2 - x/2 + 2$$

where x is the output of station A. The Electricity Company has to find that value of x for which this total cost function takes its minimum value.

In other situations we may be interested in the maximum rather than the minimum value of a function. This would certainly be the case, for example, with a monopoly coal producer who has to work out how much coal to sell to maximise his profits. Too much and the price will fall, too little and the revenue will fall. The company, in effect, has to find the maximum value of a function (profit) of one variable (coal production). We will derive an explicit form for this function in Section 1.3, where we also show how the solutions for the monopolistic and competitive situations differ.

Both types of problem, maximisation and minimisation of a function of one variable, can usually be solved by applying the differential calculus, and in the next section we shall

1.2 Stationary Points for Functions of One Variable

In this section we will consider functions that are defined for all real values of the single variable x. We will also assume them to be sufficiently 'well-behaved' to possess continuous first derivatives for all finite x—corners, cusps and discontinuities are therefore excluded. We will denote such a function by f and its values by $f(x)$.

To find the maximum value that f can take, we first require the concept of a *stationary point*, namely a finite value of x at which the derivative (gradient) of f is zero.

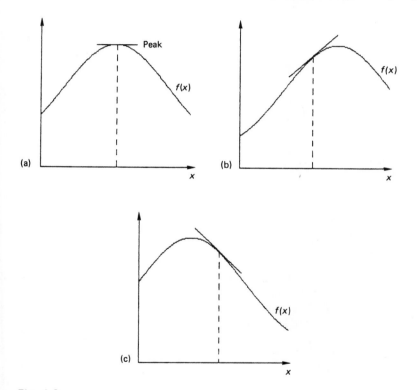

Fig. 1.2

If the maximum value of f is taken at a finite value[2] of x then this must be a stationary point (Fig. 1.2a). If the derivative were not zero at the maximum value (Figs 1.2b, 1.2c) then there would be a point nearby where f takes a larger value—and this would be a contradiction. If, conversely, x is a stationary point of f then it is not necessarily true that the maximum value of f is taken at x. Counter examples are shown in Figs 1.3a and 1.3b. Even if a stationary point does correspond to a 'peak' (Fig. 1.2a) it may still not yield the (overall) maximum value of f since, for example, there may be a peak with a higher value of f elsewhere (Fig. 1.3c).

Thus if the maximum value is taken at a finite point then it is taken at its highest peak, identified by sorting through the stationary points for the one where the value of f is largest. (In some situations there will be more than one such point.) However the

4 *The basics: unconstrained optimisation*

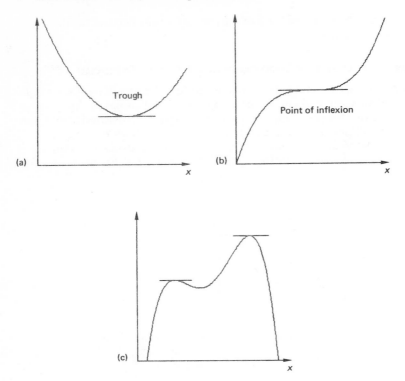

Fig. 1.3

maximum of f could be taken at its boundary at infinity. This will certainly be the case if the value taken by the function at infinity is greater than its largest value at any stationary point. Examples in Fig. 1.4 illustrate this possibility. In exceptional circumstances the limit at infinity will not exist. Example 3 suggests how to deal with one such situation.

These observations suggest the particular algorithm for finding the maximum value of a function f displayed in Procedure Box 1.1.

> **PROCEDURE 1.1**
> *To find the maximum value of a function of one variable*
> 1 Find all stationary points by solving the equation: $df/dx = 0$.
> 2 Evaluate f at each of these stationary points.
> 3 Evaluate $f(\infty)$ and $f(-\infty)$ as the limits of $f(x)$ as x tends to ∞ and $-\infty$ respectively.
> 4 Identify the largest of the values found in steps 2 and 3. This is the **maximum value** taken by the function and is denoted by: $\max f(x)$.

This algorithm can easily be adapted to find the minimum value taken by a function since it is achieved either at that stationary point which corresponds to the lowest 'trough' (Fig. 1.5a) or at infinity (Fig. 1.5b). We therefore carry out the first three steps of the

1.2 Stationary points for functions of one variable 5

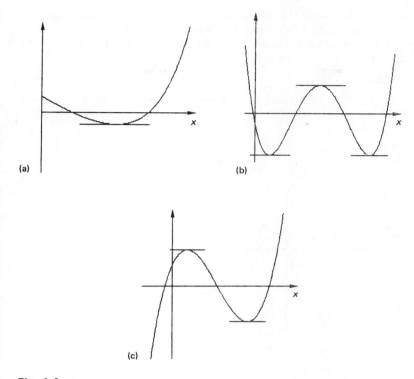

Fig. 1.4

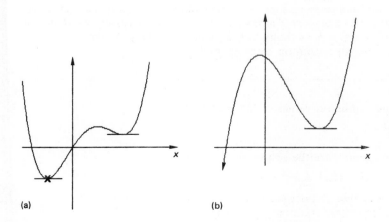

Fig. 1.5

algorithm as stated and then, in the fourth step, pick out the smallest rather than the largest of the values in steps 2 and 3. This value is denoted by: $\min f(x)$.

Alternatively we can always turn a minimisation problem into a maximisation problem by using the relationship:

$$\min f(x) = -(\max(-f(x)))$$

corresponding to the property illustrated in Fig. 1.6 that a maximum for $f(x)$ is a minimum for $-f(x)$.

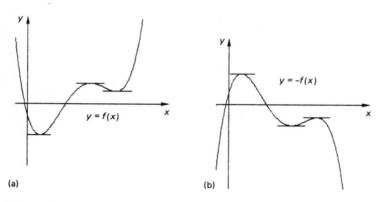

Fig. 1.6

Perhaps we should also mention at this stage the more mathematical terminology for some of the colloquialisms introduced in the development of the algorithm. We shall use the term *local maximum* synonymously with the term 'peak' to denote a point $(x, f(x))$ on the graph of f if x is a stationary point about which there is an interval[3] in which the values of the function are no greater than at x. Further from the stationary point f could of course take larger values (Fig. 1.4). If the maximum value of the function is taken at a particular 'peak' then that peak is said to be both a local and overall maximum. For a 'trough' the mathematical term is *local minimum*, and for stationary points that define neither 'peaks' nor 'troughs' we shall use the term *point of inflexion*. Finally, when necessary we shall refer to f, the function being optimised, as the *objective function* of the problem.

We will now try out this algorithm on some examples.

Example 1
Find the maximum and minimum values of the function:

$$f(x) = x^4 + 4x^3 + 1.$$

Step 1 The stationary points are the solutions of the equation:

$$df/dx = 4x^3 + 12x^2 = 4x^2(x+3) = 0.$$

There are clearly two distinct solutions: $x = 0$ and $x = -3$.

Step 2 The values of f at these stationary points are:

$$f(0) = 1; \quad f(-3) = -26.$$

1.2 Stationary points for functions of one variable 7

Step 3 As x tends to infinity, the highest power, x^4, dominates and hence f tends to ∞ in both directions.
(For a discussion on how to determine the limit at infinity refer to Section D of the Appendix.)

Step 4 The largest of the values obtained in Steps 2 and 3 is ∞, giving:

$$\max f(x) = \infty \text{ at } x = \pm \infty.$$

(i.e. at both $x = +\infty$ and $x = -\infty$.)

The smallest of these values is -26 and so:

$$\min f(x) = -26 \text{ at } x = -3.$$

We can go further, if we wish, and identify the nature of the other stationary point by using the information gained from the algorithm to sketch the graph of the function. We first plot the known points and gradients (Fig. 1.7a) and then join them smoothly in such a way that no new stationary points are generated (Fig. 1.7b). From the sketch it is clear that the stationary point at $x = 0$ defines a point of inflexion.

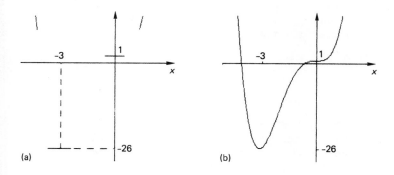

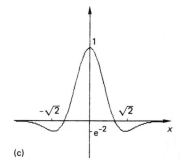

Fig. 1.7

Example 2
Find the maximum and minimum values of:

$$f(x) = (1 - x^2) \exp(-x^2).$$

8   The basics: unconstrained optimisation

Step 1   The stationary points are the solutions of the equation:
$$df/dx = 2x(x^2 - 2)\exp(-x^2) = 0,$$
and are given by $x = 0, \pm\sqrt{2}$.

Step 2   At the stationary points f takes the values:
$$f(\pm\sqrt{2}) = -\exp(-2); \quad f(0) = 1.$$

Step 3   As x tends to infinity, the rapidly decaying exponential $\exp(-x^2)$ dominates the increasing x^2 term and hence f, with $\exp(-x^2)$, tends to zero at infinity in both directions.

Step 4   The largest of the values obtained in the two previous steps is 1, attained at $x = 0$. Hence
$$\max f(x) = 1 \text{ at } x = 0.$$
The smallest of the values is $-\exp(-2)$ and hence
$$\min f(x) = -\exp(-2) \text{ at } x = \pm\sqrt{2}.$$

For completeness, the graph of the function is sketched in Fig. 1.7c.

Example 3
Find the maximum and minimum values of:
$$f(x) = 3\cos x + 4\sin x.$$

Step 1   The stationary point equation is
$$df/dx = -3\sin x + 4\cos x = 0,$$
with solution given by $\tan x = 4/3$. In the interval $-\pi < x \leqslant \pi$ there are two solutions θ and $\theta - \pi$ (Fig. 1.8a), with θ defined as in Fig. 1.8b. Since $\tan x$ repeats its values after every interval of 2π, there must be a 'double infinity' of stationary points, located at: $x = (\theta + 2k\pi), (\theta - \pi + 2l\pi)$, where k and l are arbitrary integers.

Step 2   Since the sine and cosine functions also repeat their values after every interval of 2π, so does our function f. Hence:
$$f(\theta + 2k\pi) = f(\theta) = 3(3/5) + 4(4/5) = 5.$$
It is also true that the sine and cosine functions (and hence f) change their signs after an interval of π, so that
$$f(\theta - \pi + 2l\pi) = f(\theta - \pi) = -f(\theta) = -5.$$
A sketch of f is shown in Fig. 1.8c.

Step 3   As we move towards infinity f must still continue oscillating (with period 2π) between 5 and -5, and so f can have no limit at infinity. The crucial question is, however, whether f can exceed in magnitude its values at the stationary points as we move to infinity, and because of the periodicity the answer must be no.

Step 4   Our conclusion, therefore, is that:
$$\max f(x) = 5 \text{ when } x = \theta + 2k\pi \text{ (k arbitrary integer)},$$
$$\min f(x) = -5 \text{ when } x = \theta - \pi + 2l\pi \text{ (l arbitrary integer)}.$$

The graph of f in Fig. 1.8c looks very much like a cosine curve shifted to the right by an angle θ and scaled up by a factor 5. That this is indeed the case can be shown by the

1.2 *Stationary points for functions of one variable* 9

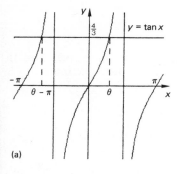
(a)

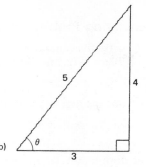
(b)

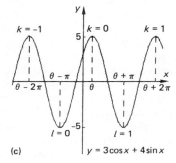
(c) $y = 3\cos x + 4\sin x$

Fig. 1.8

following argument using the addition rule for the cosine function:

$$f(x) = 3\cos x + 4\sin x$$
$$= 5(3/5 \cos x + 4/5 \sin x)$$
$$= 5(\cos \theta \cos x + \sin \theta \sin x)$$
$$= 5\cos(x - \theta).$$

Exercise Set 1
This is an appropriate point for you to try your hand at a few examples. Solutions to the starred questions will be found at the end of the book.

Find the maximum and minimum values of the following functions $f(x)$:

(i)* $x^2/2 - x/2 + 2$ (ii) $x^3 - 3x^2 + 3$

(iii)* $x + \exp(-2x)$ (iv) $(x^2 + 3/4)\exp(-x)$

(v)* $\ln(1 + x^2)/(1 + x^2)$ (vi) $2\exp(x) + \exp(-x)$.

In each case sketch the graph of $f(x)$.

1.3 Applications

A The Electricity Generating Problem Revisited

We are now in a position to solve the generating problem modelled in the introduction. In mathematical terms the problem is to find the minimum value of the function:

$$f(x) = x^2/2 - x/2 + 2.$$

The solution is obtained from the algorithm in the following steps.

1. Since $df/dx = x - 1/2 = 0$, there is just one stationary point at $x = 1/2$.
2. $f(1/2) = 15/8$.
3. $f(\pm \infty) = \infty$, since the dominant term in f has even power and a positive coefficient.
4. So $\min f(x) = 15/8$ at $x = 1/2$.

Power station A should therefore produce 0.5 megawatts and power station B 1.5 megawatts. This result is depicted in Fig. 1.9, where we have superimposed the cost functions for the two stations. At the optimal allocation the two cost curves have equal gradients. In Economic terminology 'optimality occurs when the marginal costs are equal'.[1] This result is perhaps a little surprising, since one might have thought that the optimal allocation would have been at the point P (Fig. 1.9), where the costs are equal and the demand is satisfied. That this is not in fact the case can be understood by imagining that the output of power station A is reduced by a small amount with a consequent increase in the other power station's output by the same amount. At P, the reduction in A's cost is greater than the increase in B's cost since the gradient of C_A is greater than that of C_B. The new allocation therefore meets demand but at a lower cost, and so is to be preferred. In fact the cost can be progressively decreased as the output of station A is decreased until the point where the gradients are equal is reached, when this argument no longer holds.

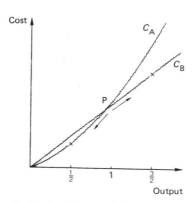

Fig. 1.9

B The Monopolistic Coal Producer

Suppose that at the turn of the century, a particular coal company owned two underground mines. It cost an amount c_1 to produce one tonne of coal from the older mine, for which the maximum (annual) output was limited to x_1 tonnes. For the second mine, the maximum output was x_2 and the unit cost c_2. The total cost of operating both mines at full capacity is therefore given by:

$$c_1 x_1 + c_2 x_2,$$

equalling the sum of the areas of the two rectangles shown on the unit cost/output diagram of Fig. 1.10a.

Suppose that since then, the company has expanded so much by capital investment and takeovers that it is effectively in a monopoly position, with a large number of mines operating. We can represent their cost structure by a step function (AB in Fig. 1.10b) giving the unit costs of the mines ordered in terms of increasing cost. As the number of mines is large we can approximate this step function by a smooth cost function, $C(x)$, (Fig. 1.10b), and the total cost of producing a given amount of coal, x, by the area under it.

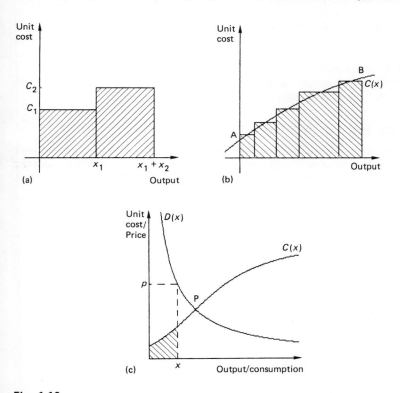

Fig. 1.10

How much coal a company can sell depends on the price asked. If the price is increased then the amount the customers buy will decrease. This behaviour can be modelled by a demand function $p = D(x)$ relating the price, p, the customers are willing to pay to the amount, x, they are willing to buy (Fig. 1.10c). If x units of coal are in fact produced and sold then the net profit to the coal producer is given by:

revenue − total cost of production,

or mathematically as:

$$xD(x) - \int_0^x C(u)du.$$

12 *The basics: unconstrained optimisation*

Let us suppose, for simplicity, that both the cost and demand functions are well approximated in the region of interest by the linear functions:

$$C(x) = 20 + 10x$$
$$D(x) = 40 - 10x,$$

where the price unit is US$ and coal output is measured in millions of tonnes. For these functions the profit is given by:

$$20x - 15x^2.$$

This function takes its maximum value[4] at its unique stationary point: $x = 2/3$.

This quantity is less than the quantity ($x = 1$) that would be produced in a competitive market, described in the general case by point P in Fig. 1.10c at the intersection of the demand and cost functions. That this is the competitive solution follows from the observation that if there were still profit to be made (i.e. for a given x the demand curve lies above the cost curve) then it would pay somebody to open another mine. Point P is the limit where individual profit reduces to zero.

In summary, we can say that under the assumptions of the model a monopoly position leads to less output and a higher price[1].

1.4 Stationary Points for Functions of Two Variables

If there were three power stations on our Caribbean island, then the optimal power allocation problem would involve two output variables, with the third determined from the demand constraint. To handle problems with two variables we will have to generalise

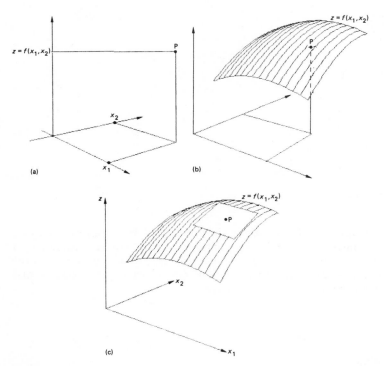

Fig. 1.11

1.4 Stationary points for functions of two variables

the concepts that proved useful in our analysis of the one variable problem, in particular, the concept of a stationary point.

We start with some geometry. A function f of the two variables x_1, x_2, can be represented geometrically as the surface with equation:

$$z = f(x_1, x_2) \tag{1.1}$$

in the three-dimensional space of the variables x_1, x_2, z. For each choice of x_1, x_2 the function f generates a value z and hence a point P with coordinates (x_1, x_2, z) in the three-dimensional space (Fig. 1.11a). As we vary x_1, x_2 we obtain a surface whose equation is (1.1). Fig. 1.11b shows a portion of such a surface.

At a point P on this surface the gradient depends on the direction θ in which we face. Its value can be obtained by taking a vertical cross-section of the surface through P in the direction θ (Fig. 1.12a). The required gradient is that of the tangent at P to the resulting cross-sectional curve. If such tangents exist for all θ and they lie in a common plane (called the *tangent plane* to the surface at P (Fig. 1.11c)), then we say that the function f is *differentiable*[5] at P.

The gradient at P in direction θ (Fig. 1.12a) is called the *directional derivative* of f at P and is denoted by $D_\theta f$. When $\theta = 0$ it is usually referred to as the 'partial derivative of f with respect to x_1' and denoted by $\partial f / \partial x_1$ (Fig. 1.12b). Similarly when $\theta = \pi/2$ (Fig. 1.12c) the directional derivative becomes the 'partial derivative of f with respect to x_2' being denoted by $\partial f / \partial x_2$. To obtain the partial derivative with respect to x_1 we differentiate f as if it were a function of the single variable x_1 with x_2 kept fixed, since all

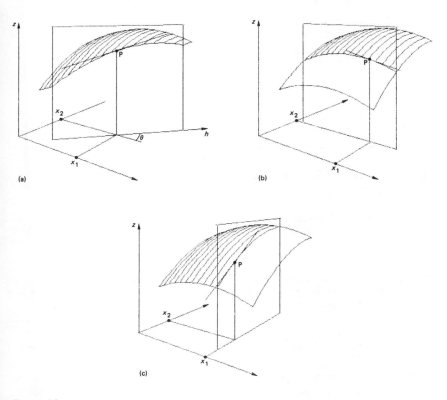

Fig. 1.12

points on the vertical plane through P with $\theta = 0$ have the same value of x_2. For the partial derivative with respect to x_2 we keep x_1 fixed in the differentiation.

From the partial derivatives we can obtain the directional derivatives in *any* direction θ using the following formula[6]:

$$D_\theta f = \partial f/\partial x_1 \cos\theta + \partial f/\partial x_2 \sin\theta \tag{1.2}$$

We are now in a position to generalise the concept of stationary point to functions of two variables. We say that a point (x_1, x_2) is a *stationary point* of a function f if the tangent plane at P: (x_1, x_2, z) is horizontal, i.e. if the directional derivatives in *all* directions at P are zero (Fig. 1.13a). From equation (1.2) we see that this happens when and only when the partial derivatives are both zero.

Before proceeding with the argument let us first find the stationary points in a simple example.

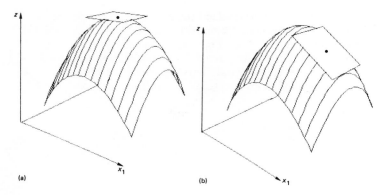

Fig. 1.13

Example 4
To find the stationary points of the function:

$$f(x_1, x_2) = -x_1^2 + x_1 - x_2^2 - 2x_2 - 1$$

we first determine the partial derivatives:

$$\partial f/\partial x_1 = -2x_1 + 1$$
$$\partial f/\partial x_2 = -2x_2 - 2.$$

(Remember to keep 'the other variable' constant in the partial differentiation!)

For both partial derivatives to be zero, we must have $x_1 = 1/2$ and $x_2 = -1$. There is therefore just one stationary point at $(x_1, x_2) = (1/2, -1)$.

If the maximum value of f is achieved at a finite point (x_1, x_2), then that point must be a stationary point, since if the tangent plane is not horizontal then there must be a point nearby where the function is greater in value (Fig. 1.13b). If, conversely, (x_1, x_2) is a stationary point then it is not necessarily the case that the maximum value of f is taken there. Counter examples are shown in Figs 1.14b, c.

1.4 Stationary points for functions of two variables

There are three basic types of stationary point (Fig. 1.14). For a local maximum ('peak') there is a region about the stationary point in which the value of the function is nowhere *greater* than its value at the stationary point. For a local minimum ('trough'), on the other hand, there is a region about the stationary point in which the function is nowhere *less* than that at the stationary point. For the third type any region about the stationary point contains some points where the function is greater than at the stationary point and some points where it is less. This is called a *saddle point*, or in geographical terms, a 'mountain pass' or 'col'.

If the maximum value of f is achieved at a finite point, then it must be at its highest local maximum, identified by sorting through the stationary points and finding the one at which the value is largest. It might be, however, that f takes its maximum value on the boundary at infinity. We have therefore to compare the maximum value of f at the stationary points with the values at infinity. Taking the limit to infinity is not as easy as in the one-dimensional case, since the limiting value in general depends on the *direction* taken—and there are an infinity of these!

These observations lead to the algorithm displayed in Procedure Box 1.2.

PROCEDURE 1.2
To find the maximum value of a function of two variables

1 **Find all stationary points by solving the equations:**

   $$\partial f/\partial x_1 = 0; \quad \partial f/\partial x_2 = 0.$$

2 **Evaluate f at each of these stationary points.**
3 **Evaluate f on the boundary at infinity.**
4 **Choose the largest of the values found in steps 2 and 3. This is the maximum value taken by the function and is denoted by: $\max f(x_1, x_2)$.**

To find the minimum value, $\min f(x_1, x_2)$, we simply pick the smallest of the values in steps 2 and 3. Alternatively, we can convert the minimum problem into a maximum problem using the identity:

$$\min f(x_1, x_2) = -(\max(-f(x_1, x_2))).$$

Let us apply this algorithm in the following two examples.

Example 5
Find the maximum and minimum values of the function:

$$f(x_1, x_2) = -x_1^2 + x_1 - x_2^2 - 2x_2 - 1.$$

1 In Example 4 we found that there is just one stationary point for this function, located at $(x_1, x_2) = (1/2, -1)$.
2 The value of f at this stationary point is given by:

   $$f(1/2, -1) = 1/4.$$

3 On completing the square:

   $$f = -(x_1 - 1/2)^2 - (x_2 + 1)^2 + 1/4$$

   it is clear that however we move to infinity at least one of the squared terms and hence f tends to $-\infty$.

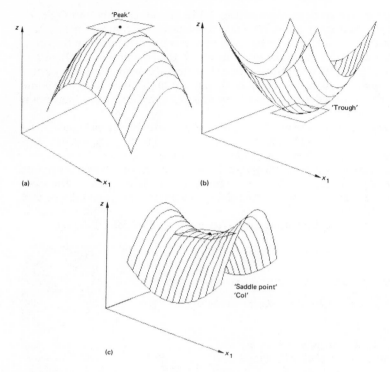

Fig. 1.14

4   From a comparison of the results of steps 2 and 3 we conclude that:
$$\max f(x_1, x_2) = 1/4 \text{ at } (1/2, -1),$$
$$\min f(x_1, x_2) = -\infty \text{ at infinity.}$$

The single stationary point must therefore define both a local and overall maximum of f.

Example 6
Find the maximum and minimum values of the function:
$$f(x_1, x_2) = \exp(x_1^2 - x_2^2).$$

1, 2   The stationary point equations have the form:
$$\partial f / \partial x_1 = 2x_1 \exp(x_1^2 - x_2^2) = 0$$
$$\partial f / \partial x_2 = -2x_2 \exp(x_1^2 - x_2^2) = 0,$$

with unique solution $x_1 = x_2 = 0$, since the exponential function can never be zero for any finite exponent. The corresponding value of f is 1.

3, 4   The difference in signs of the squared terms in the exponent of f suggests that the behaviour at infinity depends on direction. This is indeed the case, for if we go to infinity along the x_1 axis (in either direction) the exponent, and hence f, tends to ∞, but along the x_2 axis the exponent tends to $-\infty$, making f tend to 0. In summary, we have:

$$\max f(x_1, x_2) = \infty \text{ at infinity,}$$
$$\min f(x_1, x_2) = 0 \text{ at infinity.}$$

The single stationary point we found in step 2 defines a saddle point, as you will confirm later.

Exercise Set 2

1 In this question, the behaviour at infinity *does not* depend on the direction in which we approach it.

Find the maximum and minimum values and all stationary points of the following functions:

(i)* $x_1^2 + 5x_2^2 - 4x_2 + 1$ (ii) $(x_1 + x_2 - 1)^2 + (2x_1 + x_2 - 1)^2 + (3x_1 + x_2 - 1)^2$
(iii)* $\exp(-x_1^2 - x_2^2)$ (iv) $x_1 \exp(-x_1^2 - x_2^2)$
(v)* $-(x_1^2 - 1)^2 - x_2^2$ (vi) $-(x_1^2 - 1)^2 - (x_2^2 - 1)^2$.

2 Repeat question 1 for the following functions whose behaviour at infinity *does* depend on the direction taken:

(i)* $x_1(1 - x_2^2)$ (ii) $x_1(x_1^2 + x_2^2 - 1)$
(iii)* $-(x_1 - x_2^2 + 1)(x_1 + x_2^2 - 1)$ (iv) $(x_1^2 - x_2^2 - 1)(x_1^2 + x_2^2 - 2)$.

In each case find the relevant information about the behaviour of f by approaching infinity along the axes.

1.5 Applications

A Linear Regression[7]

If we plot the data from a repeated experiment to test the validity of a supposed linear relationship between the variables x and y we shall not usually find the data points lying exactly on a straight line even if the supposition is correct (Fig. 1.15a). The main reasons for this 'scatter' are measurement errors and changes in the background conditions of the experiment (temperature, humidity, etc.). To derive the 'line of best fit' to the data we could, of course, draw in a line 'by eye'—a rather subjective process. Fortunately there are more satisfactory ways of finding the equation of this line—the regression line. One of these, the 'method of least squares', chooses that line which minimises the sum of squares of the 'scatters' about the line. Precisely, suppose at the ith of n experiments x takes the value x_i (without error) and y is observed to take the value y_i (possibly with some error). If we assume the exact underlying relationship is:

$$y = ux + v \tag{1.3}$$

where u and v are the as yet undetermined gradient and intercept, then the 'error', e_i, in the ith experiment (Fig. 1.15a) is:

$$e_i = y_i - ux_i - v.$$

For simplicity, let us suppose that just three experiments are performed with results:

$$(x_1, y_1) = (1, 1); \quad (x_2, y_2) = (2, 2); \quad (x_3, y_3) = (3, 4),$$

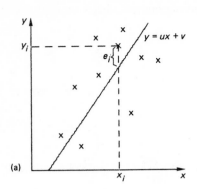 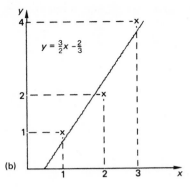

Fig. 1.15

then, $e_1 = 1 - u - v$, $e_2 = 2 - 2u - v$, $e_3 = 4 - 3u - v$, so that the sum of squares of the errors is given by:

$$f(u, v) = (1 - u - v)^2 + (2 - 2u - v)^2 + (4 - 3u - v)^2.$$

In least squares regression we choose u and v to minimise f.[8]

Solution
1   The equations for the stationary points

$$\partial f/\partial u = 2(u + v - 1) + 4(2u + v - 2) + 6(3u + v - 4)$$
$$= 2(14u + 6v - 17) = 0,$$
$$\partial f/\partial v = 2(u + v - 1) + 2(2u + v - 2) + 2(3u + v - 4)$$
$$= 2(6u + 3v - 7) = 0$$

have the unique solution $u = 3/2$, $v = -2/3$.

2   $f(3/2, -2/3) = 1/6$.
3   f tends to ∞ in all directions. (As the line moves increasingly away from its optimal position, the errors must increase without bound.)
4   Min $f = 1/6$ at $u = 3/2$, $v = -2/3$ and hence the 'best' straight line is given by (Fig. 1.15b):

$$y = 3x/2 - 2/3.$$

In the general case with n pairs of values (x_i, y_i), $(i = 1 \ldots n)$, the sum of squared errors on the assumption of a linear relationship of the form (1.3) is then:

$$f(u, v) = \sum_{i=1}^{n} (y_i - ux_i - v)^2$$

where x_i, y_i are *constants*, and Σ denotes the summation operator which is reviewed in Section B of the Appendix. This function has a unique stationary point given by:

$$\partial f/\partial u = -2 \sum_{i=1}^{n} x_i(y_i - ux_i - v) = 0$$

$$\partial f/\partial v = -2 \sum_{i=1}^{n} (y_i - ux_i - v) = 0.$$

Rearranging these expressions we obtain the following simultaneous equations for u and v:

$$s_{xx}u + s_x v = s_{xy}$$
$$s_x u + nv = s_y,$$

where $s_x = \sum_{i=1}^{n} x_i$; $s_y = \sum_{i=1}^{n} y_i$; $s_{xx} = \sum_{i=1}^{n} x_i^2$; $s_{xy} = \sum_{i=1}^{n} x_i y_i$.

These are two linear equations for u and v and their solution is easily found to be:

$$u = (ns_{xy} - s_x s_y)/(ns_{xx} - s_x^2); \quad v = (s_{xx} s_y - s_x s_{xy})/(ns_{xx} - s_x^2).$$

This solution yields the overall minimum, since the summed square form of f implies that f tends to infinity in all directions.

B A Road Building Problem

Let us suppose that a road is to be built between town A and town B, 30 miles to the East and 50 miles to the North of A. In between A and B is a mountain range, shown shaded in Fig. 1.16. The cost of building a road on the plain is £1 million per mile, and over the mountains £α million per mile (with $\alpha > 1$). In analysing which route to take between A and B to minimise total cost, the contractors have reduced the problem to one of building the road in three straight sections (Fig. 1.16). All that remains to be calculated are the points at which the road enters and leaves the mountain region. These points are parametrised by the variables x_1, x_2 identified in Fig. 1.16. In terms of x_1, x_2 the total cost of building the road is given by:

$$f(x_1, x_2) = d_1 + \alpha d_2 + d_3$$

where $d_1 = (x_1^2 + 100)^{1/2}$, $d_2 = ((x_1 + x_2 - 50)^2 + 100)^{1/2}$ and $d_3 = (x_2^2 + 100)^{1/2}$ in units of £1 million. The minimum occurs at the stationary point given by the equations:

$$\partial f/\partial x_1 = x_1/d_1 + \alpha(x_1 + x_2 - 50)/d_2 = 0$$
$$\partial f/\partial x_2 = \alpha(x_1 + x_2 - 50)/d_2 + x_2/d_3 = 0.$$

If we subtract the two equations and square out the terms, we obtain the result $x_1 = x_2$ (which is otherwise obvious from the symmetry of the problem!). Substituting back for x_2

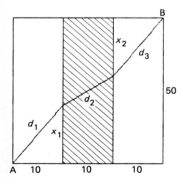

Fig. 1.16

in, say, the first of the equations yields:

$x_1/d_1 = \alpha(50 - 2x_1)/d_2.$

When $\alpha^2 = 1.6$, for example, the unique solution of this equation is given by $x_1 = 20$ with, as a consequence, $x_2 = 20$, also.

Had it not been for symmetry this problem would have been much more difficult to solve; indeed we should probably have had to resort to numerical methods.

1.6 Contour Maps: An Introduction

To find out more about the behaviour of a function of one variable, in particular the nature of its non-optimal stationary points, we used the information gained from Procedure 1.1 to sketch the graph of the function. For a two variable function the analogous process would be to sketch its surface from the information gained from using Procedure 1.2—not an easy task! An alternative approach is to do what mapmakers and weather forecasters do—represent the surface by its contour map.

If we project the set of points at the same height on the surface:

$$z = f(x_1, x_2) \tag{1.4}$$

onto the x_1, x_2 plane (Fig. 1.17a), we obtain a curve called a *contour* of the surface (1.4). If the common height of these points is c then the equation of the contour is simply: $c = f(x_1, x_2)$. As we vary c we obtain a family of curves (Fig. 1.17b) that constitutes a *contour map* of the surface (Fig. 1.17c).

To gain experience in sketching contour maps let us start with some examples based on familiar planar curves—lines, circles, ellipses, parabolae and hyperbolae, reviewed in Section F of the Appendix.

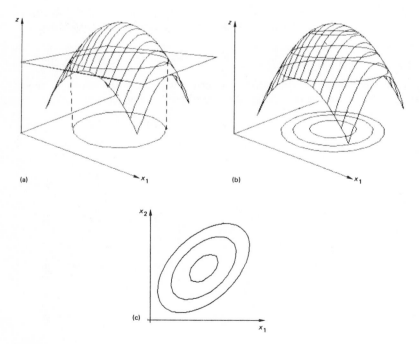

Fig. 1.17

1.6 Contour maps: an introduction

Example 7
The equation for the contours of the surface

$$z = 2x_1 + x_2 + 3 \tag{1.5}$$

is

$$c = 2x_1 + x_2 + 3,$$

i.e. $\quad 2x_1 + x_2 = c - 3.$

For each c this is the equation of a line with gradient -2 and intercept $c - 3$ on the x_2 axis (Fig. 1.18a). The set of contours for different values of c is therefore the set of parallel lines with gradient -2 (Fig. 1.18c). From the intercepts on the axes (Fig. 1.18b), we deduce that the greater the value of c the further to the right the contour line lies. Hence climbing up the surface (1.5) means moving in this direction, a direction we mark on the contour map with arrows (Fig. 1.18c). On a geographical map the contour heights are marked against the contours. We shall not usually need this amount of information, and so we will be content with the qualitative information given by the arrow structure.

Example 8
The contour equation for the surface:

$$z = x_1 - x_2^2$$

is, on rearrangement, given by:

$$x_1 - c = x_2^2.$$

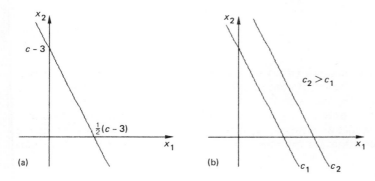

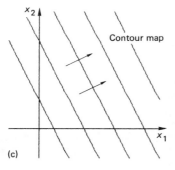

Fig. 1.18

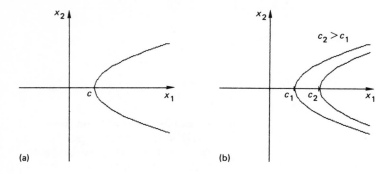

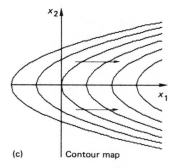

(c) Contour map

Fig. 1.19

This is the equation of a parabola with its axis coincident with the x_1 axis, intersecting that axis at $x_1 = c$ (Fig. 1.19a). The set of contours for different values of c is therefore the family of parabolae with the x_1 axis as common axis (Fig. 1.19b). The greater the value of c the further to the right the contour parabola lies and hence climbing the surface entails moving in this direction, a fact we represent by arrows in Fig. 1.19c. If we move along the x_1 axis in the positive direction, for example, we are continually gaining height without limit.

Example 9
For the surface:
$$z = -x_1^2 + x_1 - x_2^2 - 2x_2 - 1$$
we obtain the contour equation:
$$x_1^2 + x_2^2 - x_1 + 2x_2 = -(c+1).$$
On 'completing the square', this equation simplifies to the form:
$$(x_1 - 1/2)^2 + (x_2 + 1)^2 = (1/4 - c)$$
defining a circle, radius $(1/4 - c)^{1/2}$, centred at $(1/2, -1)$. For example, if $c = 0$ the radius is $1/2$ and if $c = 1/4$ the radius is zero (Fig. 1.20a). Clearly c cannot be greater than $1/4$. As we decrease c from $1/4$ we obtain a family of concentric circles of increasing radius, forming the contour map shown in Fig. 1.20b. The height therefore progressively decreases from its

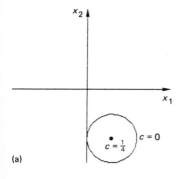
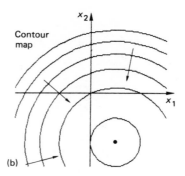

(a) (b)

Fig. 1.20

maximum value at the centre $(1/2, -1)$ as the distance from the centre increases. The arrows indicating increasing height must therefore point towards the centre (Fig. 1.20b).

Exercise Set 3
Sketch[9] the contour map for each of the following surfaces:

(i)* $z = 2x_1 - x_2 + 3$ (ii) . $z = -2x_1 + x_2 + 3$

(iii)* $z = x_1 + x_2^2 + 1$ (iv) $z = x_1^2 - x_2 + 1$

(v)* $z = x_1^2 + 2x_1 + x_2^2 + 3$ (vi) $z = x_1^2 - x_1 + x_2^2 + x_2 - 3$

(vii)* $z = \exp(x_1 x_2)$ (viii) $z = \exp(x_1^2 - x_2^2)$

(ix)* $z = x_1^2 + 2x_2 - x_2^2 + 1$ (x) $z = 2x_1 - x_1^2 + 2x_2 + x_2^2 - 3$.

Hint: in (vii) and (viii) take the logarithm of the contour equation.

1.7 Exotic Contours[10]

Sketching the contour map of a function with several stationary points is usually quite difficult due to the complicated nature of the contour equation. The strategy we shall adopt in this section to cope with this complexity involves the following four stages.

(a) Where feasible, sketch the contours that pass through the stationary points whose natures have not yet been determined. The expectation is that these special contours do have a sufficiently simple analytic form. This is the case in our first example below where the quartic contour equation factorises into two quadratic equations. In the second example this is not the case and so we will be content to determine the shape of the contour near the stationary point by approximating the contour equation.

(b) Deduce the nature of each of these stationary points by determining whether the function increases or decreases as we move into those regions defined by the contour through the stationary point.

(c) Sketch the contours close to the stationary points by noting that near an 'isolated'[11] local maximum or minimum the contours are usually roughly elliptical in shape (Figs 1.21a, b), and for a saddle point of the type shown in Fig. 1.21c they are usually roughly hyperbolic with the two contour sections through the point as common asymptotes.

(d) The final step is to sketch the complete contour map by smoothly piecing together the contours and contour sections determined in stages (a) and (c).

24   *The basics: unconstrained optimisation*

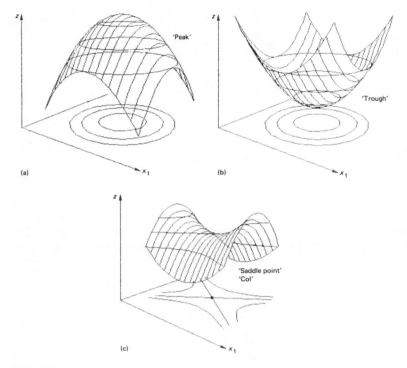

Fig. 1.21

Example 10
Let us consider the function:
$$f = -(x_1 - x_2^2 + 1)(x_1 + x_2^2 - 1)$$

(a) In problem 2(iii) of Exercise Set 2 we found this to have three stationary points, at $(0, 1)$, $(0, -1)$, $(0, 0)$.

None of these points is optimal since the function takes its maximum and minimum values at infinity. The first two stationary points share a common contour with equation:
$$-(x_1 - x_2^2 + 1)(x_1 + x_2^2 - 1) = 0,$$
giving:
$$(x_1 - x_2^2 + 1) = 0 \quad \text{or} \quad (x_1 + x_2^2 - 1) = 0.$$

This contour therefore consists of two parabolae which intersect in a saddle-like structure at these stationary points (Fig. 1.22a).
(b) To prove[12] that they are in fact saddle points we have to show that the contour divides the area about the stationary point into regions where the function is higher than the stationary point and regions where it is lower. To do this we note that the zero contour divides the plane into five regions in each of which f carries the same sign at all points. To find this sign, all we need to do is evaluate f at any point in the chosen region. For example, in the middle region A (Fig. 1.22a) f is positive since $f(0, 0) = 1$. In the region B, however, the sign is negative since $f(2, 0) = -3$. The signs for the other regions are given in Fig. 1.22a. The stationary points are now seen to be saddle points since movement away

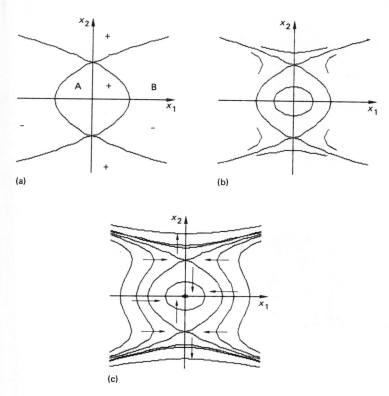

Fig. 1.22

from them parallel to the x_1 axis results in negative values of f, whereas movement parallel to the x_2 axis results in positive values.

The third stationary point is located at the origin, trapped between the parabolae of the zero contour. We can in fact deduce its nature without recourse to the contour equation from the following argument. There must be a maximum to the positive values taken by f in this bounded region, and this maximum must be taken at a stationary point. Since there is only one stationary point in this region it must be this local maximum.

(c, d) The contours about the stationary point at the origin are roughly elliptical and, about the other two stationary points, hyperbolic (Fig. 1.22b). From these contours we construct the contour map of Fig. 1.22c.

Example 11
(a) As a second example let us consider the function:

$$f(x_1, x_2) = -(x_1^2 - 1)^2 - x_2^2,$$

which was found to have three stationary points, at $(1, 0)$, $(-1, 0)$ and $(0, 0)$ (problem 1(v), Exercise Set 2). The first two are local maxima because they take the maximum value of f. The third point is not optimal. The equation of the contour through it is given by:

$$-(x_1^2 - 1)^2 - x_2^2 = -1,$$

i.e. $\quad -x_1^4 + 2x_1^2 - x_2^2 = 0.$

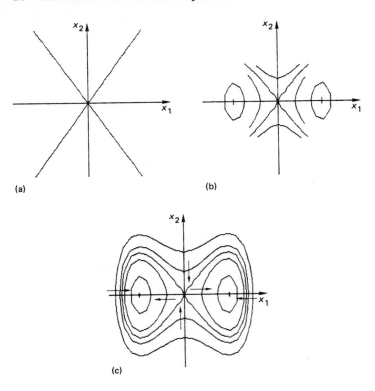

Fig. 1.23

This is rather a nasty equation, so we will be content to examine its shape near the stationary point at the origin. We can simplify the equation by dropping the x_1^4 term which becomes insignificant compared with x_1^2 close enough to the origin. The equation then becomes:

$$(\sqrt{2}x_1 - x_2)(\sqrt{2}x_1 + x_2) = 0 \tag{1.6}$$

which represents a pair of lines dividing the area about the origin into four regions (Fig. 1.23a).
(b) This again suggests a saddle point, but to be absolutely sure, we have to check the function values in its neighbourhood. We first note that the left-hand side of equation (1.6) gives an approximation to the change in the value of f from its value at the stationary point since:

$$f(x_1, x_2) - f(0, 0) = (-x_1^4 + 2x_1^2 - x_2^2 - 1) - (-1) \simeq 2x_1^2 - x_2^2.$$

We can easily check that the sign of this left-hand side alternates in the four regions defined by equation (1.6), hence we have a saddle point.
(c, d) The contour map of the function can now be completed (Fig. 1.23c) by merging the ellipses about the local maxima with the hyperbolae at the saddle (Fig. 1.23b). The contour through the origin is now seen to be in the shape of a figure eight enclosing the two maxima.

Exercise Set 4
By first finding the stationary points and the behaviour at infinity sketch the contour maps for the functions:

(i)* $x_1(1-x_2^2)$ (ii) $(x_1^2-1)(x_2^2-1)$
(iii)* $\sin x_1 \sin x_2$ (iv) $x_1(x_1^2+x_2^2-1)$
(v)* $(x_1^2-x_2^2)(x_1^2+x_2^2-1)$ (vi) $-(x_1^2-1)^2-(x_2^2-1)^2$

and identify the nature of all stationary points. (Where appropriate use the results obtained for Exercise Set 2.)

Problem Set 1

1 This first problem is a revision question on the material in this chapter. Its parts are designed to be straightforward, but not necessarily so simple as those in the Exercise Sets.

(a) Find all the stationary points and the maximum and minimum values of the following functions of one variable:

(i)* $(x^2+(1-2x)^2)^{1/2}$ (ii) $(1+x^2)^{1/2}+((2-x)^2+1)^{1/2}$
(iii)* $x+\cos x$ (iv) $x^2(1-x^2)(1+x^2)$
(v)* $\tan^{-1} x + 2/(1+x^2)$ (vi) $x(1+x^2)^{1/2}+\ln(x+(1+x^2)^{1/2})$.

In each case sketch the graph of the function.

(b) Find all the stationary points and the maximum and minimum values of the following functions of two variables:

(i)* $x_1 x_2 \exp(-x_1^2-x_2^2)$ (ii) $x_1 x_2/(1+x_1^2+x_2^2)^2$
(iii)* $x_1 x_2(x_1^2+x_2^2-1)$ (iv) $-x_2^2+(\cos x_1 - 1)^2$
(v)* $((x_1-2)^2+x_2^2-9)((x_1+2)^2+x_2^2-9)$ (vi) $-(x_1^2+2x_2^2-1)(2x_1^2+x_2^2-1)$.

(c) Sketch the contour maps for the following functions:

(i)* $x_1^2+x_2^2-4x_2+4$ (ii) $x_1^2-2x_1-x_2^2+x_2-1$
(iii)* $2x_1^2-4x_1-x_2$ (iv) $2x_1^2+5x_2^2+3x_2+6$
(v)* $x_1^2-2x_2^2+5x_1-1$ (vi) $x_1 x_2 - 2x_1 + x_2 - 3$.

(*Hint*: where appropriate complete the square.)

2 For those who enjoy sketching contours for functions with multiple stationary points, sketch the contour maps for the functions in question 1(b) and identify the nature of all stationary points.

3 To examine the variety of saddle points that can occur, sketch the contour maps for the following functions:

(i)* $x_1 x_2(x_1^2-x_2^2)$ (ii) $x_1 x_2(2x_1^2-x_2^2)(x_1^2-2x_2^2)$
(iii)* $x_1(x_1-x_2^2)$ (iv) $-(x_1^2+x_2^2-1)(x_1^2+2x_2^2-1)$.

4 In all the situations discussed so far the stationary points have always been isolated but this is not always the case. Find the stationary points and maximum and minimum values of the following functions:

(i)* $(x_1+x_2)^2$ (ii) $(x_1-x_2)^3$
(iii)* $x_1^2 x_2^2$ (iv) $(x_1-x_2^2)^2$
(v)* $(x_1^2-x_2^2-1)^2$ (vi) $(x_1^2+x_2^2)\exp(-x_1^2-x_2^2)$.

5* (a) (i) Find the stationary points of the function

$$(x) = x^3 - 3ax$$

when $a = -1, 0, 1$.

(ii) Discuss how the shape of the graph of $f(x)$ changes as a increases in value from -1 to 1.

(b) Sketch the contour map of the function:

$$f(x_1, x_2) = x_1^3 - 3x_1 x_2$$

and relate it to the discussion in part (a) by making the identification $x_1 \to x$, $x_2 \to a$.

6 In this question we use the concept of directional derivative to derive some useful results.

(a) Using the relation:

$$D_\theta f = \partial f / \partial x_1 \cos \theta + \partial f / \partial x_2 \sin \theta \qquad (1.2)$$

find the direction in which the gradient at a point on the surface $z = f$ is zero (i.e. the direction of the contour).

(b) Find the directions in which the gradient takes its maximum and minimum values. (These are the directions of steepest ascent and descent.)

(*Hint*: consider $D_\theta f$ to be a function of the one variable θ.)

(c) Show that the directions identified in (a) and (b) are at right angles to one another.

(*Hint*: prove and use the result that two directions θ_1, θ_2 are at right angles if $\tan \theta_1 \tan \theta_2 = -1$.)

(d) From equation (1.2) show that if we change position by varying the x_1, x_2 coordinates by small amounts dx_1, dx_2 respectively, then the value of f changes by df, given approximately by:

$$df \simeq (\partial f / \partial x_1)(dx_1) + (\partial f / \partial x_2)(dx_2) \qquad (1.7)$$

7* In this question we indicate how to derive property (1.2).

(a) Verify that the plane:

$$z - a_3 = b_1(x_1 - a_1) + b_2(x_2 - a_2)$$

passes through the point P: $(x_1, x_2, z) = (a_1, a_2, a_3)$ for any choice of parameters b_1, b_2.

(b) By taking vertical cross-sections through P show that $b_1 = \partial f / \partial x_1$, $b_2 = \partial f / \partial x_2$ if this plane is the tangent plane to the surface $z = f(x_1, x_2)$ at P.

(c) By taking a vertical cross-section through the tangent plane at P in direction θ derive equation (1.2).

(*Hint*: let $x_1 - a_1 = h \cos \theta$, $x_2 - a_2 = h \sin \theta$ and refer to Fig. 1.12a.)

References

1. For a discussion on the historical development of Optimisation Theory you may wish to refer to: J. F. McCloskey and F. N. Trefethen (Eds), *Operations Research for Management*, vol. 1, John Hopkins Press or M. J. Lighthill (Ed), *Newer Uses of Mathematics*, Penguin (1968).

    For further reading on the Economic models that we will be discussing in this chapter you should consult: R. G. Lipsey, *An Introduction to Positive Economics*, Wiedenfeld and Nicholson (1966) or P. A. Samuelson, *Economics—An Introductory Analysis*, McGraw-Hill (1961).

2. For conciseness of presentation we are supposing that f is also defined at ∞ and $-\infty$, with values obtained as the limit as x tends to ∞ and $-\infty$. In some cases these limits may not exist (see Example 3) but we will in the main avoid this complication.

3   An interval about $x = a$ is the set of values of x defined by $a - \epsilon < x < a + \epsilon$ where ϵ is some positive constant.
4   At infinity the quadratic term dominates and since its coefficient is negative the function must tend to $-\infty$ (refer to Section D of the Appendix).
5   We shall only consider functions f which are continuously differentiable at all points.
6   This will be proved in the Problem Set at the end of the chapter.
7   See Chapter 21 of G. M. Clarke and D. Cooke, *A Basic Course in Statistics*, 2nd edition, Edward Arnold (1983).
8   The independent variables of the problem are no longer labelled x_1, x_2 but u and v. The particular choice of labels is irrelevant.
9   To start with, you may wish to choose specific contours to draw—after some practice you will find you can roughly sketch in representative contours.
10 The techniques discussed in this section will only be used in Chapter 4. They are included especially for those who appreciate graphical analysis and would like to extend their expertise in this direction.
11 The other possibility is for a 'continuum' of stationary points—see the Problem Set.
12 Intersecting sections of a contour do not necessarily imply a saddle point. Consider the function $f = (x_1^2 - x_2^2)^2$ whose zero contour consists of two lines $x_1 = x_2$ and $x_1 = -x_2$, intersecting at the origin. This cannot be a saddle point because the function can never be negative.

2
Optimisation with Equality Constraints

2.1 Introduction

In setting up an optimisation problem the decision-maker has to identify those variables under his or her control that are relevant to the optimisation problem. It is often the case that these 'decision variables' cannot be chosen independently—indeed there may be functional relations between them that limit the decision-maker's freedom.

This is the experience of an Engineer in a Chemical Company who has been asked to design an open rectangular tank from a given amount of surplus steel sheet to hold as much of a new chemical product as possible. Choosing a square base with sides of length x_1 and height x_2, the problem is to maximise the volume $x_1^2 x_2$. Now x_1 and x_2 cannot be chosen independently since the total surface area, $x_1^2 + 4x_1 x_2$, is a given quantity A, say. The design problem is therefore to:

maximise $x_1^2 x_2$

subject to the constraint

$x_1^2 + 4x_1 x_2 = A$.

Similarly in the Electricity Generating Problem discussed previously the planner in Head Office has to decide at what outputs x_1, x_2 the two stations should operate to minimise the total cost of meeting a demand of, say, D megawatts. If the cost functions of the two stations are defined by $C_1(x_1)$ and $C_2(x_2)$, then the planner's problem is to:

minimise $C_1(x_1) + C_2(x_2)$

subject to the demand constraint:

$x_1 + x_2 = D$.

Both these problems are examples of an optimisation problem in two variables subject to a single constraint. In general we have:

maximise $f(x_1, x_2)$ or minimise $f(x_1, x_2)$
subject to $g(x_1, x_2) = 0$ subject to $g(x_1, x_2) = 0$.

In the first of our examples we have the identification $f = x_1^2 x_2$ and $g = x_1^2 + 4x_1 x_2 - A$, while in the second $f = C_1(x_1) + C_2(x_2)$ and $g = x_1 + x_2 - D$. The obvious way to solve such problems is to use the constraint to eliminate one of the variables from the objective function. The problem is thereby reduced from one in two variables and one constraint to one in one variable and seemingly no constraints. Indeed this is how we solved the Electricity Generating Problem in Chapter 1. Difficulties, however, can be encountered in using this substitution method. It might not be possible, for example, to extract from the constraint function g one variable as an explicit (analytic) function of the other. Try, for

example:
$$g(x_1, x_2) = x_1 \exp x_2 + x_2 \sin x_1 - 1.$$
Even if it is possible, the explicit function might be so unwieldy that the resulting one variable optimisation problem becomes very awkward to solve. In addition, there are other traps for the unwary as we shall see in the following problem:

$$\text{maximise } -(x_1^2 + x_2^2) \tag{2.1a}$$
$$\text{subject to } (x_1 - 2)^2 + x_2^2 = 1 \tag{2.1b}$$

where $f = -(x_1^2 + x_2^2)$ and $g = (x_1 - 2)^2 + x_2^2 - 1$.

From the constraint equation we find that $x_2^2 = 1 - (x_1 - 2)^2$, and hence on substitution the objective function becomes: $f = 3 - 4x_1$, a linear function in x_1 with no stationary points. Its maximum would appear to be infinite, corresponding to x_1 tending to $-\infty$. But it is easy to see that the objective function (2.1a) can *never* be greater than zero! The reason for this apparent contradiction is that constraint (2.1b) does not allow x_1 to take all real values: in fact x_1 is restricted to the range $1 \leqslant x_1 \leqslant 3$. It is, therefore, *not* generally true that the one variable problem obtained as a result of substitution has no constraints.

In this example the difficulty was easily spotted but, as you can imagine, this is not always the case. An alternative method that avoids these difficulties is the 'method of undetermined multipliers' discovered by the 18th century Italian mathematician, Lagrange. This method is immensely powerful and forms the basis for most of the theory in this and subsequent chapters.

2.2 Constrained Stationary Points

Before explaining and justifying the Lagrange method for solving the problem:

$$\text{maximise } f(x_1, x_2) \tag{2.2a}$$
$$\text{subject to } g(x_1, x_2) = 0 \tag{2.2b}$$

we shall try to give some insight into its geometry. Let us consider again the surface:

$$z = f(x_1, x_2) \tag{2.2a}$$

in the three-dimensional space of the variables x_1, x_2, z. Without the constraint we can move freely over this surface (Fig. 2.1a) searching out any stationary points, and in particular the highest one[1]. If, however, the constraint is imposed we are restricted to moving along a path on that surface. The path is defined as the intersection of the surface (equation (2.2a)) with the vertical surface[2]:

$$g(x_1, x_2) = 0 \tag{2.2b}$$

The maximum value of f in the problem (2.2) (i.e. the problem defined by equations (2.2a) and (2.2b)) is the maximum height reached on the path (point B in Fig. 2.1a) and is normally less than the height of the highest point on the surface (2.2a)—point B_0.

A more compact geometric representation is provided by the contour map of the surface (2.2a) obtained by projecting that surface onto the x_1, x_2 plane. The constraint is superimposed as the curve in this plane with equation $g(x_1, x_2) = 0$ (Fig. 2.1b).

As we move along the constraint curve from A towards B we climb the surface crossing contours at increasingly higher levels. Past the point B, however, we are descending, recrossing contours that we crossed on our ascent. Point B, the highest point on the path, is an example of a *constrained stationary point*, defined provisionally as a point at which a contour of f touches the constraint curve. A constrained stationary point does not

32 Optimisation with equality constraints

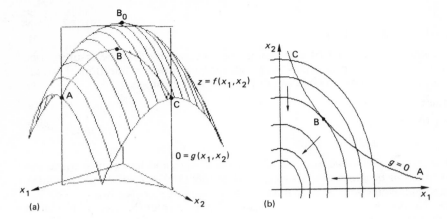

Fig. 2.1

necessarily correspond to a (local) constrained maximum as at B—it could be a (local) constrained minimum as in Fig. 2.2a, or it could be a 'point of inflexion' as in Fig. 2.2b.

As an example, let us find the constrained stationary points of the problem we discussed

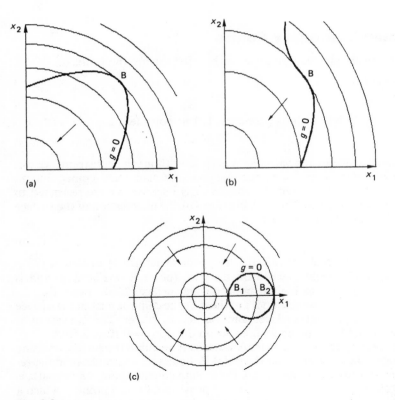

Fig. 2.2

in the introduction:

maximise $-(x_1^2+x_2^2)$
subject to $(x_1-2)^2+x_2^2=1$.

The contours of the objective function are the circles centred at the origin (Fig. 2.2c), where the unconstrained maximum is attained. The constraint path is the circle centred at (2, 0) with radius 1. There are two constrained stationary points (where the constraint path and a contour touch), one at $B_1(1, 0)$ and the other at $B_2(3, 0)$. Using the information given by the arrows, it is clear that B_1 is the constrained maximum and B_2 the constrained minimum.

Exercise Set 1
In each of the following cases sketch the contour map and superimpose the constraint path for the problem (2.2), where:

(i)* $f = x_1 + x_2$; $g = x_1^2 + x_2^2 - 1$
(ii) $f = -(x_1^2 + x_2^2)$; $g = x_1 + x_2 - 1$
(iii)* $f = -((x_1 + (x_2 + 1/2)^2)/2 - 1/8)$; $g = x_1 + x_2 - 2$
(iv) $f = x_1 - x_1^2 - x_2^2 - x_2 + 3$; $g = x_1 + x_2^2 + 1$.

In each case identify those points on the map where the constraint path touches a contour, and deduce the nature of these constrained stationary points.

2.3 The Lagrange Multiplier Method

To derive the Lagrange method for solving problem (2.2), we have to translate the geometric condition for a constrained stationary point into an analytic form. To do this we use the result established in Chapter 1 (Problem Set 1, Question 6) that the gradient of the tangent to a contour of a function f is given by:

$$\tan \theta = -\frac{\partial f}{\partial x_1} \Big/ \frac{\partial f}{\partial x_2}, \qquad (2.3)$$

where θ is the angle defined in Fig. 2.3a. Further, if we consider the constraint path as a contour of the surface:

$$z = g(x_1, x_2),$$

then a tangent to this path has gradient:

$$\tan \phi = -\frac{\partial g}{\partial x_1} \Big/ \frac{\partial g}{\partial x_2} \qquad \text{(see Fig. 2.3b)}.$$

The condition for the path and a contour to touch is that their tangents have equal gradients at the point of contact[3], i.e.

$$-\frac{\partial f}{\partial x_1} \Big/ \frac{\partial f}{\partial x_2} = -\frac{\partial g}{\partial x_1} \Big/ \frac{\partial g}{\partial x_2} \qquad (2.4)$$

This condition, together with the condition that the point in question must lie on the constraint curve (2.2b):

$$g(x_1, x_2) = 0$$

defines two equations for the constrained stationary points.

34 Optimisation with equality constraints

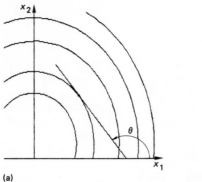
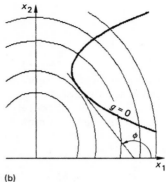

(a) (b)

Fig. 2.3

With a little manipulation, these equations can be rewritten in a form that will prove to be more useful, especially when we generalise the problem to more than two variables and more than one constraint. The steps in this manipulation are as follows.

1 Rearrange equation (2.4) to give:
$$\frac{\partial f/\partial x_1}{\partial g/\partial x_1} = \frac{\partial f/\partial x_2}{\partial g/\partial x_2}.$$

2 Equate each of these ratios to an as yet undetermined parameter:
$$\frac{\partial f/\partial x_1}{\partial g/\partial x_1} = \lambda; \quad \frac{\partial f/\partial x_2}{\partial g/\partial x_2} = \lambda.$$

3 Multiply out to give:
$$\partial f/\partial x_1 - \lambda \partial g/\partial x_1 = 0; \quad \partial f/\partial x_2 - \lambda \partial g/\partial x_2 = 0. \tag{2.5}$$

4 Define the function:
$$L = f(x_1, x_2) - \lambda g(x_1, x_2),$$

then the equations (2.5) can be written as:
$$\partial L/\partial x_1 = 0; \quad \partial L/\partial x_2 = 0 \tag{2.6}$$

These two equations, together with the constraint equation (2.2b) are an alternative set of equations for determining the constrained stationary points. They provide three equations for the three unknowns x_1, x_2 and λ.

The function L is called the *Lagrangean*, λ the *Lagrange multiplier* and the equations (2.6) the *Lagrange equations* for the problem.

Having found the constrained stationary points using the Lagrange equations, we then have to determine which (if any) of them solve problem (2.2). This can be done, as in the unconstrained case, by evaluating the objective function at the constrained stationary points and comparing the largest of these values with its values at points on the constraint curve at infinity, if such points exist. These steps are summarised in Procedure 2.1.

2.3 The Lagrange multiplier method

PROCEDURE 2.1
The Lagrange Method for solving Problem (2.2)
1. **Construct the Lagrangean:**

    $L = f(x_1, x_2) - \lambda g(x_1, x_2).$

2. **Find the constrained stationary points by solving the equations:**

    $$\frac{\partial L}{\partial x_1} = 0; \quad \frac{\partial L}{\partial x_2} = 0; \quad g(x_1, x_2) = 0 \qquad (2.7)$$

3. Evaluate f at these points and, where appropriate, at points on the constraint curve at infinity.
4. The largest of the values found in step 3 is the solution to problem (2.2).

In the corresponding minimisation problem:

   minimise $f(x_1, x_2)$
   subject to $g(x_1, x_2) = 0$,

the algorithm still applies provided we replace 'largest' by 'smallest' in step 4. Alternatively we can translate the problem into a maximisation problem using the usual identity:

   $\min f = -\max(-f).$

Example 1
Consider the problem:

   maximise $-(x_1 + (x_2 + 1/2)^2/2 - 1/8)$
   subject to $x_1 + x_2 = 2$.

In the last exercise set we found the contours of the objective function to be the family of parabolae with the line $x_2 = -1/2$ as their common axis (Fig. 2.4a). The constraint path is a line with gradient -1 which has only one point of tangency with a contour.

To find the coordinates of this point let us apply the Lagrange algorithm.

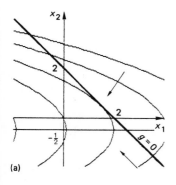
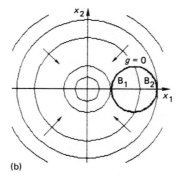

Fig. 2.4

36 *Optimisation with equality constraints*

1 The Lagrangean is:
$$L = -(x_1 + (x_2 + 1/2)^2/2 - 1/8) - \lambda(x_1 + x_2 - 2).$$

2 The three stationary point equations are:
$$\partial L/\partial x_1 = -1 - \lambda = 0$$
$$\partial L/\partial x_2 = -1/2 - x_2 - \lambda = 0$$
and
$$x_1 + x_2 = 2.$$

The first equation gives us $\lambda = -1$, the second $x_2 = 1/2$, and the third $x_1 = 3/2$. There is therefore just one constrained stationary point at $(3/2, 1/2)$.

3 At this point we have $f(3/2, 1/2) = -15/8$.
The constraint path unfortunately contains points at infinity, so we have to work out what happens to f as we move towards infinity along the constraint. It is clear, however, from the contour map (Fig. 2.4a) that f progressively decreases in value to $-\infty$ as we move towards infinity in either direction.

4 The maximum value is therefore achieved at the single constrained stationary point $(3/2, 1/2)$ with $f = -15/8$.

Note: This is just the two power station problem we discussed in Chapter 1. There it was solved by substitution after being transformed into a minimisation problem.

Example 2

As an example of a problem with a bounded constraint region (i.e. one that does not contain points at infinity), let us consider again:

maximise $-(x_1^2 + x_2^2)$
subject to $(x_1 - 2)^2 + x_2^2 = 1.$

We discussed this problem in Section 2.2 where we observed that the contours of the objective function are the family of circles centred at the origin (the unconstrained maximum) and the constraint path is a circle centred at $(2, 0)$ with radius 1 (Fig. 2.4b). The maximum and minimum points B_1 and B_2 can be determined by symmetry as $(1, 0)$ and $(3, 0)$ respectively.

We will now verify this result using the Lagrange method.

1 $L = -(x_1^2 + x_2^2) - \lambda((x_1 - 2)^2 + x_2^2 - 1).$
2 $\partial L/\partial x_1 = -2x_1 - 2\lambda(x_1 - 2) = -2x_1(1 + \lambda) + 4\lambda = 0.$
   $\partial L/\partial x_2 = -2x_2 - 2\lambda x_2 = -2x_2(\lambda + 1) = 0.$
   $(x_1 - 2)^2 + x_2^2 = 1.$

From the second equation we see that $x_2 = 0$ or $\lambda = -1$.
If $x_2 = 0$ then the third equation gives $x_1 = 3$ or 1, and the first $\lambda = -3$ or 1 (respectively).
If we take $\lambda = -1$ then the first equation gives us the contradiction $-4 = 0$.
There are therefore just two constrained stationary points at $(3, 0)$ and $(1, 0)$.

3 At these points $f(3, 0) = -9$ and $f(1, 0) = -1$.
4 Since the constraint path contains no points at infinity, the maximum value is taken at $(1, 0)$ with $f = -1$, as we previously observed.

Exercise Set 2

Use Procedure 2.1 to solve problem (2.2), where:
(i)* $f = x_1 + x_2;$ $g = x_1^2 + x_2^2 - 1$
(ii) $f = -(x_1^2 + x_2^2);$ $g = x_1 + x_2 - 1$
(iii)* $f = -(x_1/2 + x_2^2/2 + x_2);$ $g = x_1 + x_2 - 1$
(iv) $f = -(x_1/2 + x_1^2 + x_2 + x_2^2/2);$ $g = x_1 + x_2 - 2$
(v)* $f = x_1^2 - x_2^2;$ $g = x_1^2 + (x_2 - 1)^2 - 1$
(vi) $f = x_1 - x_1^2 - x_2^2 - x_2 + 3;$ $g = x_1 + x_2^2 + 1.$

In each case check your results by sketching a contour map or using the one already obtained in Exercise Set 1.

2.4 Singularities

The contour map for the problem:

maximise $x_1^2 - x_2^2$
subject to $x_1^2 + (x_2 - 1)^2 = 1$

consists of the family of hyperbolae with common asymptotes $x_1 = x_2$, $x_1 = -x_2$ and the constraint curve is the unit circle centred at (0, 1). There are three points of tangency, P_1, P_2 and P_3 (Fig. 2.5a) with coordinates ($\sqrt{3}/2$, 1/2), ($-\sqrt{3}/2$, 1/2) and (0, 2). However, in the Lagrange analysis of the problem in the last Exercise Set, we found a fourth point, the origin (0, 0). This happens to be a local maximum for the constrained problem since the height initially decreases as we move away from the origin along the constraint path in either direction. The reason why there is no tangency between the contour and the constraint path at the origin is that the contour there does not possess a tangent. This is because both partial derivatives $\partial f/\partial x_1$, $\partial f/\partial x_2$ are zero and hence the gradient (equation (2.3)) is undefined. This singularity shows up in the solution algebra through the value of

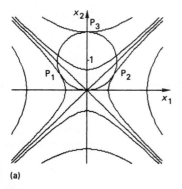
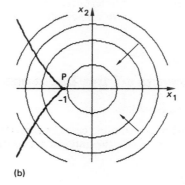

(a)   (b)

Fig. 2.5

38 *Optimisation with equality constraints*

the Lagrange multiplier. In fact if

$$\frac{\partial L}{\partial x_1} = \frac{\partial f}{\partial x_1} - \lambda \frac{\partial g}{\partial x_1} = 0; \quad \frac{\partial L}{\partial x_2} = \frac{\partial f}{\partial x_2} - \lambda \frac{\partial g}{\partial x_2} = 0 \tag{2.8}$$

and

$$\frac{\partial f}{\partial x_1} = 0; \quad \frac{\partial f}{\partial x_2} = 0$$

then it must be that $\lambda = 0$ (unless the partial derivatives of g are also both zero).

It could also be the case that the contour has a tangent but that the constraint path does not. From equations (2.8) we see that the effect of this is to make λ infinite (since we are assuming that at least one of the partial derivatives of f is non-zero). As an example consider the following problem:

$$\begin{aligned} \text{maximise} \quad & -(x_1^2 + x_2^2) \\ \text{subject to} \quad & (x_1 + 1)^3 + x_2^2 = 0. \end{aligned}$$

The maximum occurs at the point P: $(-1, 0)$ in Fig. 2.5b, where the constraint path comes closest to the unconstrained maximum at the origin. At P the path has a cusp with no tangent. This point is identified by Procedure 2.1 as follows:

$$L = -(x_1^2 + x_2^2) - \lambda((x_1 + 1)^3 + x_2^2)$$

$$\frac{\partial L}{\partial x_1} = -2x_1 - 3\lambda(x_1 + 1)^2 = 0 \text{ giving } \lambda = -2x_1/(3(x_1 + 1)^2)$$

$$\frac{\partial L}{\partial x_2} = -2x_2 - 2\lambda x_2 = -2x_2(1 + \lambda) = 0$$

$$(x_1 + 1)^3 + x_2^2 = 0.$$

One solution of the second equation is $x_2 = 0$, implying from the other two equations that $x_1 = -1$ and $\lambda = \infty$. The other solution, corresponding to $\lambda = -1$, leads to a complex value for x_1 and can therefore be discarded.

There is a third singular case mentioned above in which neither the contour nor the constraint path possess tangents at a particular point. In this case the Lagrange multiplier λ can take any value (see equations (2.8)). To include all these singularities we generalise our definition of a constrained stationary point to be *any* point corresponding to a solution of the Lagrange system of equations (2.7).

In summary, we can say that our algorithm for solving problem (2.2) still applies when one or both tangents do not exist. Such singularities are present when λ takes the value zero or infinity[4] or remains undetermined. In these cases there is a coincidence of the constrained stationary point with an unconstrained stationary point of the f function, the g function or both.

2.5 Other Aspects of the Lagrange Method

(a) The Lagrange Surface

The Lagrange algorithm can in most cases be considered as a two-stage process in which we first find for a given value of λ the unconstrained stationary points of the Lagrangean L satisfying:

$$\frac{\partial L}{\partial x_1} = 0; \quad \frac{\partial L}{\partial x_2} = 0 \tag{2.9}$$

2.5 Other aspects of the Lagrange method

and then adjust λ by applying the constraint condition:

$$g(x_1, x_2) = 0,$$

so that the unconstrained stationary points of L coincide with the constrained stationary points of the original problem. Forming L by absorbing the constraint function g frees us from the need to impose the constraint until the end of the analysis.

To understand geometrically how this is possible let us take a particular case—a quadratic objective function f with a single maximum, constrained by a linear function g. The constrained problem is, as we observed previously, one of finding the highest point on the constraint path $g = 0$ lying on the surface S_0, with equation $z = f(x_1, x_2)$ (Fig. 2.6a). This point can be located using the following argument.

(i) For each value of λ the Lagrangean $L(x_1, x_2; \lambda)$ also defines a surface, S_λ (Fig. 2.6b), with equation

$$z = L(x_1, x_2; \lambda).$$

This surface coincides with the surface S_0 when $\lambda = 0$, since, by the definition of L, it is clear that:

$$L(x_1, x_2; 0) = f(x_1, x_2).$$

(ii) As the value of λ is changed from zero S_λ moves from coincidence with S_0 in such a way that the constraint path still lies on S_λ whatever the value of λ. This follows from the fact

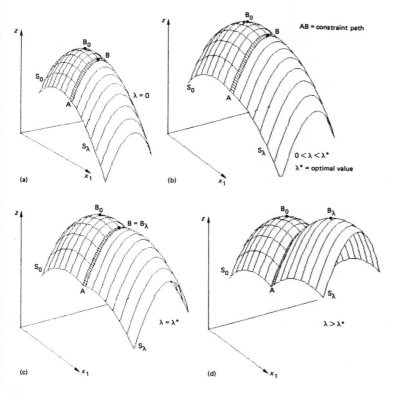

Fig. 2.6

that if x_1 and x_2 are such that $g(x_1, x_2) = 0$ then by the definition of L

$$L(x_1, x_2; \lambda) = f(x_1, x_2).$$

The curve of intersection of the surfaces S_0 and S_λ therefore includes all points on the constraint path (Fig. 2.6b).

(iii) The solution of the equations (2.9) gives the location of the maximum of S_λ as a function of its parameter λ. At the optimal value of λ, determined by imposing the constraint condition, the surface S_λ is positioned so that the constraint path actually goes over the maximum point (Fig. 2.6c). The unconstrained stationary point of L therefore coincides with the constrained stationary point of the original problem.

(iv) For values of λ beyond the optimal value, the maximum point of S_λ rises (Fig. 2.6d) and the constraint path no longer passes over it. The optimum is therefore a maximum with respect to x_1 and x_2 but a *minimum* with respect to λ. This means that the Lagrangean possesses a saddle point if we consider L to be a function of the *three* variables x_1, x_2 and λ. This important structural result will form the basis of our discussion on duality in Chapter 4.

(b) Constraint Sensitivity

In some problems the constraint function will contain a parameter and we will be interested in the effect of its variation on the optimal value of the objective function.

As an example, let us consider another electricity generation example with demand, D, taken to be a parameter:

$$\text{minimise} \quad x_1/2 + x_1^2 + x_2 + x_2^2/2$$
$$\text{subject to} \quad x_1 + x_2 = D.$$

Straightforward application of Procedure 2.1 yields:

$$x_1 = D/3 + 1/6; \quad x_2 = 2D/3 - 1/6; \quad \lambda = 2D/3 + 5/6,$$

and the minimum cost is given (after some algebra) by:

$$\min f = D^2/3 + 5D/6 - 1/24 \qquad \text{(Fig. 2.7a)} \tag{2.10}$$

For $D = 2$ we regain the result obtained previously (problem (iv) of Exercise Set 2), that:

$$x_1 = 5/6; \quad x_2 = 7/6; \quad \lambda = 13/6; \quad \text{and} \quad \min f = 71/24.$$

If we denote $\min f$ (as defined by equation (2.10)) by f^*, then a measure of the sensitivity of f^* to changes in D is the derivative df^*/dD. From equation (2.10) we calculate this derivative as

$$df^*/dD = 2D/3 + 5/6.$$

If, for example, we increase D by 0.1 when $D = 2$ (Fig. 2.7a) then the corresponding change in f^* is approximately

$$0.1 \, df^*/dD = 0.1(4/3 + 5/6) \simeq 0.22.$$

We note that the sensitivity df^*/dD is in this case equal to the optimal value of the Lagrange multiplier. This is no accident—it is a general result, as we shall now show for the two-variable case. Suppose the constraint function $g(x_1, x_2)$ contains a parameter b in the simple form:

$$g(x_1, x_2) = \hat{g}(x_1, x_2) - b,$$

so that the constraint can be expressed as:

$$\hat{g}(x_1, x_2) = b.$$

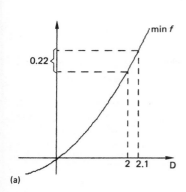
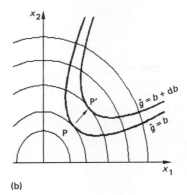

Fig. 2.7

If we increase b by a small amount, db, then the optimal point P: (x_1, x_2) (Fig. 2.7b) moves to a new optimal point P': $(x_1 + dx_1, x_2 + dx_2)$. The changes db, dx_1, dx_2 are related by:

$$db \simeq \frac{\partial \hat{g}}{\partial x_1} dx_1 + \frac{\partial \hat{g}}{\partial x_2} dx_2 \tag{2.11}$$

using the result (1.7) established in Question 6 of Problem Set 1. In the move the optimal value f^* of the objective function f is changed by an amount:

$$df^* \simeq \frac{\partial f^*}{\partial x_1} dx_1 + \frac{\partial f^*}{\partial x_2} dx_2 \tag{2.12}$$

But since $\partial f^*/\partial x_1 = \lambda^*(\partial g/\partial x_1) = \lambda^*(\partial \hat{g}/\partial x_1)$ and $\partial f^*/\partial x_2 = \lambda^*(\partial g/\partial x_2) = \lambda^*(\partial \hat{g}/\partial x_2)$ at P, we conclude from equations (2.11) and (2.12) that

$$df^* \simeq \lambda^* db$$

with λ^* the optimal value of λ. In the limit as db tends to zero we obtain the exact relationship:

$$df^*/db = \lambda^*.$$

2.6 Generalisation of the Lagrange Method

Since most optimisation problems of interest involve more than two variables and several constraints we have to generalise the Lagrange method to cover these cases. Fortunately the generalisation is quite straightforward—we just add an extra Lagrange multiplier for each additional constraint and differentiate partially with respect to each variable in turn. Suppose our problem is one in n variables and m constraints, so that it can be written as:

maximise $f(x_1, x_2, \ldots, x_n)$
subject to $g_1(x_1, x_2, \ldots, x_n) = 0$
$g_2(x_1, x_2, \ldots, x_n) = 0$
. . . .
. . . .
$g_m(x_1, x_2, \ldots, x_n) = 0,$

42 Optimisation with equality constraints

or, equivalently, in the more compact form:

$$\text{maximise} \quad f(x)$$
$$\text{subject to} \quad g_j(x) = 0 \quad j = 1, 2, \ldots, m \tag{2.13}$$

where x stands for the set of n variables $x_1, x_2, \ldots, x_n$ and the constraints are ordered by index j. To solve this problem we use the algorithm given in Procedure Box 2.2.

PROCEDURE 2.2
The Lagrange Method for solving Problem (2.13)

1 Construct the Lagrangean
$$L = f(x) - \lambda_1 g_1(x) - \lambda_2 g_2(x) - \ldots - \lambda_m g_m(x)$$
$$= f(x) - \sum_{i=1}^{m} \lambda_i g_i(x).$$

2 **Find the constrained stationary points as solutions of the equations**[5]:
$$\partial L/\partial x_1 = \partial L/\partial x_2 = \ldots = \partial L/\partial x_n = 0$$
and $g_1(x) = g_2(x) = \ldots = g_m(x) = 0$.

(These are $n+m$ equations for the $n+m$ variables $x_1, x_2, \ldots, x_n, \lambda_1, \lambda_2, \ldots, \lambda_m$.)

3 Evaluate f at these points and, where appropriate, at points at infinity satisfying the constraints.

4 **The largest of the values found in step 3 is the required solution to problem (2.13).**

Again, for the equivalent minimisation problem we can use this Procedure, picking out the smallest rather than the largest value in step 4 or we can transform the problem into one of maximisation using the identity: $\min f = -(\max(-f))$.

Example 3

As a first example in the use of this Procedure let us look at the following 3-variable, 2-constraint problem:

$$\text{maximise} \quad -(x_1^2 + x_2^2 + x_3^2)$$
$$\text{subject to} \quad x_1 + x_2 + x_3 = 0$$
$$x_1 + 2x_2 + 3x_3 = 1.$$

Here $g_1 = x_1 + x_2 + x_3$ and $g_2 = x_1 + 2x_2 + 3x_3 - 1$.

1 $L = -(x_1^2 + x_2^2 + x_3^2) - \lambda_1(x_1 + x_2 + x_3) - \lambda_2(x_1 + 2x_2 + 3x_3 - 1).$
2 The stationary point equations are:

$$\partial L/\partial x_1 = -2x_1 - \lambda_1 - \lambda_2 = 0$$
$$\partial L/\partial x_2 = -2x_2 - \lambda_1 - 2\lambda_2 = 0$$
$$\partial L/\partial x_3 = -2x_3 - \lambda_1 - 3\lambda_2 = 0.$$

These equations enable us to express the x variables in terms of the multipliers as:

$$x_1 = -(\lambda_1 + \lambda_2)/2; \quad x_2 = -(\lambda_1 + 2\lambda_2)/2; \quad x_3 = -(\lambda_1 + 3\lambda_2)/2 \tag{2.14}$$

2.6 *Generalisation of the Lagrange method* 43

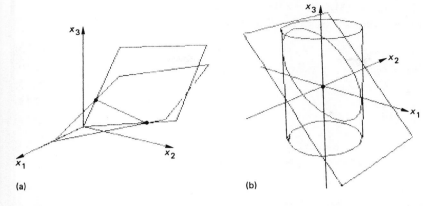

Fig. 2.8

Substitution into the constraint equations then yields the following equations for the multipliers:
$$-\lambda_1 - 2\lambda_2 = 0; \quad -3\lambda_1 - 7\lambda_2 = 1$$
with solution: $\lambda_1 = 2; \quad \lambda_2 = -1$.

Hence from equations (2.14) we conclude that $(-1/2, 0, 1/2)$ is the only constrained stationary point.

3   $f(-1/2, 0, 1/2) = -1/2$.

There are points at infinity on the constraint, since they define two planes in the space of variables x_1, x_2, x_3 (Fig. 2.8a). However, in *all* directions to infinity (including the directions allowed by the constraints) the objective function tends to $-\infty$.

4   We conclude that the maximum value is taken at $(-1/2, 0, 1/2)$ with $f = -1/2$.

Solving the set of equations generated by Procedure 2.2 can be difficult. One strategy, which we adopted in the last example, is to first solve the Lagrange equations for the x variables as a function of the multipliers and then determine these multipliers by substitution in the constraint equations. This worked in the last example because the equations were linear in the xs and the λs. When this is not the case, for example when a constraint is quadratic rather than linear, the strategy is liable to lead to non-linear equations for the λs and the possibility of degeneracies in the equations for the xs. To avoid this sort of algebraic difficulty—this book is about optimisation rather than algebra—we will only consider those problems where the equations have special features that can be exploited to simplify the solution procedure. For example, in the next problem one of the equations can be factored into linear terms. We can then explore the implications of each factor being zero.

Example 4

$$\text{Maximise} \quad x_1^2 + 2x_2^2 + 3x_3^2$$
$$\text{subject to} \quad x_1^2 + x_2^2 = 25$$
$$x_1 + x_3 = 2.$$

44   *Optimisation with equality constraints*

1   $L = x_1^2 + 2x_2^2 + 3x_3^2 - \lambda_1(x_1^2 + x_2^2 - 25) - \lambda_2(x_1 + x_3 - 2)$.
2   The constrained stationary point conditions are:

$$\partial L/\partial x_1 = 2x_1 - 2\lambda_1 x_1 - \lambda_2 = 2x_1(1 - \lambda_1) - \lambda_2 = 0 \tag{2.15}$$
$$\partial L/\partial x_2 = 4x_2 - 2\lambda_1 x_2 = 2x_2(2 - \lambda_1) = 0 \tag{2.16}$$
$$\partial L/\partial x_3 = 6x_3 - \lambda_2 = 0 \tag{2.17}$$
$$\text{and} \quad x_1^2 + x_2^2 = 25 \tag{2.18}$$
$$x_1 + x_3 = 2 \tag{2.19}$$

From equation (2.16) we have $x_2 = 0$ or $\lambda_1 = 2$.
Consider the first alternative: $x_2 = 0$. The other equations yield:

2.18: $x_1 = 5$ or -5
2.19: $x_3 = -3$ or 7 (respectively)
2.17: $\lambda_2 = -18$ or 42
2.15: $\lambda_1 = 1 - \lambda_2/(2x_1) = 14/5$ or $26/5$.

We conclude that there are two constrained stationary points $(5, 0, -3)$ and $(-5, 0, 7)$.

Now consider the second alternative: $\lambda_1 = 2$. The other equations now yield:

2.15: $\lambda_2 = -2x_1$
2.17: $\lambda_2 = 6x_3$. Together these give $x_1 = -3x_3$.
2.19: $x_3 = -1$. Hence $x_1 = 3$ and $\lambda_2 = -6$.
2.18: $x_2^2 = 25 - 9 = 16$, i.e. $x_2 = \pm 4$.

There are therefore constrained stationary points at $(3, 4, -1)$ and $(3, -4, -1)$:
3   Evaluation of f at the stationary points yields:

$$f(5, 0, -3) = 52; \quad f(-5, 0, 8) = 172; \quad f(3, 4, -1) = 44; \quad f(3, -4, -1) = 44.$$

There are no points at infinity that satisfy the constraints as can be seen from the following argument. From equation (2.18) it is clear that neither x_1 nor x_2 may be greater than five in magnitude, and since $x_3 = 2 - x_1$ from equation (2.19) we conclude that x_3 must also be finite. The constraint curve in the space of the variables x_1, x_2, x_3 is in fact an ellipse, being the intersection of a cylinder (2.18) with a plane (2.19) (see Fig. 2.8b).
4   We therefore conclude that the maximum value occurs at $(-5, 0, 8)$ with the objective function taking the value 172.

Example 5
One important use of Procedure 2.2 is in handling problems containing a set of parameters and involving an arbitrary number of variables. As an example let us consider:

$$\text{maximise} \quad \sum_{k=1}^{n} a_k x_k$$
$$\text{subject to} \quad \sum_{k=1}^{n} x_k^2 = 1,$$

where there are n variables and the coefficients a_k in f are unspecified. The algorithm yields the following results.

1. $L = \sum_k a_k x_k - \lambda \left(\sum_k x_k^2 - 1 \right)$
   $= (a_1 x_1 + a_2 x_2 + \ldots + a_n x_n) - \lambda(x_1^2 + x_2^2 + \ldots + x_n^2 - 1).$

2. $\partial L/\partial x_1 = a_1 - 2\lambda x_1 = 0$, i.e. $x_1 = a_1/2\lambda$
   $\partial L/\partial x_2 = a_2 - 2\lambda x_2 = 0$, i.e. $x_2 = a_2/2\lambda$ (2.20)
   $\ldots \ldots$
   $\partial L/\partial x_n = a_n - 2\lambda x_n = 0$, i.e. $x_n = a_n/2\lambda.$

Substituting these values for the variables in the single constraint we obtain:

$$\sum_k x_k^2 = (x_1^2 + x_2^2 + \ldots + x_n^2) = (a_1^2 + a_2^2 + \ldots + a_n^2)/4\lambda^2 = 1.$$

i.e. $\left(\sum_k a_k^2 \right) \Big/ 4\lambda^2 = 1$ or $2\lambda = \pm \sqrt{\sum_k a_k^2}.$

Substituting back in the equations (2.20) we find that:

$$x_k = \pm a_k \Big/ \sqrt{\sum_k a_k^2} \qquad k = 1, 2, \ldots, n.$$

There are therefore two constrained stationary points.

3. At these points f has the values $\pm \sqrt{\sum_k a_k^2}$ (respectively).

4. There are no points at infinity—the constraint surface being a 'sphere'—and hence the maximum is taken at $x_k = a_k / \sqrt{\sum_k a_k^2}$, and the minimum at $x_k = -a_k / \sqrt{\sum_k a_k^2}$.

Put another way, we can say that the objective function $\sum_k a_k x_k$ is bounded above by $\sqrt{\sum_k a_k^2}$ and below by $-\sqrt{\sum_k a_k^2}$, i.e.

$$-\sqrt{\sum_k a_k^2} \leq \sum_k a_k x_k \leq \sqrt{\sum_k a_k^2}$$

for all x_k such that $\sum_k x_k^2 = 1$ (2.21)

We can do away with the condition (2.21) by introducing new variables b_k such that $x_k = b_k / \sqrt{\sum_k b_k^2}$. Substitution in the inequalities then leads to:

$$-\sqrt{\sum_k a_k^2} \sqrt{\sum_k b_k^2} \leq \sum_k a_k b_k \leq \sqrt{\sum_k a_k^2} \sqrt{\sum_k b_k^2} \qquad (2.22)$$

This is the important and very useful *Schwarz Inequality* that holds for any set of numbers a_k, b_k. It is a generalisation of a well-known property of vectors in a plane. The scalar product[6] of two row vectors $\mathbf{v} = (a_1, a_2)$, $\mathbf{w} = (b_1, b_2)$ is defined as:

$\mathbf{v}\mathbf{w}^T = a_1 b_1 + a_2 b_2.$

46 *Optimisation with equality constraints*

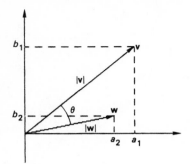

Fig. 2.9

Its value can be shown to be equal to

$$|\mathbf{v}||\mathbf{w}|\cos\theta$$

where $|\mathbf{v}|$, $|\mathbf{w}|$ denote the lengths of the corresponding geometric vectors (Fig. 2.9) and θ the angle between them. Since $-1 \leq \cos\theta \leq 1$ we must have

$$-|\mathbf{v}||\mathbf{w}| \leq \mathbf{v}\mathbf{w}^T \leq |\mathbf{v}||\mathbf{w}| \qquad (2.23)$$

Clearly $|\mathbf{v}| = \sqrt{a_1^2 + a_2^2}$ and $|\mathbf{w}| = \sqrt{b_1^2 + b_2^2}$ and so equation (2.23) is a special case of equation (2.22) with $n = 2$.

Exercise Set 3

1 Solve the problems:

(i)* maximise $x_1 - x_2 + x_3$
  subject to $x_1^2 + x_2^2 + x_3^2 = 1$

(ii) maximise $-(x_1^2 + 2x_2^2 + 3x_3^2)$
  subject to $x_1 + x_2 + x_3 = 1$
  $2x_1 - x_2 = 3.$

(iii)* maximise $x_2 x_3 - 5(x_1 - 900)^2$
  subject to $x_1 + x_2 = 1000,$
  $x_2 = 250 - x_3/20$

(iv) minimise $ax_1 + x_2/2 + x_2^2/2 + 3x_3/4 + x_3^2/4$
  subject to $x_1 + x_2 + x_3 = 3$
  in the two cases $a = 1$ and $a = 1/4$

2 Solve the following n variable problems:

(i)* maximise $-\sum_{i=1}^{n} a_i x_i^2$ with $a_i > 0$ for all i,

  subject to $\sum_{i=1}^{n} x_i = 1$

(ii) minimise $\sum_{i=1}^{n} x_i^2$

subject to $\sum_{i=1}^{n} a_i x_i = 1,$

$\sum_{i=1}^{n} x_i = 0.$

2.7 Applications

A University Admission Policy

Consider the problem faced by a University Administrator who has the task of maximising revenue from student fees. The University is obliged by the Government to charge 'home' students a fixed fee, but can charge what it likes for 'overseas' students. It would seem to be in the interest of the University, therefore, to enrol as many overseas students as possible— but if it enrols too many the Government will impose financial penalties! The Administrator therefore has to balance the higher revenue from overseas students with the heavy penalties that might be incurred. Precisely, his job is to work out the values of three variables—the number of home students, the number of overseas students and the fees to charge these overseas students—in order to maximise net revenue. These variables, however, cannot be varied independently since they have to satisfy certain constraints. For example the total number of students is limited by the teaching resources of the University and the number of overseas students wishing to enrol presumably depends on the level of their fees.

To model this problem mathematically, let us suppose the number of home students admitted is denoted by x_1, the number of overseas students by x_2, and the fee charged to overseas students by x_3. Suppose the Government wants 900 of the total 1000 places available at the University to be taken by home students and exacts a penalty of £$5(x_1 - 900)^2$ if the number deviates from that figure, then the net income for the University is:

$$f = R - 5(x_1 - 900)^2 + x_2 x_3 \tag{2.24}$$

where R is the fixed Government grant that includes the home student fees. Further, we shall suppose that there are enough students for the University to fill all its places, i.e.

$$x_1 + x_2 = 1000 \tag{2.25}$$

and that the relationship between the number of overseas students admitted and the fee charged is well approximated by the linear relationship:

$$x_2 = 250 - x_3/20 \tag{2.26}$$

The problem is therefore to maximise equation (2.24) subject to equations (2.25) and (2.26).

The solution is obtained as follows:

$$L = R - 5(x_1 - 900)^2 + x_2 x_3 - \lambda_1(x_2 - 250 + x_3/20) - \lambda_2(x_1 + x_2 - 1000).$$
$$\partial L/\partial x_1 = -10(x_1 - 900) - \lambda_2 = 0$$
$$\partial L/\partial x_2 = x_3 - \lambda_1 - \lambda_2 = 0 \tag{2.27}$$
$$\partial L/\partial x_3 = x_2 - \lambda_1/20 = 0$$

with:
$$x_1 + x_2 = 1000$$
$$x_2 = 250 - x_3/20.$$

The first set of equations gives:
$$x_1 = 900 - \lambda_2/10; \quad x_2 = \lambda_1/20; \quad x_3 = \lambda_1 + \lambda_2.$$

Substituting these into the constraints gives:
$$100 = \lambda_1/20 - \lambda_2/10; \quad 250 = \lambda_1/10 + \lambda_2/20$$

and hence $\lambda_1 = 2400$ and $\lambda_2 = 200$. If we now substitute these back into equations (2.27) we obtain the single constrained stationary point: $x_1 = 880$, $x_2 = 120$, $x_3 = 2600$. This is the maximum, since at infinity in the directions dictated by the constraints the objective function tends to $-\infty$. That is, the University should admit 880 home students together with 120 overseas students, each charged £2600.

B Electricity Generation

We have already discussed several examples of this problem, involving two and sometimes three power stations. Let us now consider the general case in which there are n power stations, the kth of which generates x_k units of electricity at cost $C_k(x_k)$. The company wishes to meet the demand D at minimum cost, that is it seeks the solution of the problem:

$$\text{minimise} \quad \sum_{k=1}^{n} C_k(x_k)$$

$$\text{subject to} \quad \sum_{k=1}^{n} x_k = D.$$

The Lagrangean, in this general case, can be written as

$$L = \sum_k C_k(x_k) - \lambda \left(\sum_k x_k - D \right)$$
$$= \sum_k C_k(x_k) + \lambda \left(D - \sum_k x_k \right) \qquad (2.28)$$

This Lagrangean has a simple interpretation if we imagine that the Electricity company has the option of buying in electricity from, say, a local industrial complex at a unit price λ if it does not wish to meet all the demand from its own stations. The Lagrangean can then be considered as the sum of the costs of producing electricity from its own stations and the cost of buying in electricity. At an arbitrary price λ some electricity will usually be bought in but at the optimal price λ^* all the electricity would be produced by the Electricity company itself. Therefore λ^* is the price at which it is worth the company producing all the required demand. This optimal price is in fact equal to the unit cost of producing extra electricity in the most efficient manner. This follows immediately from the sensitivity result:

$$df^*/dD = \lambda^*$$

established in Section 2.5 for the two variable case, but which can be shown to be true for any number of variables. If demand is increased by a small amount dD then the minimum cost is increased by an amount df^* where

$$df^* \simeq \lambda^* dD.$$

In the economic literature λ^* is called the *shadow price* and is equal to the *marginal* cost of producing electricity efficiently.[7]

The Lagrange structure also suggests a means of decentralising the running of the company. This can be seen by rearranging the Lagrangean in the form:

$$-L = \sum_k (\lambda x_k - C_k(x_k)) - \lambda D.$$

The kth term in the sum is the profit made by the kth power station when we take λ to be the price at which the Electricity company buys electricity from its stations for final sale to its customers. Maximising $-L$ with respect to x_k corresponds to the manager of the kth station choosing output to maximise the profit of that station. To decentralise, therefore, the company sets the following rule.

> It will buy at price λ (fixed by itself) as much electricity as the station managers calculate will yield maximum profits for their individual stations.

In this way Head Office is released from the day-to-day running of the stations, and its only problem is to adjust λ so that the output will meet demand.[7]

C Commodity Dealing[8]

After lengthy negotiations the purchasing department of a large food processing company has signed contracts with three wheat producing countries C_1, C_2, C_3 to supply wheat in 12 month's time. There was much haggling over the contract prices, but it was eventually agreed that each contract price should be the local price of wheat at the time of delivery. If the prices were certain then clearly the company would have signed a contract only with the country with the lowest price. Since, however, it can only forecast the price ($p_1 p_2 p_3$ per tonne respectively) the company has decided to 'hedge its bets', just in case one or more of the forecasts are badly wrong, by buying from all three countries. To calculate the proportion x_i to buy from country C_i the company first decided on the average wheat price $\hat{p}$ to budget for in its current corporate plan. These proportions would then have to satisfy

$$x_1 p_1 + x_2 p_2 + x_3 p_3 = \hat{p}$$

together with the identity

$$x_1 + x_2 + x_3 = 1.$$

Being conservatively minded the company wishes to choose the values for x_1, x_2, x_3 to minimise the uncertainty that this average price $\hat{p}$ is wildly out of line with the actual market price in 12 month's time. An obvious measure of risk is the *variance* of the forecasted average price, which can be expressed as[9]

$$\sum_i x_i^2 \sigma_i^2$$

where σ_i^2 is the variance of the forecast p_i, if we assume that the weather conditions and other factors affecting the harvests in the three countries are effectively unrelated. The company's problem therefore reduces to:

minimise $\sum_i x_i^2 \sigma_i^2$

subject to $\sum_i x_i p_i = \hat{p}$

and $\sum_i x_i = 1.$

As a particular example let us take $(p_1, p_2, p_3) = (10, 15, 20)$ and $(\sigma_1^2, \sigma_2^2, \sigma_3^2) = (15, 10, 5)$, then the problem becomes:

$$\text{minimise} \quad 15x_1^2 + 10x_2^2 + 5x_3^2$$
$$\text{subject to} \quad 10x_1 + 15x_2 + 20x_3 = \hat{p}$$
$$x_1 + x_2 + x_3 = 1$$

for which the Lagrange algorithm yields the solution[10]:

$$x_1 = -0.08\hat{p} + 1.58; \quad x_2 = -0.03\hat{p} + 0.83; \quad x_3 = 0.12\hat{p} - 1.42 \quad \text{and}$$
$$\lambda_1 = 0.37\hat{p} - 6.17; \quad \lambda_2 = -6.17\hat{p} + 109.17.$$

As a check we note from the optimal proportions that an increase in the price $\hat{p}$ that the company is prepared to pay increases the proportion of the high priced wheat (from country C_3) at the expense of the other two suppliers.

Using the sensitivity relationship:

$$\lambda_1^* = dR^*/d\hat{p},$$

where R^* is the minimum risk, we conclude by integration that this minimum risk as a function of expected price is given by:

$$R^* = 0.19\hat{p}^2 - 6.2\hat{p} + \alpha,$$

where α is a constant equalling 55, determined for example by substitution into the objective function. The budgeted price $\hat{p}$ that would give minimum risk is given by $dR^*/d\hat{p} = \lambda_1^* = 0$, and has value 16.8 (Fig. 2.10).

It would not be in the interests of the company to budget for a price above 16.8 since the uncertainty is increasing the higher the price paid. It is therefore only the section of the curve to the left of the minimum point that is of interest. Over this section the company can buy reduced uncertainty by increasing the budgeted price.

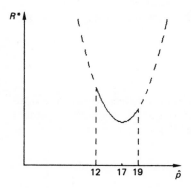

Fig. 2.10

Problem Set 2

1 (a) Solve problem (2.2) using Procedure 2.1 when:

(i)* $f = x_1 x_2, \quad g = x_1^2 + 2x_2^2 - 1$

(ii) $f = -(x_1 - 1)^2 - (x_2 - 1/2)^2, \quad g = x_1 x_2$

(iii)* $f = x_1^2 + 2x_1x_2 + 2x_2^2$, $g = x_1^2 + x_2^2 - 1$
(iv) $f = x_1^2 + 2x_1x_2 + x_2^2/2$, $g = x_1^2 + x_2^2 - 1$.

In the first two cases check your answer by sketching a contour map.
(b) Solve problem (2.13) using Procedure 2.2 when:
(i)* $f = -(3x_1^2 + 2x_2^2 + x_3^2)$,
$g_1 = x_1 + x_2 - 1$, $g_2 = 2x_2 + x_3 - 1$
(ii) $f = -(x_1 + x_2)^2 - (x_2 + x_3)^2 - (x_3 + x_1)^2$,
$g_1 = x_1 + x_2 + x_3 - 1$, $g_2 = x_1 - x_2 + x_3 - 1$
(iii)* $f = x_1^2 + 2x_1x_2 + 2x_2 + 3x_3^2$,
$g = x_1^2 + x_2^2 + x_3^2 - 1$
(iv) $f = x_1^2 + 2x_1x_2 + x_2^2/2$,
$g = x_1^2 + x_2^2 + x_3^2 - 1$.

2 (a) (i) Solve the problem:

minimise $-(x_1 - x_3)^2 \pm (x_2 - x_4)^2$
subject to $x_1 + x_2 = 1$, $(x_3 - 3)^2 + (x_4 - 3)^2 = 1$

by the Lagrange method.
(ii) Explain how this optimisation problem finds the shortest distance between the line $x + y = 1$ and the circle $(x - 3)^2 + (y - 3)^2 = 1$ in the x, y plane.
(b) By solving the appropriate optimisation problem find the shortest distance between the line $x + y = 1$ and the parabola $x + y^2 = 0$.

3 In studying the behaviour of the surface S_λ we restricted attention to the case in which the relevant stationary point of S_λ remained a peak as λ passed through its optimal value. This is not always the case as the following problems show.
(a)* (i) Solve the problem:

maximise $-(x_1^2 + x_2^2)/2$
subject to $1 - x_1^2 + x_2^2 = 0$

by the Lagrange method.
(ii) Sketch the contour map for the Lagrangean of this problem when λ equals 0, $\lambda^*/2$, λ^*, $2\lambda^*$, where λ^* is its optimal value.
(iii) Discuss the changes in the contour map as λ varies from 0 to $2\lambda^*$.
(b) Repeat part (a) for the problem:

maximise $-(x_1 + 4x_2^2)$
subject to $x_1^2 - 4x_2^2 + 1 = 0$.

4 In this question we analyse some situations involving singularities.
(a)* Solve the problem:

maximise $-(x_1 - a)^2 - (x_2 - b)^2$
subject to $x_1(x_1 - x_2^2) = 0$,

by the Lagrange method, when (a, b) equals:
(i) $(-1, 0)$, (ii) $(0, 0)$, (iii) $(0, 1)$.

Check your results by sketching the contour maps.

(b) Repeat part (a) for the problem:

$$\text{minimise} \quad (x_1 + a)^2 + x_2^2$$
$$\text{subject to} \quad (x_1^2 + x_2^2 - 1)^2 = 0,$$

when $a = 2, 1, 0$.

5 (a)* The contours of the function:

$$f = ax_1^2 + 2bx_1x_2 + cx_2^2 \qquad (a, b, c \text{ not all zero}) \tag{2.29}$$

form either a family of ellipses or hyperbolae centred at the origin. To determine which, we find the maximum and minimum values of f subject to the unit circle constraint $x_1^2 + x_2^2 = 1$. If both values have the same sign then the family consists of ellipses, if not then hyperbolae. The pair of axes common to the members of the family intersects the unit circle at the optimal points. Use these facts to sketch the contours for the functions f in question 1(a) (iii)*, (iv).
(b) Find conditions on the coefficients a, b, c in equation (2.29) for its contours to be (i) elliptical, (ii) hyperbolic.
(c) If you are familiar with the concept of the eigenvalues of a matrix, relate the constrained minimum and maximum values of f to the eigenvalues of the matrix

$$A = \begin{pmatrix} a & b \\ b & c \end{pmatrix}$$

and the optimal points to its eigenvectors.

6 For the following problems verify the sensitivity result $\partial f^*/\partial b = \lambda^*$, when f^*, λ^* are the optimal values of the objective function and multiplier:

(i)*   maximise $-(x_1^2 + x_2^2)$, subject to $b = x_1 + x_2$
(ii)   maximise $-(x_1^2 + x_2^2)$, subject to $x_1 + 2x_2 = b$
(iii)* maximise $x_1 + x_2$, subject to $x_1^2 + x_2^2 = b$
(iv)   maximise $-(x_1^2 + x_2^2)$, subject to $(x_1 - 2)^2 + x_2^2 = b$.

References

1 For the moment we shall ignore the possibility that the maximum of f occurs at infinity.
2 If the point P with coordinates $(x_1, x_2, 0)$ satisfies $g(x_1, x_2) = 0$ then so does any point (x_1, x_2, z) on the vertical line through P, since the equation $g = 0$ does not involve z. These vertical lines generate a vertical surface.
3 If $\partial f/\partial x_1 \neq 0$ and $\partial f/\partial x_2 = 0$ then the tangent is vertical. The case in which both derivatives are zero will be considered later.
4 To avoid infinite multipliers we could use the technique of employing a second multiplier, λ_0, such that

$$L = \lambda_0 f - \lambda g$$

with, for example, the constraint $\lambda_0 + \lambda = 1$ to define a scale. In this case, instead of $\lambda = \infty$ we would use $\lambda_0 = 0$.
5 By $\partial L/\partial x_i$ we mean the derivative of L with respect to x_i keeping all other variables fixed.
6 See Section E of the Appendix for a discussion of scalar products.
7 See G. M. Heal, *The Theory of Economic Planning*, North Holland (1973).
8 See H. Markowitz, *Portfolio Selection*, Monograph 16, Cowles Foundation for Research, Yale University.
9 See, for example, Section 9.4 of *A Basic Course in Statistics*, by G. M. Clarke and D. Cooke, Edward Arnold (1983).

10   We have for simplicity ignored the fact that the proportions must be non-negative. These conditions are satisfied if $12.5 \leq \hat{p} \leq 18.5$ (Fig. 2.10). Fortunately only values in this range will be of interest to us in the subsequent analysis. However, in general for a problem of this type we would have to properly take into account these non-negativity constraints, using for example the methods discussed in the next chapter.

3
Inequality Constraints

3.1 Introduction

In Chapter 2 we explored the implication of a decision-maker's freedom being limited by the existence of equality constraints between the variables identified in the problem. That freedom could also be limited by restrictions on the range of the values that each variable can take. It may be that for variable x:

$$a \leqslant x \leqslant b \tag{3.1}$$

where a is its *lower* bound and b its *upper* bound. For example, in the Chemical Tank Design Problem of Section 2.1, suppose that the tank had to be placed in a recess in a corner of the factory, the width and depth of this recess being w and its height h (Fig. 3.1a). To fit in this recess, the width and height of the tank must satisfy the constraints:

$$0 \leqslant x_1 \leqslant w, \quad 0 \leqslant x_2 \leqslant h.$$

Similarly in the Two Station Electricity Problem the (necessarily) non-negative output of each station is limited by its design capacity, i.e.

$$0 \leqslant x_1 \leqslant d_1, \quad 0 \leqslant x_2 \leqslant d_2.$$

More generally, there could be inequality constraints that involve more than one of the decision variables. For example, in the tank problem the roof of the recess could be sloping (Fig. 3.1b). If it slopes from a height h at an angle of $45°$ then the dimensions of the tank must satisfy:

$$x_1 + x_2 \leqslant h.$$

The constraint imposed by the limited amount of sheet steel available should also be considered as an inequality constraint:

$$x_1^2 + 4x_1 x_2 \leqslant A$$

since the engineer may no longer be able to use up all the sheet metal, although, obviously, he will use up as much as he can consistent with the other constraints. In summary, we have

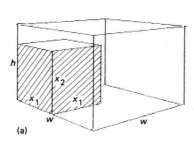
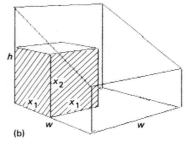

(a)   (b)

Fig. 3.1

as a complete specification of the Tank Problem, the following inequality constrained optimisation problem:

maximise $x_1^2 x_2$
subject to $0 \leqslant x_1 \leqslant w$
$0 \leqslant x_2 \leqslant h$
$x_1 + x_2 \leqslant h$
$x_1^2 + 4x_1 x_2 \leqslant A.$

We can also express the Electricity Problem as an inequality constrained optimisation problem by observing that the Company is obliged to at least meet demand, i.e.

$x_1 + x_2 \geqslant D.$

It could produce more than demanded, although we know from common sense that at the optimum the equality will hold. A completely specified model for the problem will therefore have the form:

minimise $C_1(x_1) + C_2(x_2)$
subject to $x_1 + x_2 \geqslant D$
$0 \leqslant x_1 \leqslant d_1$
$0 \leqslant x_2 \leqslant d_2.$

It was the absence of the non-negativity conditions $x_1 \geqslant 0, x_2 \geqslant 0$ in part (iii) of Exercise Set 2 in the last chapter that led to the unacceptable result $x_1 = 3/2, x_2 = -1/2$.

In this chapter we show how to solve an optimisation problem subject to a set of inequality constraints. We start with the simplest case, the one-variable problem with upper and lower bounds (3.1). The procedure for solving this problem requires only a simple modification to the one-variable algorithm discussed in Section 1.2. We determine the largest (or smallest) value of the objective function at the stationary points satisfying the inequalities and then compare this with the value of the function at the boundaries, whether finite or infinite.

Similarly in the two-variable case the inequality constraints define a region whose boundary no longer lies entirely at infinity, as it did in Chapter 1. The optimum occurs either inside the boundary or on the boundary, so we can split the problem into two separate sub-problems. Inside the boundary the problem is one of unconstrained optimisation and can be solved by the methods of Chapter 1, while on the boundary we have an equality constrained optimisation problem which can be solved by the Lagrange algorithm discussed in Chapter 2. In the second half of the chapter the method is generalised to handle an arbitrary number of variables and constraints. In so doing we will obtain the well-known Complementary Slackness and Kuhn–Tucker conditions, whose meaning is illustrated in the analysis of a transportation problem involving the efficient distribution of a primary resource.

3.2 The One-Variable Case

In the problem:

maximise $f(x)$
subject to $a \leqslant x \leqslant b,$ (3.2)

the value to be determined is the maximum of the values of f when x is restricted to the interval $a \leqslant x \leqslant b$. The maximum will occur either at a stationary point inside the permitted interval (Fig. 3.2a) or at a boundary point (Figs 3.2b, c). To find the maximum,

56 Inequality constraints

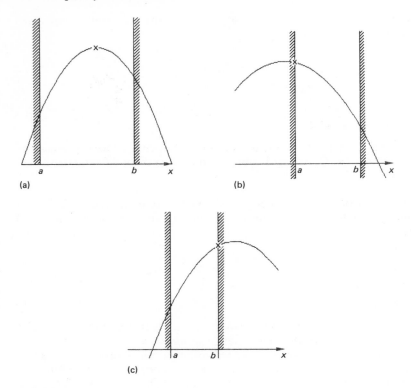

Fig. 3.2

therefore, we pick the largest of the values of f at the stationary points in the interval and at the boundaries. The algorithm is shown in Procedure Box 3.1.

PROCEDURE 3.1
To solve Problem (3.2)

1. Find all the stationary points by solving the equation:
   $$df/dx = 0.$$
2. Evaluate f at each of the stationary points that lie in the permitted interval $a \leqslant x \leqslant b$.
3. Evaluate f at the boundaries $x = a$ and $x = b$.
4. The largest of the values found in steps 2 and 3 is the required constrained maximum value.

As before, for the minimum problem we choose the smallest rather than the largest value in step 4, or translate the problem into one of maximisation using the identity:
$$\min f = -\max(-f).$$

Example 1
Let us use the algorithm to solve the problem:

maximise $-x^2 + 3x + 1$
subject to $0 \leqslant x \leqslant 1$.

1. The equation $df/dx = -2x + 3 = 0$ has solution $x = 3/2$, which is outside the permitted region.
2. There are no stationary points in the permitted interval.
3. $f(0) = 1$; $f(1) = 3$.
4. Hence, for $0 \leqslant x \leqslant 1$ max $f = 3$ at $x = 1$ (Fig. 3.3a).

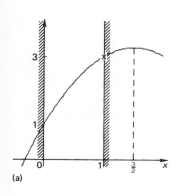
(a)

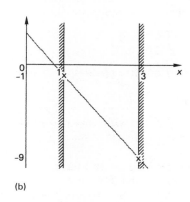
(b)

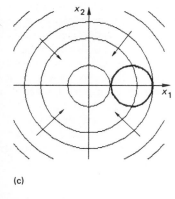
(c)

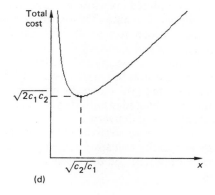
(d)

Fig. 3.3

Example 2
The steps in solving the problem:

maximise $3 - 4x$
subject to $1 \leqslant x \leqslant 3$.

are:

1. $df/dx = -4 \neq 0$—hence there are no stationary points.
2. Not applicable.

58 Inequality constraints

3   $f(1) = -1;\ f(3) = -9.$
4   Hence over the interval $1 \leqslant x \leqslant 3$ max $f = -1$ at $x = 1$ (Fig. 3.3b).

Note: this is the problem that resulted from the elimination of x_2 in the equality constrained two-variable problem:

maximise $-(x_1^2 + x_2^2)$
subject to $(x_1 - 2)^2 + x_2^2 = 1$

discussed in Section 2.1 (Fig. 3.3c). The answer obtained here agrees with that obtained by the Lagrange method.

Example 3
In the final example of this section we minimise a function over a semi-infinite interval.

minimise $c_1 x + c_2/x$
subject to $x \geqslant 0$ (i.e. $0 \leqslant x < \infty$)

where c_1, c_2 are two positive constants.

The Procedure gives:

1   $df/dx = c_1 - c_2/x^2 = 0$ which has two solutions $x = \pm\sqrt{c_2/c_1}$, but only the positive root lies in the permitted interval.
2   $f(\sqrt{c_2/c_1}) = 2\sqrt{c_1 c_2}$.
3   $f(0) = \infty$ because of the second term.
    $f(\infty) = \infty$ because of the first term.
4   Hence over the interval $x \geqslant 0$, min $f = 2\sqrt{c_1 c_2}$ at $x = \sqrt{c_2/c_1}$ (Fig. 3.3d).

Example 3 is a simplified version of the well-known Procurement–Inventory Problem[1] in Operational Research. This is the problem faced, for example, by a steel stockholding company which has to decide how frequently to order steel from the producers to satisfy a steady demand from its customers. If the steel is delivered in bulk at infrequent intervals then the company is faced with large storage costs, whereas if the steel is delivered in smaller quantities but more frequently then the storage costs are less but the handling costs increase due to the increased number of deliveries. To model this problem let x denote the amount delivered in each shipment and d the amount sold to the customers each month. The frequency of delivery is therefore d/x and hence the total handling costs are $c_h d/x$, where c_h is the (assumed) fixed handling cost per delivery. The storage costs are given by $c_s x/2$, where c_s is the unit storage cost per month, the factor $1/2$ allowing for the fact that on average only half the delivered amount is being stored. The total cost is therefore given by:

$c_1 x + c_2/x$

where $c_1 = c_s/2$ and $c_2 = c_h d$. The optimal delivery we found to be given by $\sqrt{(2c_h d/c_s)}$ and the minimum cost by $\sqrt{(2c_s c_h d)}$.

Exercise Set 1
Solve problem (3.2) using Procedure 2.1 when:

(a)* $f = 2x - x^2$ and (a, b) equals
        (i) $(-1, 0)$, (ii) $(0, 2)$, (iii) $(2, \infty)$

(b) $f = x/(1+x^2)$ and (a, b) equals
  (i) $(-\infty, -1)$, (ii) $(-2, 0)$, (iii) $(0, 2)$
(c)* $f = 2xc - x^2$, $(a, b) = (-1, 1)$ and c equals
  (i) 3/2, (ii) 1, (iii) $-1/2$, (iv) $-3/2$
(d) $f = (x+c)/(1+(x+c)^2)$, $(a, b) = (-1, 1)$ and c equals
  (i) $-5/2$, (ii) -1, (iii) 1/2, (iv) 1.

3.3 The Two-Variable Case: A Single Inequality Constraint

First let us consider the problem where we

$$\text{maximise} \quad f(x_1, x_2) \tag{3.3a}$$
$$\text{subject to} \quad g(x_1, x_2) \leqslant 0 \tag{3.3b}$$

The points (x_1, x_2) which satisfy

$$g(x_1, x_2) = 0$$

usually define a curve in the x_1, x_2 plane which acts as a boundary[2] between a region where $g > 0$ at all points and a region where $g < 0$ at all points (Fig. 3.4a). To determine whether

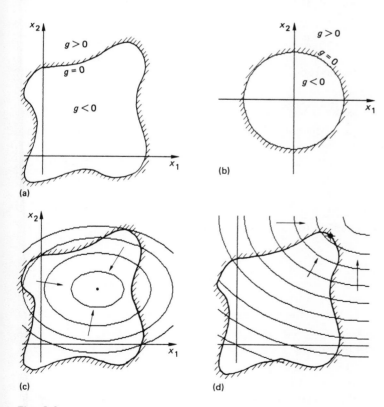

Fig. 3.4

60 Inequality constraints

a particular region corresponds to $g > 0$ or $g < 0$ all we need to do is determine the sign of g at *any* point in that region, since this sign must be common to all its points.

For example, if

$$g(x_1, x_2) = x_1^2 + x_2^2 - 1 \tag{3.4}$$

then $g = 0$ is the unit circle centred at the origin (Fig. 3.4b). Inside the circle g is negative since at the origin $g = -1$. Outside the circle g is positive since, for example, $g = 3$ at $(2, 0)$.

The inequality constraint (equation (3.3b)) restricts us to a region called the *feasible region*, consisting of those points satisfying $g < 0$ (forming the *interior* of the feasible region) and those points satisfying $g = 0$ (forming the *boundary* of the feasible region). If the optimum occurs in the interior (Fig. 3.4c) then it occurs at an unconstrained stationary point with the tangent plane at that point horizontal to the surface $z = f(x_1, x_2)$, i.e.

$$\partial f/\partial x_1 = \partial f/\partial x_2 = 0 \tag{3.5}$$

or at a point at infinity.

If it occurs on the boundary (in which case the constraint is said to be *binding*), then the point must be either a constrained stationary point (Fig. 3.4d), i.e.

$$\partial L/\partial x_1 = \partial L/\partial x_2 = 0; \quad g(x_1, x_2) = 0 \tag{3.6}$$

where $L = f - \lambda g$, or lie at infinity.

To solve our inequality constrained problem therefore we need to search through all unconstrained and constrained stationary points (and, where necessary, points at infinity) to locate the maximum of f. Let us see how this search procedure works out in the following example.

Example 4

The inequality constraint for the problem

$$\text{maximise} \quad x_1 - x_2^2 - 1$$
$$\text{subject to} \quad x_1^2 + x_2^2 \leq 1 \quad \text{(i.e. } g = x_1^2 + x_2^2 - 1 \leq 0\text{)}$$

restricts us to points on or inside the unit circle centred at the origin—hence there are no points at infinity to consider.

(a) The Interior.
Since

$$\partial f/\partial x_1 = 1 \neq 0; \quad \partial f/\partial x_2 = -2x_2 = 0$$

there are no unconstrained stationary points.

(b) The Boundary.

$$L = f - \lambda g = x_1 - x_2^2 - 1 - \lambda(x_1^2 + x_2^2 - 1).$$

Hence $\partial L/\partial x_1 = 1 - 2\lambda x_1 = 0;\quad \partial L/\partial x_2 = -2x_2 - 2\lambda x_2 = 0;\quad x_1^2 + x_2^2 = 1$.

The second equation yields $x_2 = 0$ or $\lambda = -1$. In the former case the other equations give $x_1 = \pm 1$ and $\lambda = \pm 1/2$. In the latter case $x_1 = -1/2$ and $x_2 = \pm\sqrt{3/2}$. There are therefore four constrained stationary points: $(1, 0)$, $(-1, 0)$, $(-1/2, \sqrt{3/2})$, $(-1/2, -\sqrt{3/2})$, with multipliers $1/2, -1/2, -1, -1$ and f values $0, -2, -9/4, -9/4$ respectively.

The maximum value of the objective function over the feasible region (i.e. where $g \leq 0$) is therefore zero, and occurs at $(1, 0)$, agreeing with the contour map shown in Fig. 3.5a. At this optimum the constraint is binding with $x_1^2 + x_2^2 = 1$.

3.3 The two-variable case: a single inequality constraint 61

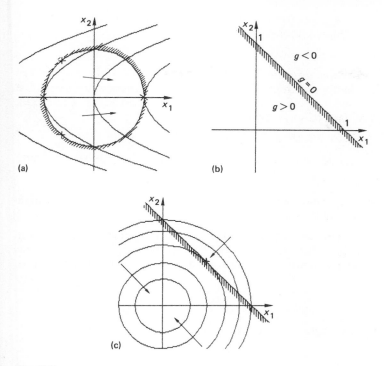

Fig. 3.5

We can simplify the solution procedure a little by introducing the Lagrangean right from the start. We use the observation that equations (3.5) are just the first two equations of (3.6) with $\lambda = 0$. The equations to be solved are therefore:

$$\partial L/\partial x_1 = 0; \quad \partial L/\partial x_2 = 0$$

and $\quad \lambda = 0$ or $g(x_1, x_2) = 0$ \hfill (3.7)

Let us use this formulation in solving the next problem.

Example 5
For the problem:

maximise $\quad -(x_1^2 + x_2^2)$
subject to $\quad 1 \leq x_1 + x_2 \quad$ (i.e. $g = 1 - x_1 - x_2 \leq 0$)

the boundary, $g = 0$, is a line through the points (1, 0) and (0, 1) (Fig. 3.5b) with the region $g < 0$ lying above and to the right of this line (since $g(0, 0) = 1$ and $g(1, 1) = -1$) and stretching to infinity.

The Lagrangean is given by:

$$L = -(x_1^2 + x_2^2) - \lambda(1 - x_1 - x_2)$$

with Lagrange equations:

$$\partial L/\partial x_1 = -2x_1 + \lambda = 0; \quad \partial L/\partial x_2 = -2x_2 + \lambda = 0 \hfill (3.8)$$

62   Inequality constraints

(a) The Interior.

With $\lambda = 0$ these equations reduce to

$$x_1 = 0, \quad x_2 = 0,$$

and hence there is just one unconstrained stationary point located at the origin. This happens to lie outside the feasible region and hence is of no interest to us.

(b) The Boundary.

On the boundary we have to solve three equations:

$$-2x_1 + \lambda = 0, \quad -2x_2 + \lambda = 0 \quad \text{and} \quad x_1 + x_2 = 1.$$

They have in fact only one solution: $x_1 = x_2 = 1/2$ and $\lambda = 1$, at which point $f = -1/2$.

(c) Infinity.

At all points at infinity, in particular those lying in the feasible region, the objective function clearly takes the value $-\infty$.

Comparing the values in (b) and (c) we see that the maximum value over the feasible region is $-1/2$ and is taken on the boundary at the point $(1/2, 1/2)$, agreeing with the contour map shown in Fig. 3.5c. The unconstrained maximum is zero at the origin, but is unattainable.

Exercise Set 2

Solve problem (3.3) using the Lagrange method when:

(i)* $f = x_1 - 2x_2$, $g = x_1^2 + x_2^2 - 1$

(ii) $f = -(x_1 + x_2)$, $g = x_2^2 - 1 - x_1$

(iii)* $f = -(x_1 - c)^2 - x_2^2$, $g = -x_1$ and $c = -1, 1$

(iv) $f = -(x_1 - c)^2 - x_2^2$, $g = x_1^2 + x_2^2 - 1$ and $c = -2, -1/2$

(v)* $f = -(x_1 + x_2/2 + x_2^2/2)$, $g = 2 - x_1 - x_2$.

In each case check your solutions by sketching a contour map.

3.4   The Two-Variable Case: Several Inequality Constraints

For the problem:

$$\text{maximise} \quad f(x_1, x_2) \tag{3.9a}$$
$$\text{subject to} \quad g_1(x_1, x_2) \leq 0 \tag{3.9b}$$
$$g_2(x_1, x_2) \leq 0 \tag{3.9c}$$

we can only reach those points (x_1, x_2) that satisfy *both* constraints (3.9b) and (3.9c). For example if $g_1 = x_2 - 1$ and $g_2 = x_1^2 - 1 - x_2$ then we must not consider points above the horizontal line $x_2 = 1$ (Fig. 3.6a) nor below the parabola $x_1^2 - 1 = x_2$ (Fig. 3.6b). We are therefore restricted to points in the feasible region shown in Fig. 3.6c. The maximum value of problem (3.9) could occur internally (an unconstrained stationary point), or on one of the boundary sections (a constrained stationary point), or at an intersection of the boundaries (a vertex).

The simplest way to carry out the search amongst these possibilities is to start with the Lagrangean which includes both constraint functions g_1 and g_2. We isolate internal points by setting both multipliers zero and each boundary section by setting the multiplier of the other constraint zero. To see how this works in practice let us consider the following example.

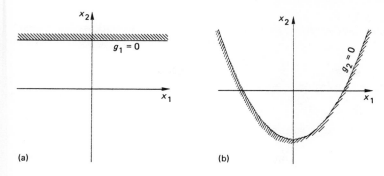

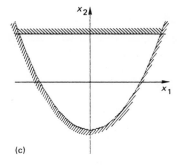

Fig. 3.6

Example 6
The 'Master' Lagrangean for the problem

   maximise $x_1^2 - x_2^2$
   subject to $x_2 \leq 1$ (i.e. $g_1 = x_2 - 1 \leq 0$)
   $x_1^2 \leq 1 + x_2$ (i.e. $g_2 = x_1^2 - 1 - x_2 \leq 0$)

is given by:
$$L = x_1^2 - x_2^2 - \lambda_1(x_2 - 1) - \lambda_2(x_1^2 - 1 - x_2)$$
with:
$$\partial L/\partial x_1 = 2x_1 - 2\lambda_2 x_1 = 0; \quad \partial L/\partial x_2 = -2x_2 - \lambda_1 + \lambda_2 = 0 \tag{3.10}$$

(a) The Interior.
Inside the boundary (Fig. 3.7a) we must have $\lambda_1 = \lambda_2 = 0$ and so equations (3.10) give the unique unconstrained stationary point $(0, 0)$ with $f = 0$.

(b) (i) Boundary Section ACB.
The other (parabolic) boundary section is removed from the Lagrangean by putting $\lambda_2 = 0$. The (constrained) stationary point equations then reduce to:

$$2x_1 = 0; \quad -2x_2 - \lambda_1 = 0; \quad x_2 - 1 = 0,$$

with unique solution $(0, 1)$ with $\lambda_1 = -2$ and $f = -1$.

64 Inequality constraints

(ii) Boundary Section ADB.
The other (linear) boundary section is removed by setting $\lambda_1 = 0$. Equations (3.10) then reduce to:

$$2x_1(1-\lambda_2) = 0; \quad -2x_2 + \lambda_2 = 0; \quad x_1^2 - 1 - x_2 = 0,$$

for which there are three solutions:

$$(0, -1), \ (\sqrt{3/2}, 1/2) \ \text{and} \ (-\sqrt{3/2}, 1/2)$$

with multipliers $-2, 1, 1$ and f values $-1, 5/4, 5/4$ respectively.

(c) Vertices A and B.
The coordinates of the vertices A and B are obtained by solving the simultaneous boundary section equations:

$$x_2 = 1; \quad x_1^2 - 1 = x_2.$$

At both vertices A: $(-\sqrt{2}, 1)$ and B: $(\sqrt{2}, 1)$ f takes the value one.

Since the constraint region does not include points at infinity we conclude, by comparing the values of f at its various stationary points and vertices, that the maximum value of f over the feasible region is 5/4 and is taken at the two points $(\sqrt{3/2}, 1/2)$ and $(-\sqrt{3/2}, 1/2)$ on the parabolic section of the boundary (Fig. 3.7b). The second constraint is therefore binding at the optimum. Our search technique generalises in an obvious fashion when there are more than two inequality constraints, as we shall see in the next example.

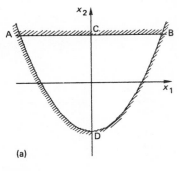
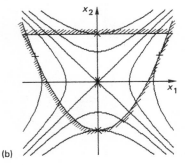

(a)   (b)

Fig. 3.7

Example 7

$$\begin{aligned} \text{maximise} \quad & -(x_1/2 + x_2 + x_2^2/2) \\ \text{subject to} \quad & x_1 + x_2 \geq 1 \\ & 0 \leq x_1 \leq 3/4 \\ & 0 \leq x_2 \leq 3/4. \end{aligned}$$

We first note in sketching the feasible region (Fig. 3.8a) that two of the constraints, $x_1 \geq 0$, $x_2 \geq 0$, are redundant in that they are automatically satisfied if the other constraints are satisfied. We can therefore exclude them from the discussion, reducing the number of constraints to three.

The first step in the solution procedure is to form the Master Lagrangean by including all relevant constraints with their multipliers:

$$L = -(x_1/2 + x_2 + x_2^2/2) - \lambda_1(1 - x_1 - x_2) - \lambda_2(x_1 - 3/4) - \lambda_3(x_2 - 3/4).$$

3.4 The two-variable case: several inequality constraints

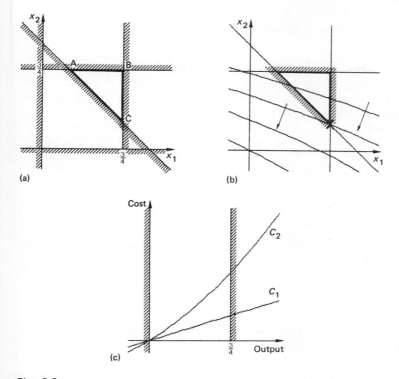

Fig. 3.8

The Lagrange equations have the form:

$$\partial L/\partial x_1 = -1/2 + \lambda_1 - \lambda_2 = 0; \quad \partial L/\partial x_2 = -1 - x_2 + \lambda_1 - \lambda_3 = 0 \qquad (3.11)$$

We now begin our systematic search of the interior of the feasible region, the boundary sections and the vertices.

(a) The Interior. ($\lambda_1 = \lambda_2 = \lambda_3 = 0$)
Equations (3.11) reduce, in this case, to the insoluble set:
$$-1/2 = 0; \quad -1 - x_2 = 0.$$

(b) (i) Boundary Section AC. ($\lambda_2 = \lambda_3 = 0$)
The equations (3.11) together with the constraint equation yield
$$-1/2 + \lambda_1 = 0; \quad -1 - x_2 + \lambda_1 = 0; \quad x_1 + x_2 = 1,$$
with solution: $\lambda_1 = 1/2, \quad x_1 = 3/2, \quad x_2 = -1/2$.
This is unacceptable since it lies outside the feasible region.
(ii) Boundary Section BC. ($\lambda_1 = \lambda_3 = 0$)
The corresponding equations:
$$-1/2 - \lambda_2 = 0; \quad -1 - x_2 = 0; \quad x_1 = 3/4$$
have the unacceptable solution:
$$\lambda_2 = -1/2, \quad x_1 = 3/4, \quad x_2 = -1.$$

66 Inequality constraints

(iii) Boundary Section AB. ($\lambda_1 = \lambda_2 = 0$)
In this case equations (3.11) have no solution.
(c) At the vertices we obtain:

$$f(1/4, 3/4) = -37/32; \quad f(3/4, 3/4) = -45/32; \quad f(3/4, 1/4) = -21/32.$$

We conclude that since there are no stationary points and no points at infinity in the feasible region the maximum is attained at a vertex—the point C:(3/4, 1/4) in Fig. 3.8a. This is consistent with the contour map shown in Fig. 3.8b.

Note that this is an example of the Power Allocation Problem mentioned in the introduction to this chapter. From the cost functions shown in Fig. 3.8c it is clear that the first power station is always cheaper to run and is therefore run at full capacity (0.75 megawatts). The second station is used to generate the remaining demand (0.25 megawatts).

Exercise Set 3
Use the Lagrange method to solve the inequality constrained maximisation problems given in the following table.

	Objective function	Constraints		
	f	g_1	g_2	g_3
(i)*	$x_1^2 + x_2^2$	$x_2 - 1$	$x_1^2 - 1 - x_2$	
(ii)	$x_1 - x_2$	$x_2 - 1$	$x_1^2 - 1 - x_2$	
(iii)*	$x_1 x_2$	$x_1^2 + x_2^2 - 2$	$1 - x_1$	
(iv)	$-(x_1^2 + x_2^2)$	$1 - x_1 - x_2$	$x_1 - 1 - x_2$	
(v)*	$(x_1 - 1/2)^2 + (x_2 - 1/2)^2$	$2 - x_1 - x_2$	$x_1 - 2$	$x_2 - 2$
(vi)	$-(x_1 - 1/2)^2 - (x_2 - 1/2)^2$	$2 - x_1 - x_2$	$x_1 - 2$	$x_2 - 2$
(vii)*	$2x_1 + x_2$	$1 - x_1$	$x_1 - 1 - x_2$	$x_2 - 1$
(viii)	$2x_1 - x_2$	$1 - x_1$	$x_1 - 1 - x_2$	$x_2 - 1$

In each case sketch the contour map.

3.5 The Sign of the Lagrange Multiplier

If we look back over the maximisation problems discussed in the last two sections we shall see that no multiplier has been negative at an optimal constrained stationary point. In Example 4, for instance, the optimal multiplier was 1/2 at (1, 0) and in Example 6 the multiplier was 1 at ($\pm\sqrt{3/2}, 1/2$).

This property, in fact, holds generally, and in non-singular situations has a simple geometric explanation. As we shall argue later in this section, a negative multiplier implies that as we move away from the constrained stationary point into the interior of the feasible region we *climb* the surface $z = f(x_1, x_2)$. Hence there are points in the interior where the function f takes values greater than at the stationary point. This can be seen to be the case at the point $(-1, 0)$ in Example 4 where the multiplier is $-1/2$. This point corresponds to a local maximum with respect to movements along the boundary, but a local minimum with respect to movements off the boundary (Fig. 3.5a).

We can generalise this negativity property to test for optimality at vertices. The position of a vertex is determined by solving the constraint equations:

$$g_1 = 0; \quad g_2 = 0.$$

If we solve, in addition, the Lagrange equations:

$$\partial L/\partial x_1 = 0; \quad \partial L/\partial x_2 = 0$$

for the multipliers and find either or both of them to be negative then the overall maximum cannot be taken at this vertex. As an illustration let us look at Example 6. At vertex A with coordinates $(-\sqrt{2}, 1)$, the Lagrange equations:

$$\partial L/\partial x_1 = 2x_1(1 - \lambda_2) = 0; \quad \partial L/\partial x_2 = -2x_2 - \lambda_1 + \lambda_2 = 0$$

yield $\lambda_1 = -1$ and $\lambda_2 = 1$. Similarly at vertex B: $(\sqrt{2}, 1)$ the multipliers are $\lambda_1 = -1$ and $\lambda_2 = 1$. Neither of these vertices corresponds to the maximum which occurs at $(\pm\sqrt{3/2}, 1/2)$.

To justify these properties in non-singular cases we give the following geometric arguments.

(a) Boundary Sections[3]

We first note that the gradient vectors

$$\nabla f \equiv (\partial f/\partial x_1, \partial f/\partial x_2); \quad \nabla g \equiv (\partial g/\partial x_1, \partial g/\partial x_2)$$

at the constrained stationary point P on the boundary section $g = 0$ are collinear, both being perpendicular to the common tangent to the contour and boundary section at that point (Fig. 3.9a). (See question 6 of Problem Set 1 for the proof.) These vectors are therefore scalar multiples of each other, the scalar multiple being in fact the Lagrange multiplier:

$$\nabla f = \lambda \nabla g$$

since this is just the Lagrange equations in vector form. If λ is positive then the vectors point in the same direction and if negative in opposite directions. But ∇g points into the 'forbidden' region $g > 0$ (shown shaded in Fig. 3.9a) since it gives the direction of steepest ascent on the surface $z = g(x_1, x_2)$. If λ is negative, therefore, ∇f must point into the feasible region $g < 0$ (Fig. 3.9b). Since ∇f indicates the direction of steepest ascent on the surface $z = f$, there must be internal points with greater f value than at the constrained stationary point P. This point, therefore, cannot provide the solution to the problem. If, however, λ is positive it still may not be that P is an optimal point since:

(i) it might be possible to increase the value of f by moving along the boundary (Fig. 3.9c),
(ii) it could be that P is a local maximum, surpassed in value by another local maximum elsewhere (Fig. 3.9d).

(N.B. Remember that the constraints must be in the form $g \leq 0$ with $L = f - \lambda g$ for this analysis to apply.)

(b) Vertices[4]

If the contour through vertex C (Fig. 3.10a) passes into the feasible region then that vertex cannot yield the optimal value for the problem since we can do better by moving away from the vertex. If the contour does not enter the feasible region about C then ∇f must point away from the feasible region, otherwise we can do better by moving away from C

68 *Inequality constraints*

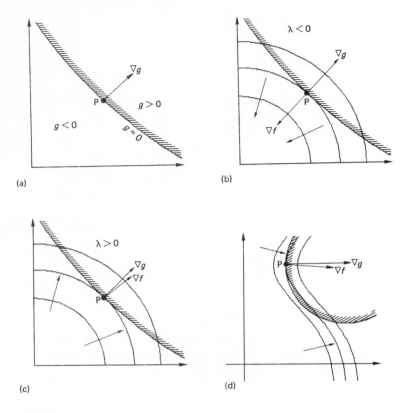

Fig. 3.9

(Fig. 3.10b). Therefore *only* the positioning of the gradient vectors shown in Fig. 3.10c can yield an optimum. The condition for this is that ∇f can be written as a linear combination of vectors ∇g_1, ∇g_2 with non-negative scalar weights, i.e.

$$\nabla f = \lambda_1 \nabla g_1 + \lambda_2 \nabla g_2, \quad \lambda_1, \lambda_2 \geq 0,$$

or, taking components and rearranging,

$$\partial f/\partial x_1 - \lambda_1 \partial g_1/\partial x_1 - \lambda_2 \partial g_2/\partial x_1 = 0$$
$$\partial f/\partial x_2 - \lambda_1 \partial g_1/\partial x_1 - \lambda_2 \partial g_2/\partial x_2 = 0.$$

But these are just the Lagrange equations $\partial L/\partial x_1 = \partial L/\partial x_2 = 0$ with non-negative multipliers. The conclusions we have reached in (a) and (b) in fact hold in all situations and lead to the rule that we should reject a constrained stationary point if any of its multipliers are negative.

If we follow through these arguments for a minimisation problem we can show that the equivalent rule is that we reject boundary or vertex solutions with positive multipliers. However, in order to keep to one sign convention we shall always rewrite a minimisation problem as a maximisation problem.

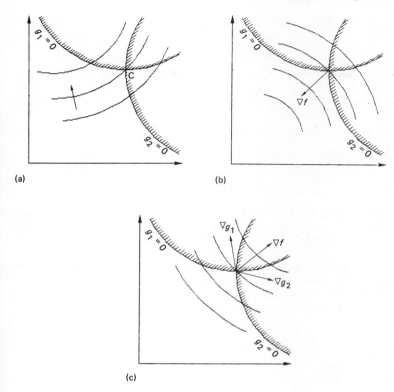

Fig. 3.10

3.6 The Complementary Slackness Conditions

The Master Lagrangean technique and the observations concerning the sign of the Lagrange multipliers generalise quite straightforwardly to the case of n variables and r inequality constraints, i.e. to the problem:

$$\begin{aligned}
\text{maximise} \quad & f(x_1, x_2, \ldots, x_n) \\
\text{subject to} \quad & g_1(x_1, x_2, \ldots, x_n) \leq 0 \\
& g_2(x_1, x_2, \ldots, x_n) \leq 0 \\
& \quad \cdots \cdots \cdots \\
& g_r(x_1, x_2, \ldots, x_n) \leq 0.
\end{aligned} \qquad (3.12)$$

The algorithm for solving this problem is listed in Procedure Box 3.2.

PROCEDURE 3.2
To solve Problem (3.12)

1 **Construct the Master Lagrangean:**

$$L = f - \sum_{j=1}^{r} \lambda_j g_j.$$

2 **Solve the set of simultaneous conditions:**

$$\partial L / \partial x_i = 0 \quad i = 1, \ldots, n \tag{3.13}$$

$$\begin{Bmatrix} \lambda_j = 0 \\ g_j < 0 \end{Bmatrix} \quad \text{or} \quad \begin{Bmatrix} g_j = 0 \\ \lambda_j \geqslant 0 \end{Bmatrix} \quad j = 1, \ldots, r \tag{3.14}$$

3 Evaluate f at all solution points of step 2 and, where appropriate, at infinity.
4 The optimal value is the largest of the values obtained in step 3.

The search through the various possible locations for the optimal points has been succinctly summarised in equations (3.14). For each constraint either we are on that boundary section or we are not; hence either $g_j = 0$ or $\lambda_j = 0$. Conditions (3.14) are termed the *Complementary Slackness Conditions* and can be written more symmetrically as:

$$\begin{Bmatrix} \lambda_j = 0 \\ \partial L / \partial \lambda_j > 0 \end{Bmatrix} \quad \text{or} \quad \begin{Bmatrix} \partial L / \partial \lambda_j = 0 \\ \lambda_j \geqslant 0 \end{Bmatrix} \quad j = 1, \ldots, r \tag{3.15}$$

with the obvious identification $g_j = - \partial L / \partial \lambda_j$.

A more condensed form of these conditions is given by:

$$\lambda_j \geqslant 0, \quad \partial L / \partial \lambda_j \geqslant 0, \quad \lambda_j \partial L / \partial \lambda_j = 0 \quad \text{for } j = 1, \ldots, r$$

stating that λ_j and $\partial L / \partial \lambda_j$ are never negative and at least one of them is zero.

To gain some understanding of the significance of these conditions let us apply them to an important type of optimisation problem, the so-called *linear program*, where the objective function and all constraint functions are linear in the variables. We met an example of such a program in Exercise Set 3 (part vii):

$$\begin{aligned} \text{maximise} \quad & 2x_1 + x_2 \\ \text{subject to} \quad & -x_1 \leqslant -1 \\ & x_2 \leqslant 1 \\ & x_1 - x_2 \leqslant 1. \end{aligned} \tag{3.16}$$

The general linear program[5] in n variables and with r constraints can be written in the form:

$$\begin{aligned} \text{maximise} \quad & \sum_{i=1}^{n} c_i x_i \\ \text{subject to} \quad & \sum_{i=1}^{n} a_{ji} x_i \leqslant b_j \quad j = 1, \ldots, r \end{aligned} \tag{3.17}$$

with Lagrangean

$$L = \sum_i c_i x_i - \sum_j \lambda_j \left(\sum_i a_{ji} x_i - b_j \right)$$

$$= \sum_i \left(c_i - \sum_j a_{ji} \lambda_j \right) x_i + \sum_j \lambda_j b_j.$$

3.6 The complementary slackness conditions

The complementary slackness conditions are given in this case by:

$$\left\{\begin{matrix} \lambda_j = 0 \\ \sum_i a_{ji} x_i \leqslant b_j \end{matrix}\right\} \quad \text{or} \quad \left\{\begin{matrix} \sum_i a_{ji} x_i = b_j \\ \lambda_j \geqslant 0 \end{matrix}\right\} \quad j = 1, \ldots, r \tag{3.18}$$

where the multipliers satisfy the equations:

$$\partial L / \partial x_i = c_i - \sum_j \lambda_j a_{ji} = 0 \quad i = 1, \ldots, n \tag{3.19}$$

One important property of a linear program with a bounded feasible region (i.e. one not containing points at infinity) is that the optimal value is taken at a vertex of that region. This was the case in (3.16) and can be established generally from the equations (3.18) and (3.19). The proof is straightforward when there is no linear dependency[6] in the equation sets and can be achieved in the following steps.

(a) Suppose that for a finite solution of (3.17) p of the r constraints are binding, i.e. the equalities hold. Now p must be less than or equal to n (if there is no linear dependence between the constraint equations) since, in general, there are no solutions to a linear set of equations with more equations than variables.
(b) From the complementary slackness condition (3.18) we can deduce that at least $r - p$ of the multipliers must be zero. Equations (3.19) reduce, therefore, to n equations for at most $p \leqslant n$ multipliers. But if there is no linear dependence there will be no solution unless $p \geqslant n$—hence $p = n$.
(c) If the n binding constraints are themselves linearly independent they intersect in a unique point defining a vertex on the boundary of the feasible region.

(In the Problem Set we examine what can happen when there is dependency.)

This result means that all we need to do in order to find an optimal point for a linear program with a bounded feasible region is to evaluate the objective function at the vertices of the feasible region and identify as optimal the one(s) with maximum value.

The number of vertices increases rapidly with the size of the problem so that an efficient method is required for searching through the vertices. One such method is based on a second important property of linear programs concerning neighbouring vertices on the boundary of the feasible region. Two vertices P and Q are said to be neighbouring if $n - 1$ of the n boundary constraints binding at P remain binding at Q. In the case $n = 2$, shown in Fig. 3.11a, points P and R are both neighbours of Q as they each share a binding constraint with Q. The second property states that if at a vertex P the objective function is

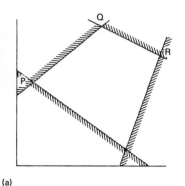
(a)

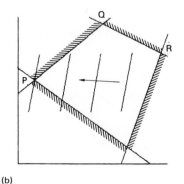
(b)

Fig. 3.11

72 Inequality constraints

not less than at any neighbouring vertex then P is an optimal point. This property is clear from a contour map in two dimensions (Fig. 3.11b), but generalises to higher dimensions.

This second property suggests the following method of search: we follow a path from vertex to vertex along the boundary edges, moving at each step to a neighbouring vertex with larger objective function value until we reach a vertex where there are no neighbouring vertices with larger objective function. This final point must then be an optimal point.

This is the essence of the so-called Simplex Algorithm proposed by Dantzig in 1947 and discussed at length in Chapter 5.

3.7 An Application: The Coal Transportation Problem

After the oil price shock of the 1970s increasing interest is being shown in alternative sources of energy, in particular coal. This is a dirty fuel to handle, and without expensive precautions is a heavy pollutant, but it is available in abundant supply. The problem is that it is not always found in places where it is needed and hence has to be transported across the globe from mine to final user.

There are three principal regions with export surplus, the USA, South Africa and Australia, and two major regions which cannot meet their own needs—Europe and the Far East.

We can represent the problem of transporting coal between these regions on the graph of Fig. 3.12a. There are three nodes representing producers and two nodes representing the consumers. The arcs between the two sets of nodes represent the transportation routes between producing and consuming regions. To each arc we associate the unit cost for

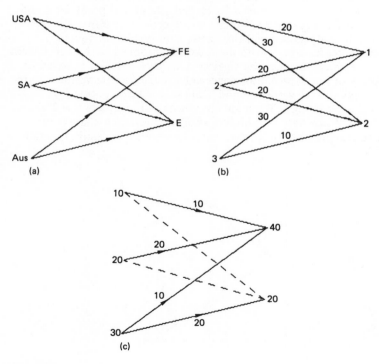

Fig. 3.12

3.7 An application: the coal transportation problem

transporting coal along that route (Fig. 3.12b). For coal from the USA to the Far East, for example, this cost is $20 per tonne.

If the flow of coal from the ith producer to the jth consumer is denoted by x_{ij}, and the unit cost for that route by c_{ij}, then the total transportation cost is given by:

$$\sum_{i=1}^{3} \sum_{j=1}^{2} c_{ij} x_{ij}.$$

Using the data from Fig. 3.12b, this is:

$$(20x_{11} + 30x_{12}) + (20x_{21} + 20x_{22}) + (30x_{31} + 10x_{32}) \text{ in \$US}.$$

The object of the analysis is to minimise this quantity subject to the following constraints.

(i) The flow x_{ij} cannot be negative, i.e.

$$x_{ij} \geq 0 \quad \text{for all values of } i \text{ and } j \tag{3.20}$$

(ii) The ith producer cannot supply more than the amount s_i available for export, i.e.

$$\sum_j x_{ij} \leq s_i \quad \text{for all } i.$$

Writing these out in full for the three producers, we have:

$$\begin{aligned} x_{11} + x_{12} &\leq s_1 & \text{for the USA} \\ x_{21} + x_{22} &\leq s_2 & \text{for South Africa} \\ x_{31} + x_{32} &\leq s_3 & \text{for Australia} \end{aligned} \tag{3.21}$$

(iii) The consumer demands d_j are to be met, so that

$$\sum_i x_{ij} \geq d_j \quad \text{for each } j$$

i.e.
$$\left. \begin{aligned} x_{11} + x_{21} + x_{31} &\geq d_1 & \text{for Europe} \\ x_{12} + x_{22} + x_{32} &\geq d_2 & \text{for the Far East} \end{aligned} \right\} \tag{3.22}$$

This problem has a linear objective function and linear constraints, and hence is a linear program. The only new feature is that the variables have two indices rather than one. This is for convenience only; a simple relabelling of the variables can reduce the number of indices to one.

Let us suppose that

$$(s_1, s_2, s_3) = (10, 20, 30) \quad \text{and} \quad (d_1, d_2) = (40, 20)$$

in units of a million tonnes of coal. Since in this case total supply equals total demand all the inequalities in equations (3.21) and (3.22) must be equalities.[7] It follows that one of the constraints must be redundant in the sense that it is implied by the others. This can be seen by adding the supply equations in (3.21) and subtracting the first demand equation in (3.22). The remainder is identical to the second demand equation.

In our linear program, therefore, we have in total 10 independent constraints, six of which must be binding at a vertex of the feasible region since there are six variables in the problem. The supply and demand constraints provide 4 of these equalities, the remaining two coming from the non-negativity constraints. Hence, in the absence of degeneracies, two of the flow variables are zero and four non-zero at any vertex. The flow depicted in Fig. 3.12c with:

$$(x_{11}, x_{12}, x_{21}, x_{22}, x_{31}, x_{32}) = (10, 0, 20, 0, 10, 20)$$

therefore corresponds to a vertex—which happens to be optimal.

74 Inequality constraints

Generalising this argument to the case in which there are n producers and m consumers we can show that $(m-1)(n-1)$ of the flow variables are zero and hence there is no trade between this number of pairs of producers and consumers in the optimal solution.

3.8 The Kuhn–Tucker Conditions

In many optimisation problems of interest n of the inequality constraints are non-negativity conditions of the form:

$$x_1 \geq 0, x_2 \geq 0, \ldots, x_n \geq 0.$$

For such a problem:

$$\begin{aligned}
\text{maximise} \quad & f(x_1, x_2, \ldots, x_n) \\
\text{subject to} \quad & g_1(x_1, x_2, \ldots, x_n) \leq 0 \\
& \quad \cdots \cdots \cdots \\
& g_m(x_1, x_2, \ldots, x_n) \leq 0 \\
& x_1 \geq 0, x_2 \geq 0, \ldots, x_n \geq 0
\end{aligned} \quad (3.23)$$

the Lagrangean can be written as:

$$L = f - \sum_{j=1}^{m} \lambda_j g_j + \sum_{i=1}^{n} \mu_i x_i = L' + \sum_{i=1}^{n} \mu_i x_i,$$

where μ_i are the multipliers for the non-negativity conditions and L' is the Lagrangean for the other constraints. (In the notation of Section 3.6, we have $m + n = n$.) The stationary point equations (3.13) become:

$$\partial L / \partial x_i = \partial L' / \partial x_i + \mu_i = 0.$$

If $x_i = 0$ then $\mu_i \geq 0$ and hence $\partial L' / \partial x_i \leq 0$.
If $x_i > 0$ then $\mu_i = 0$ and hence $\partial L' / \partial x_i = 0$.

With the further observation that $\partial L / \partial \lambda_j = \partial L' / \partial \lambda_j$, we can rewrite conditions (3.13) and (3.15) in terms of the 'reduced' Lagrangean L' as:

$$\left\{ \begin{array}{c} x_i = 0 \\ \partial L'/\partial x_i \leq 0 \end{array} \right\} \quad \text{or} \quad \left\{ \begin{array}{c} \partial L'/\partial x_i = 0 \\ x_i > 0 \end{array} \right\} \quad i = 1, \ldots, n \quad (3.24)$$

and

$$\left\{ \begin{array}{c} \lambda_j = 0 \\ \partial L'/\partial \lambda_j > 0 \end{array} \right\} \quad \text{or} \quad \left\{ \begin{array}{c} \partial L'/\partial \lambda_j = 0 \\ \lambda_j \geq 0 \end{array} \right\} \quad j = 1, \ldots, m \quad (3.25)$$

The relationships (3.24) have a simple geometric interpretation. If the optimal point is on the boundary at $x_i = 0$ then $\partial L'/\partial x_i$ need not be zero provided it is not positive (Fig. 3.13a). If, however, $x_i > 0$ then for optimality it must be that $\partial L'/\partial x_i = 0$ (Fig. 3.13b). The two relationships (3.24) and (3.25) are called the *Kuhn–Tucker conditions* after their originators. An alternative, and more compact form for these conditions is:

$$x_i \geq 0, \ \partial L'/\partial x_i \leq 0 \quad \text{and} \quad x_i \partial L'/\partial x_i = 0, \quad i = 1, \ldots, n \quad (3.24a)$$

and

$$\lambda_j \geq 0, \ \partial L'/\partial \lambda_j \geq 0 \quad \text{and} \quad \lambda_j \partial L'/\partial \lambda_j = 0, \quad j = 1, \ldots, m \quad (3.25a)$$

These conditions show an almost complete symmetry between the x variables and multipliers, the only asymmetry appearing in the inequalities on the partial derivatives of L'. This structure will be elaborated upon in the next chapter.

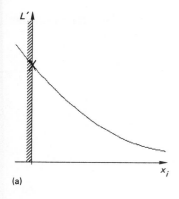 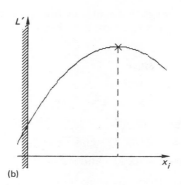

Fig. 3.13

To gain some insight into these conditions let us apply them to the Coal Transportation Problem with n producers and m consumers. The reduced Lagrangean in this case is given by:

$$L' = -\sum_i \sum_j c_{ij} x_{ij} - \sum_i \lambda_i \left(\sum_j x_{ij} - s_i \right) - \sum_j v_j \left(d_j - \sum_i x_{ij} \right)$$

where λ_i and v_j denote two sets of Lagrange multipliers, one set for the supply constraints and the other for the demand constraints. The Kuhn–Tucker conditions have the form:

(i) $\quad \left\{ \begin{array}{c} x_{ij} = 0 \\ v_j \leqslant \lambda_i + c_{ij} \end{array} \right\}$ or $\left\{ \begin{array}{c} v_j = \lambda_i + c_{ij} \\ x_{ij} > 0 \end{array} \right\}$ (3.24b)

and

(ii) $\quad \left\{ \begin{array}{c} \lambda_i = 0 \\ \sum_j x_{ij} < s_i \end{array} \right\}$ or $\left\{ \begin{array}{c} \sum_j x_{ij} = s_i \\ \lambda_i \geqslant 0 \end{array} \right\}$ (3.25b)

and

(iii) $\quad \left\{ \begin{array}{c} v_j = 0 \\ d_j < \sum_i x_{ij} \end{array} \right\}$ or $\left\{ \begin{array}{c} d_j = \sum_i x_{ij} \\ v_j \geqslant 0 \end{array} \right\}$ (3.25c)

These rather complicated-looking expressions have a simple economic interpretation if we suppose that λ_i is the price paid to the ith producer and v_j is the price paid by the jth consumer. Let us look at each set of conditions in turn.

(i) There is flow of coal from i to j (i.e. $x_{ij} \neq 0$) if the price paid by the consumer less the transportation cost is equal to the price demanded by the producer. If it is less then the producer is not willing to supply and hence $x_{ij} = 0$.

(ii) The export price λ_i will normally remain at a positive value if there is sufficient demand to buy up all the production s_i. If not $\left(\text{i.e. } \sum_j x_{ij} < s_i \right)$ then the price drops to zero.

(iii) If there is a glut of coal with more coal being shipped to a country than is required then according to Kuhn–Tucker the import price will drop to zero.

The last two sets of conditions show that the standard transportation problem does not take into account the fact that supply and demand usually depend on price. If, for example, the price drops demand is likely to go up and supply down. Inclusion of these feedback

76 Inequality constraints

effects yields the so-called generalised transportation problem[8] which in its simplest form is quadratic in its objective function and linear in its constraints.

Problem Set 3

1 Solve the problem (3.2) when:
(a)* $f = x^3 - 3x$ and $(a, b) = (-2, -1/2), (-1/2, 1/2), (1/2, 2)$
(b) $f = x^2(x^2 - 1)$ and $(a, b) = (-1/2, 0), (-1/2, 3/4), (0, 3/2)$
(c)* $f = x^3 - 3cx$, $(a, b) = (-1, 1)$ and $c = -1, 1/4, 9/4$
(d) $f = x^2(x^2 - c)$, $(a, b) = (0, 1)$ and $c = -1, 1, 2$.

2 Solve the following problems using Lagrange's method and check your answers by sketching a contour map:
(a)* maximise $x_1 + x_2$, subject to $(x_1 - 1)^2 + x_2^2 \leq 4$, $(x_1 + 1)^2 + x_2^2 \leq 4$
(b) maximise $x_1 + x_2$, subject to $x_2 \geq -1 + x_1^2$, $x_2 \leq 1 - x_1^2$
(c)* maximise $x_1 + x_2$, subject to $x_1^2 \leq x_2^2$, $x_1^2 + x_2^2 \leq 1$
(d) maximise $-(x_1 + x_2^2)$, subject to $(x_1 - x_2)^2 \leq 1$, $x_1^2 + x_2^2 \leq 1$.

3 (a)* Solve the problem

$$\text{maximise} \quad -(x_1 - a)^2/2 - (x_2 - a)^2/2$$
$$\text{subject to} \quad x_1 \leq 2, \; x_2 \leq 2, \; x_1 + x_2 \geq 2$$

when $a = 0, 1, 3$, using a contour map. Check your answer by using Lagrange's method. Determine the optimal value of the objective function as a function of the parameter a.
(b) Repeat part (a) for the problem:

$$\text{maximise} \quad -(x_1 - 2a)^2 - (x_2 + a)^2$$
$$\text{subject to} \quad (x_1 + x_2)^2 \leq 1$$
$$(x_1 - x_2)^2 \leq 1$$

when $a = -1, -1/4, 1$.

4 In this question we look at some of the singularities that can occur at vertices of the feasible region.
For each of the following problems sketch the contour map and then find the Lagrange multipliers at the optimal point.

(i)* maximise $-x_1$
      subject to $x_2 \geq 0$, $x_2 \leq x_1^3/3$.
(ii) maximise x_1
      subject to $x_1^2 + (x_2 - 1)^2 \leq 1$, $x_1^2 + (x_2 + 1)^2 \leq 1$.
(iii)* maximise $-x_1^2 - (x_2 + 1)^2$
      subject to $x_1 \geq 0$, $x_2 \geq 0$, $x_1 \leq x_2$.
(iv) maximise $-(x_1 - 1)^2 - x_2^2$
      subject to $x_1 \leq 0$, $x_1 \leq x_2^2$.
(v)* maximise $x_1 + x_2$
      subject to $x_1^2 + x_2^2 \leq 0$, $x_1 \geq 0$.
(vi) maximise $x_1 + x_2$
      subject to $x_1^2 \leq 0$, $x_1^2 + x_2^2 \leq 1$.

5 (a) Solve the following linear programs by sketching their contour maps:
   (i)*   maximise $x_1 + x_2$
          subject to $x_2 \leq 1, x_1 \leq 1, x_1 + x_2 \geq 1$
   (ii)   maximise x_2
          subject to $x_1 \geq 0, x_2 \geq 0, x_1 + x_2 \leq 1$
   (iii)* maximise $x_1 - 2x_2$
          subject to $2 \geq x_1 + x_2 \geq 1, 1 \leq x_1 - x_2 \leq 2$
   (iv)   maximise $2x_2 - x_1$
          subject to $0 \leq x_2 \leq 1, x_1 + x_2 \geq 1, x_1 - x_2 \leq 2.$

In each case determine the values of the multipliers at the optimal vertices.

(b) In Section 3.6 we referred to some of the possible dependencies that could occur in the Lagrange equations for a linear program. Their geometric significance is illustrated by the following examples.

Repeat part (a) for the problems:
   (i)*   maximise $x_1 - x_2$
          subject to $1 \geq x_1 \geq 0, x_2 \geq 0, x_1 + x_2 \leq 1$
   (ii)   maximise $x_1 + x_2$
          subject to $x_1 \geq 0, x_2 \geq 0, x_1 + x_2 \leq 1$
   (iii)* maximise x_2
          subject to $x_1 + x_2 \leq 1, x_1 + x_2 \geq 1, x_1 \geq 0$
   (iv)   maximise $x_1 + x_2$
          subject to $1 \geq x_1 \geq 0, x_2 \geq 0, x_1 + x_2 \leq 1.$

6 (a) Solve the Chemical Tank Problem:

   maximise $x_1^2 x_2$
   subject to $x_1^2 + 4x_1 x_2 \leq 1, x_1 \geq 0, x_2 \geq 0$

using Lagrange's method. Check your solution by drawing a contour map.
(b) Repeat part (a) with the additional constraint $x_1 \leq 1/2$.
(c) Repeat part (b) with the additional constraint $x_1 + x_2 \leq 3/4$.

7 Some optimisation problems have a mixture of equality and inequality constraints. The only difference in the procedure in this case is that the equality constraints are, by definition, always binding, and so their multipliers are not sign-constrained (cf. Chapters 2 and 3).

(a) Solve the following problems using Lagrange's method:
   (i)* maximise $x_1 + x_2$ subject to $x_1^2 + x_2^2 = 1$ and $x_1 \geq 0$
   (ii) maximise $x_2 + (x_1 - 1)^2$ subject to $x_1 = x_2$ and $x_2^2 \leq 1$.
(b)* Write out the Kuhn–Tucker conditions for the problem:

   maximise $f(x)$
   subject to $g_j(x) = 0$ for $j = 1, \ldots, m$
              $x \geq 0.$

References

[1] See L. A. Johnson and D. C. Montgomery, *Operations Research in Production Planning, Scheduling and Inventory Control*, Wiley (1974).

2 Not always. Consider $(x_1 - x_2)^2$. In exceptional cases it may be that $g = 0$ does not define a curve, e.g. $g = x_1^2 + x_2^2 + 1$.
3 Refer to Section E of the Appendix for a brief introduction to gradient vectors.
4 The singularities that can occur at a vertex are examined in the Problem Set.
5 In Chapter 5 we shall use an alternative but equivalent general form for a linear program.
6 Refer to Section A of the Appendix for an introduction to linear dependence.
7 Summing equations (3.21) and (3.22) we see that $d_1 + d_2 \leq \sum_i \sum_j x_{ij} \leq s_1 + s_2 + s_3$, so that for supply to equal demand it follows that each constraint must be an equality.
8 See P. Samuelson, 'Spatial Price Equilibrium and Linear Programming', *American Economic Review*, XLII (1952), pp. 284–303.

4
Duality

4.1 Introduction

In Chapter 2 we discussed in some detail the Electricity Generating Problem[1]:

$$\text{maximise} \quad -\sum_i C_i(x_i)$$
$$\text{subject to} \quad \sum_i x_i \geq D \tag{4.1}$$

and gave a possible interpretation for its Lagrange multipliers and Lagrangean:

$$L = -\left(\sum_i C_i(x_i) + \lambda\left(D - \sum_i x_i\right)\right).$$

We imagined a scenario in which the electricity company need not generate all the electricity itself but could buy in the extra electricity at price λ per unit. The total cost to the company of meeting demand from its own and outside resources is then $-L$. We can develop the story further by supposing that a second (holding) company has been given the following privileges:

(i) it chooses the price λ;
(ii) it supplies all the additional electricity that is required to meet demand;
(iii) it supplies all the inputs (e.g. fuel, labour) required by the electricity company in generating its electricity.

As far as the holding company is concerned $-L$ is the total income made from its activities, since all the costs of the electricity company are income for the holding company. The objective of this company therefore is to control the price λ so as to maximise its income, $-L$, i.e. to minimise the Lagrangean, L. This is in contrast to the electricity company which adjusts the outputs of its power stations $(x_1, \ldots, x_n)$ to minimise its total costs, $-L$, i.e. maximise the Lagrangean, L. The solution that will satisfy both companies is given by:

$$\min_{\lambda \geq 0} \max_{x_1 \ldots x_n} L(x_1, \ldots, x_n; \lambda)$$

or more compactly by:

$$\min_{\lambda \geq 0} \max_{\mathbf{x}} L(\mathbf{x}; \lambda).$$

Because of its min/max structure this problem is called the *Saddle Point Problem* for the original (primal) problem (4.1). The multiplier is now considered on an equal footing with the x variables, being a decision variable in its own right. This structure was foreshadowed, you may remember, in our discussion in Section 2.5 of Chapter 2. The *peak* in x-space reached its *minimum* position with variation in λ at the optimal value of λ.

Let us continue our discussion a little further by supposing that the holding company has set a price λ (not necessarily optimal) and the electricity company has scheduled its plants to operate at minimum cost with respect to this price. The income to the holding

company in this case is:
$$\min_x (-L(x;\lambda)) = -\max_x L(x;\lambda) \equiv -h(\lambda),$$
a function only of the price λ. In order to determine the *optimal* price the holding company has to solve the one-variable optimisation problem
$$\text{maximise}_{\lambda \geq 0} (-h(\lambda)), \quad \text{i.e. minimise}_{\lambda \geq 0} h(\lambda).$$
This problem is said to be the *dual* of the *primal* problem (4.1).

As a numerical example let us consider again the problem (Fig. 4.1a)
$$\text{maximise} \quad -(x_1/2 + x_1^2 + x_2 + x_2^2/2)$$
$$\text{subject to} \quad x_1 + x_2 \geq 2$$
whose Lagrangean is given by:
$$L = -(x_1/2 + x_1^2 + x_2 + x_2^2/2) - \lambda(2 - x_1 - x_2).$$
To obtain $h(\lambda)$ we have to maximise L with respect to x_1, x_2 for given λ, i.e. we have to solve the equations:
$$\partial L/\partial x_1 = -1/2 - 2x_1 + \lambda = 0$$
$$\partial L/\partial x_2 = -1 - x_2 + \lambda = 0.$$
The solution:
$$x_1 = (\lambda - 1/2)/2, \quad x_2 = \lambda - 1 \tag{4.2}$$
indicates that the electricity company should generate the quantities given by equations (4.2) from its two stations and buy in an amount $2 - x_1 - x_2 = 13/4 - 3\lambda/2$ to make up the demand of two units. The function $h(\lambda)$ can now be obtained as
$$h(\lambda) = \max_x L(x_1, x_2; \lambda) = L((\lambda - 1/2)/2, \lambda - 1; \lambda) = 3\lambda^2/4 - 13\lambda/4 + 9/16.$$
The optimal price, $\lambda = 13/6$, is obtained by maximising $-h(\lambda)$ (Fig. 4.1b). At this price the electricity company is meeting all the demand with the holding company making all its profits from supplying the inputs to the electricity company.

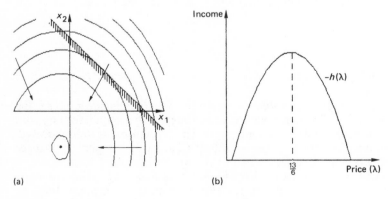

(a) (b)

Fig. 4.1

For the general constrained optimisation problem:

maximise $f(x)$

subject to $g_j(x) \leq 0$ for $j = 1, \ldots, r$ (4.3)

we can always construct a *dual* problem:

$$\underset{\lambda \geq 0}{\text{minimise}} \; h(\lambda)$$

where $h(\lambda) = \underset{x}{\max} \, L(x; \lambda)$ and L is the Lagrangean for the *primal* problem (4.3). By $\lambda \geq 0$ we mean that each multiplier is non-negative. For most cases of interest the optimal values of the primal and dual problems are equal to the optimal value of the *Saddle Point Problem*:

$$\underset{\lambda \geq 0}{\text{minimise}} \; \underset{x}{\text{maximum}} \; L(x; \lambda)$$

defined on the extended set of primal and dual variables. Intuitively the primal and dual problems can be considered as being generated by different cross-sections of the saddle point surface. This structure is summarised in Fig. 4.2.

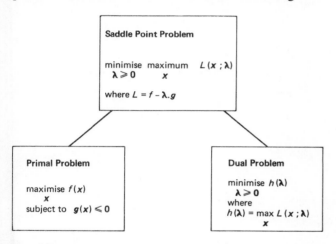

Fig. 4.2

4.2 Some Simple Examples

To gain experience in finding duals we will first look at the simplest non-trivial case where the primal problem involves just one variable and one inequality constraint.

When we solved one-variable primal problems in Chapter 3 we did not bother with Lagrange multipliers because the solution procedure was simpler by more direct methods. We could, however, have used them and we shall certainly use them now.

Example 1

Consider the primal problem:

maximise $-x^2 + 2$

subject to $x \leq -1$

with optimal value 1 on the boundary at $x = -1$ (Fig. 4.3a).

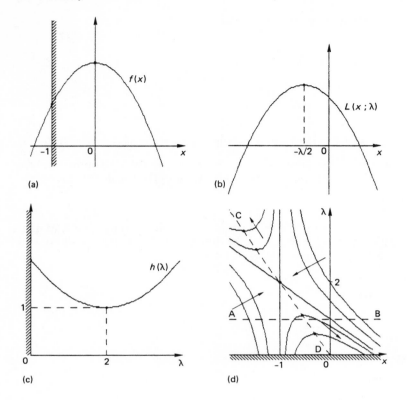

Fig. 4.3

Its Lagrangean is given by (Fig. 4.3b):

$$L(x; \lambda) = -x^2 + 2 - \lambda(x + 1)$$

and hence the solution of the *unconstrained* problem:

$$\underset{x}{\text{maximise}}\ L(x; \lambda)$$

with λ fixed at some non-negative value is reached at its unique stationary point given by:

$$dL/dx = -2x - \lambda = 0, \quad \text{i.e. } x = -\lambda/2.$$

(Remember, the Lagrangean is not constrained by the inequality $x \leqslant -1$.) The dual objective function is therefore:

$$h(\lambda) = L(-\lambda/2; \lambda) = \lambda^2/4 - \lambda + 2$$

with the graph shown in Fig. 4.3c. Its minimum is achieved at $\lambda = 2$ with $h(2) = 1$, equalling the optimal value of the primal problem.

We can check directly that the Lagrangean has a saddle point at $x = -1$, $\lambda = 2$ by sketching its contour map. The equation for these contours:

$$L = -x^2 + 2 - \lambda(x + 1) = c$$

i.e. $\quad (x + 1)(x + \lambda - 1) = 1 - c$

defines a family of hyperbolae with the lines

$$x = -1 \quad \text{and} \quad x = 1 - \lambda$$

(from the contour $c = 1$) as common asymptotes (Fig. 4.3d). The intersection of these asymptotes at $x = -1$, $\lambda = 2$ defines a saddle point, as a simple sign analysis shows.

The value of the Lagrangean at this point is the common value of the primal and dual problems, namely 1. For completeness we note that the graph in Fig. 4.3b is the cross-section AB of the Lagrangean surface for fixed λ, and the graph in Fig. 4.3c is the cross-section CD of the Lagrangean surface through the maximum points in the fixed λ cross-sections (Fig. 4.3d).

Example 2

The optimal value for the problem

$$\begin{aligned} \text{maximise} \quad & -x^2 + 2 \\ \text{subject to} \quad & x \leqslant 1 \end{aligned}$$

is 2 and is taken internally at $x = 0$ (Fig. 4.4a). The corresponding Lagrangean has the form

$$L(x; \lambda) = -x^2 + 2 - \lambda(x - 1) = -(x + \lambda/2)^2 + \lambda^2/4 + \lambda + 2$$

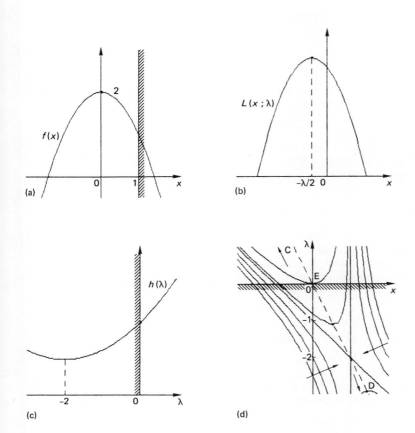

Fig. 4.4

with its maximum located at $x = -\lambda/2$ (Fig. 4.4b). Hence
$$h(\lambda) = L(-\lambda/2; \lambda) = \lambda^2/4 + \lambda + 2 = (\lambda + 2)^2/4 + 1 \qquad (\lambda \geqslant 0).$$
The minimum value of $h(\lambda)$ is achieved at the boundary $\lambda = 0$ since the trough of $h(\lambda)$ lies in the negative region (Fig. 4.4c). (The value $\lambda = 0$ is consistent with the fact that the constraint in the primal problem is not binding.)

The contour equation for the Lagrangean is given by
$$-x^2 + 2 - \lambda(x-1) = c$$
i.e. $\qquad (x-1)(x+1+\lambda) = 1-c.$

The contours are a family of hyperbolae with the lines $x = 1$, $x + \lambda + 1 = 0$ as asymptotes (Fig. 4.4d). The intersection $(x, \lambda) = (1, -2)$ is a saddle point which is inaccessible because of the non-negativity constraint on λ. The optimum occurs at the origin E, as close to the saddle point as possible on the curve CD of maximum points (with respect to x). At E the value of the Lagrangean is equal to the common optimal value of both the primal and dual problems. We shall refer to a point such as E as a *boundary saddle point*, if it has a constrained minimum with respect to λ and a constrained or unconstrained maximum with respect to x.

Example 3
Some surprises can occur when determining the dual problem as we shall see when we analyse the following primal problem:

maximise $\quad x + 3$
subject to $\quad x \leqslant 1$.

The maximum value of its linear objective function is 4 and is taken on the boundary at $x = 1$ (Fig. 4.5a). Its Lagrangean (Fig. 4.5b) is also linear (in x) and is given by:
$$L = (x+3) - \lambda(x-1) = x(1-\lambda) + \lambda + 3.$$
If $\lambda \neq 1$ the maximum value of L with respect to x is ∞, taken at $x = +\infty$ for $\lambda < 1$ or $x = -\infty$ for $\lambda > 1$, since x is no longer bounded. When $\lambda = 1$, however, the graph of L is horizontal and its maximum value is 4, the constant value of L. Hence
$$h(\lambda) = 4 \text{ when } \lambda = 1$$
$$ = \infty \text{ when } \lambda \neq 1, \lambda \geqslant 0.$$
The dual objective function is therefore a somewhat discontinuous function! (Fig. 4.5c). Even so, we find a sensible value for the minimum of $h(\lambda)$, i.e. 4 at $\lambda = 1$, an optimal value equal to that of the primal problem. The Lagrangean also retains the saddle point structure as can be seen by sketching its contours:
$$(x+3) - \lambda(x-1) = c$$
i.e. $\qquad (x-1)(1-\lambda) = c - 4.$

The saddle point is located at $x = \lambda = 1$ and has height 4 (Fig. 4.5d).

Example 4
In some cases it is convenient and in others essential that not all constraints are incorporated in the Lagrangean. As an example consider the primal problem:

maximise $\quad \sqrt{x}$
subject to $\quad x \geqslant 0, x \leqslant 1$

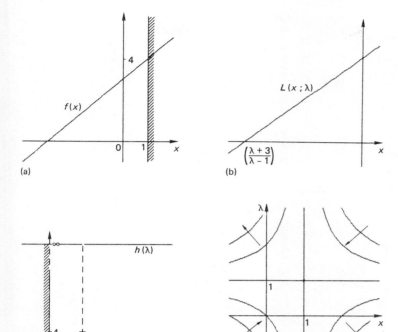

Fig. 4.5

which has optimal value 1 at $x = 1$ (Fig. 4.6a). We cannot free ourselves of the non-negativity constraint by putting it in the Lagrangean as $\sqrt{x}$ and hence the Lagrangean is only defined for non-negative x. However, inserting the second constraint we obtain

$$L = \sqrt{x} - \lambda(x - 1).$$

The maximum of this Lagrangean subject to $x \geq 0$ (Fig. 4.6b) is achieved at an internal point given by:

$$\partial L/\partial x = 1/(2\sqrt{x}) - \lambda = 0, \quad \text{i.e. } x = 1/(4\lambda^2).$$

Hence $h(\lambda) = L(1/(4\lambda^2); \lambda) = 1/(4\lambda) + \lambda$.

The dual problem:

   minimise $h(\lambda)$
   $\lambda \geq 0$

has the optimal value 1 at $\lambda = 1/2$ (Fig. 4.6c). We can check that the Lagrangean has a saddle point at $(x, \lambda) = (1, 1/2)$ by sketching the contour map (Fig. 4.6d). The value at this saddle point is again equal to the common optimal value of the primal and dual problems.

Before attempting the Exercise Set we summarise in Procedure Box 4.1 the method used to examine the primal–dual structure for one variable problems.

86 *Duality*

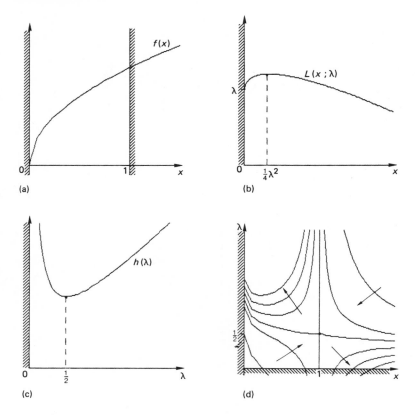

Fig. 4.6

PROCEDURE 4.1
Primal-Dual Analysis (for one variable primal and dual problems)

1 Solve the primal problem geometrically (or by Procedure 3.1).
2 Construct a Lagrangean function by including one of the constraints from the primal problem.
3 Maximise the Lagrangean with respect to x subject to the remaining constraints and with λ kept fixed.
4 Construct the dual problem by identifying its objective function as the optimal value of L in stage 3.
5 Solve the dual problem geometrically (or by Procedure 3.1).
6 Check whether:
   (i) the optimal values of the dual and primal problems are equal,
   (ii) the Lagrangean has an internal or boundary saddle point at the point defined by the solution of the primal and dual problems,
   (iii) the value of the Lagrangean at its saddle point (if it exists) is equal to the optimal values of the primal and dual problems.

Exercise Set 1

1 Apply Procedure 4.1 to the following problems:
(i)* maximise $-x^2$, subject to $x \geq 1$
(ii) maximise x, subject to $x^2 \leq 1$
(iii)* maximise x^2, subject to $x^2 \leq 1$
(iv) maximise $1-x$, subject to $x \geq 1$.

In each case check the existence of a saddle point by sketching the contours of L.

2 Repeat question 1 for the problems:
(i)* maximise $x+3$, subject to $0 \leq x \leq 1$
(ii) maximise $-(x+1)^2$, subject to $0 \leq x \leq 1$
(iii)* maximise $\ln(x+1)$, subject to $0 \leq x \leq 1$
(iv) maximise $-1/(x+1)$, subject to $0 \leq x \leq 1$.

Do not include the non-negativity conditions in the Lagrangean.

3 Repeat question 1 for the following problems, treating ε as a positive parameter:
(i)* maximise $x+1$, subject to $x^2 \leq \varepsilon^2$
(ii) maximise $1 - \varepsilon x$, subject to $x \geq 0$.

4.3 Higher Dimensional Examples

In this section we determine the duals of several two variable primal problems. The number of variables in their duals will of course depend on the number of constraints that are included in the Lagrangean.

Example 5

The solution of the problem:

$$\text{maximise} \quad -(x_1^2 + x_2^2)$$
$$\text{subject to} \quad x_1 \geq 1, x_2 \geq 1$$

can be found from its contour map shown in Fig. 4.7a. The unconstrained maximum lies at

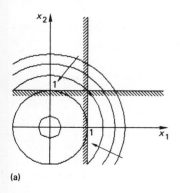
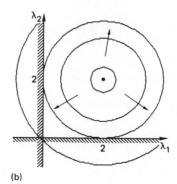

(a) (b)

Fig. 4.7

88   Duality

the origin outside the feasible region, while the constrained maximum lies, by symmetry, at the boundary corner with $x_1 = x_2 = 1$, taking the value -2.

The unconstrained Lagrangean:

$$L = -(x_1^2 + x_2^2) - \lambda_1(1 - x_1) - \lambda_2(1 - x_2)$$

for this problem has a unique stationary point determined from the equations:

$$\partial L/\partial x_1 = -2x_1 + \lambda_1 = 0, \quad \text{i.e. } x_1 = \lambda_1/2$$
$$\partial L/\partial x_2 = -2x_2 + \lambda_2 = 0, \quad \text{i.e. } x_2 = \lambda_2/2.$$

(We can check that this is a maximum by examining the behaviour at infinity.)

At the maximum point we obtain:

$$h(\lambda_1, \lambda_2) = L(\lambda_1/2, \lambda_2/2; \lambda_1, \lambda_2) = (\lambda_1^2 + \lambda^2)/4 - (\lambda_1 + \lambda_2)$$
$$= \tfrac{1}{4}(\lambda_1 - 2)^2 + \tfrac{1}{4}(\lambda_2 - 2)^2 - 2.$$

The contours of h are therefore the family of concentric circles centred at $(2, 2)$ as shown in Fig. 4.7b. The minimum is taken at this common centre and the optimal value of the dual is -2, the same as for the primal.

Example 6

The discontinuity behaviour we noted in the dual function h in Example 3 can also occur in higher dimensional primal problems. Consider for example the problem:

maximise $-(x_1 + x_2/2 + x_2^2/2)$
subject to $x_1 \geq 0, x_2 \geq 0, x_1 + x_2 \geq 2.$

In Exercise Set 2 of Chapter 3 we found the solution to be at $(3/2, 1/2)$ with the optimal value $-15/8$ (Fig. 4.8a).

To reduce the dimensions of the dual problem to manageable proportions we will separate out the non-negativity conditions and put only the third constraint into the Lagrangean:

$$L = -(x_1 + x_2/2 + x_2^2/2) - \lambda(2 - x_1 - x_2) \tag{4.4}$$
$$= (\lambda - 1)x_1 + (\lambda - 1/2)x_2 - x_2^2/2 - 2\lambda.$$

We have to maximise L with respect to *non-negative* variables x_1, x_2 with λ considered a parameter.

We will do it in two stages: firstly with respect to x_1 with x_2 fixed, and then with respect to x_2. With x_2 considered fixed, L is in fact linear in x_1 with gradient $\lambda - 1$. There are now two possibilities to consider.

(i) If $\lambda > 1$ then the gradient is positive and the maximum of L is achieved at infinity (Fig. 4.8b).
(ii) If $\lambda \leq 1$ the gradient is non-positive and the maximum of L with respect to x_1 is taken at the lower boundary[2] $x_1 = 0$, i.e.

$$\max_{x_1 \geq 0} L(x_1, x_2; \lambda) = L(0, x_2; \lambda) = (\lambda - 1/2)x_2 - x_2^2/2 - 2\lambda. \tag{4.4}$$

We now maximise with respect to x_2, i.e.

$$\underset{x_2 \geq 0}{\text{maximise }} L(0, x_2; \lambda) \quad (0 \leq \lambda \leq 1).$$

4.3 Higher dimensional examples

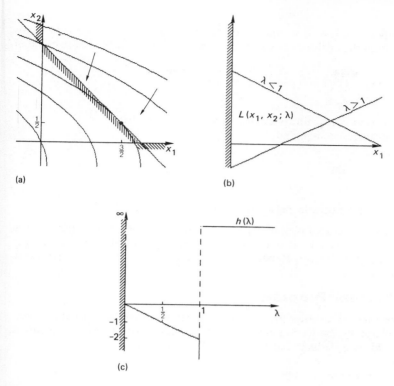

Fig. 4.8

From equation (4.4) it is clear that the solution is achieved at

$x_2 = \lambda - 1/2$ if $1/2 \leq \lambda \leq 1$

or $x_2 = 0$ if $0 \leq \lambda \leq 1/2$.

The dual objective function is therefore given by:

$$\begin{aligned} h(\lambda) &= L(0, 0; \lambda) = -2\lambda && \text{when } 0 \leq \lambda \leq 1/2 \\ &= L(0, \lambda - 1/2; \lambda) = (\lambda - 1/2)^2/2 - 2\lambda && \text{when } 1/2 \leq \lambda \leq 1 \\ &= \infty && \text{when } 1 < \lambda. \end{aligned}$$

Hence the graph of h consists of three distinct sections (Fig. 4.8c) with a discontinuity at $\lambda = 1$. Since the minimum at h cannot occur when h is infinite we can ignore all corresponding values of λ. The dual problem is then:

minimise $h(\lambda)$

subject to $0 \leq \lambda \leq 1$.

The minimum in fact occurs at $\lambda = 1$ and the dual optimal value is $-15/8$, equalling that of the primal optimum.

Exercise Set 2

1 Solve geometrically the following primal problems and then find and solve their duals geometrically. Check that the optimal values of the primal and dual problems are equal in each case.

(a)*   maximise $-(x_1+1)^2/2 - (x_2-1)^2/2$
       subject to $x_1 \geq 0, x_2 \leq 2$

when (i)* only the second constraint is included in the Lagrangean,
     (ii) both constraints are included.

(b) Repeat part (a) for:

   maximise $-(x_2 + x_1^2)$
   subject to $x_1 \geq 0, x_1 - x_2 \leq 0$.

In each case discuss the relationship between the one and two variable duals.

2* If the primal problem contains equality constraints then the only change in the procedure for constructing the dual is that the corresponding multipliers are no longer required to be non-negative. (Why?) Repeat question 1 when the second constraint is regarded as an equality.

4.4 The Dual Linear Program

A linear program (i.e. an optimisation problem with linear objective function and linear constraints) has the property that its dual is also linear. We first illustrate this result with a particular example and then deduce that it is generally true.

Example 7
Consider the linear program:

   maximise $x_1 - x_2$
   subject to $x_1 + x_2 \leq 3/2$,
                $2x_1 + x_2 \leq 2$,
                $x_1 \geq 0, x_2 \geq 0$.

The optimum, determined from the contour map in Fig. 4.9a, lies at the point $(1, 0)$, with value 1.

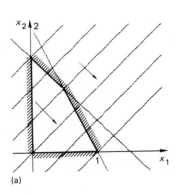

(a)   (b)

Fig. 4.9

In finding the dual we shall only include the first two constraints in the Lagrangean. We could insert all four constraints (and obtain a perfectly good dual), but in not doing so we shall preserve the symmetry between the primal and dual variables evident in the Kuhn–Tucker conditions.

The Lagrangean for this problem is given by:

$$L = x_1 - x_2 - \lambda_1(x_1 + x_2 - 3/2) - \lambda_2(2x_1 + x_2 - 2)$$
$$= (1 - \lambda_1 - 2\lambda_2)x_1 + (-1 - \lambda_1 - \lambda_2)x_2 + 3\lambda_1/2 + 2\lambda_2, \quad \text{with } x_1, x_2 \geq 0.$$

We are to maximise this expression with respect to the non-negative x variables with the multipliers treated as parameters. We will do this (as in Example 6) in two stages—maximising first with respect to x_1 and then with respect to x_2.

(a) (i) L is linear in x_1. If its gradient is positive,

i.e. $1 - \lambda_1 - 2\lambda_2 > 0$

then L has an infinite valued maximum at $x = \infty$.

(ii) If the gradient is not positive then the maximum is achieved at the lower boundary $x_1 = 0$ and so in this case:

$$\max_{x_1 \geq 0} L = L(0, x_2; \lambda_1, \lambda_2) = (-1 - \lambda_1 - \lambda_2)x_2 + 3\lambda_1/2 + 2\lambda_2.$$

(b) (i) We now have to solve the problem

$$\text{maximise } L(0, x_2; \lambda_1, \lambda_2) \quad \text{for } 1 - \lambda_1 - 2\lambda_2 \leq 0$$
$$x_2 \geq 0$$

where the objective function is linear in x_2.
If its gradient is positive,

i.e. $(-1 - \lambda_1 - \lambda_2) > 0$

then the maximum is infinite and taken when $x_2 = \infty$.

(ii) If the gradient is not positive then the maximum with respect to x_2 is taken at the lower boundary $x_2 = 0$. Hence

$$\max_{x_2 \geq 0} L(0, x_2; \lambda_1, \lambda_2) = L(0, 0; \lambda_1, \lambda_2) = 3\lambda_1/2 + 2\lambda_2$$

in this case.

In summary we conclude that

$$h(\lambda_1, \lambda_2) = 3\lambda_1/2 + 2\lambda_2 \quad \text{if} \quad 1 - \lambda_1 - 2\lambda_2 \leq 0 \quad \text{and} \quad -1 - \lambda_1 - \lambda_2 \leq 0$$
$$= \infty \qquad\qquad\qquad \text{otherwise.}$$

Our dual problem can therefore be written as:

minimise $\quad 3\lambda_1/2 + 2\lambda_2$
subject to $\quad \lambda_1 + 2\lambda_2 \geq 1$
$\qquad\qquad \lambda_1 + \lambda_2 \geq -1$
$\qquad\qquad \lambda_1 \geq 0, \lambda_2 \geq 0,$

where we have eliminated those regions in which h is infinite by imposing the first two constraints. This is clearly a linear program for the dual variables λ_1, λ_2. It is two-dimensional and hence can be solved graphically (Fig. 4.9b). The second constraint is in fact redundant, and the minimum is achieved at the vertex $(\lambda_1, \lambda_2) = (0, 1/2)$. The dual optimal value is 1, equalling that of the primal problem. We also note that $\lambda_1 = 0$ corresponds to the fact that in the optimal primal solution the first constraint is not binding.

92 Duality

The proof that the dual of the general linear program:

$$\text{maximise} \quad \sum_{i=1}^{n} c_i x_i$$

$$\text{subject to} \quad \sum_{i=1}^{n} a_{ji} x_i \leq b_j, \quad j = 1, \ldots, m \quad (4.5)$$

$$x_i \geq 0, \quad i = 1, \ldots, n$$

is also a linear program is very similar to that given for the particular case above. We first form the Lagrangean

$$L = \sum_{i=1}^{n} c_i x_i - \sum_{j=1}^{m} \lambda_j \left(\sum_{i=1}^{n} a_{ji} x_i - b_j \right)$$

$$= \sum_i \left(c_i - \sum_j \lambda_j a_{ji} \right) x_i + \sum_j \lambda_j b_j \quad (4.6)$$

without including the non-negativity conditions. We now maximise L with respect to the n non-negative x variables with the multipliers kept as parameters and we do this in n steps, optimising with respect to each x variable in turn. Since L is linear in all these variables the maximum will be infinite unless all the gradients are non-positive, i.e.

$$\sum_j \lambda_j a_{ji} \geq c_i \quad i = 1, \ldots, n \quad (4.7)$$

When all these conditions hold then the maximum of L is achieved when, in particular, all the x variables are zero and hence from equation (4.6) its optimal value is

$$h(\lambda) = L(0; \lambda) = \sum_j \lambda_j b_j$$

and ∞ otherwise. The dual problem is therefore:

$$\text{minimise} \quad \sum_j \lambda_j b_j$$

$$\text{subject to} \quad \sum_j \lambda_j a_{ji} \geq c_i \quad i = 1, \ldots, n \quad (4.8)$$

$$\lambda_j \geq 0 \quad j = 1, \ldots, m$$

provided there is at least one point satisfying these inequality conditions. Otherwise the value of the dual objective function is always infinite for $\lambda \geq 0$. This result provides us with a formula for the dual of any primal linear program.

For example, comparing the primal problem of Example 7 with (4.5) we find that:

$$c_1 = 1, \quad c_2 = -1, \quad b_1 = 3/2, \quad b_2 = 2,$$
$$a_{11} = 1, \quad a_{12} = 1, \quad a_{21} = 2, \quad a_{22} = 1,$$

and hence from (4.8) the corresponding dual is

$$\text{minimise} \quad 3\lambda_1/2 + 2\lambda_2$$
$$\text{subject to} \quad \lambda_1 + 2\lambda_2 \geq 1$$
$$\lambda_1 + \lambda_2 \geq -1$$
$$\lambda_1 \geq 0, \lambda_2 \geq 0.$$

4.4 The dual linear program

The primal/dual relationship can be seen more clearly if we use matrix notation. The primal problem can be written

for our example as:

minimise $(1 \quad -1)\begin{pmatrix} x_1 \\ x_2 \end{pmatrix}$

subject to $\begin{pmatrix} 1 & 1 \\ 2 & 1 \end{pmatrix}\begin{pmatrix} x_1 \\ x_2 \end{pmatrix} \leqslant \begin{pmatrix} 3/2 \\ 2 \end{pmatrix}$

and $\begin{pmatrix} x_1 \\ x_2 \end{pmatrix} \geqslant \begin{pmatrix} 0 \\ 0 \end{pmatrix}$

and generally as:

minimise $\mathbf{c}^T \mathbf{x}$

subject to $A\mathbf{x} \leqslant \mathbf{b}$

and $\mathbf{x} \geqslant \mathbf{0}$

where we use the vector inequality to represent an inequality for *each* corresponding pair of elements.

The dual problem becomes

for our example:

maximise $(3/2 \quad 2)\begin{pmatrix} \lambda_1 \\ \lambda_2 \end{pmatrix}$

subject to $\begin{pmatrix} 1 & 2 \\ 1 & 1 \end{pmatrix}\begin{pmatrix} \lambda_1 \\ \lambda_2 \end{pmatrix} \geqslant \begin{pmatrix} 1 \\ -1 \end{pmatrix}$

and $\begin{pmatrix} \lambda_1 \\ \lambda_2 \end{pmatrix} \geqslant \begin{pmatrix} 0 \\ 0 \end{pmatrix}$

and generally:

maximise $\mathbf{b}^T \boldsymbol{\lambda}$

subject to $A^T \boldsymbol{\lambda} \geqslant \mathbf{c}$

and $\boldsymbol{\lambda} \geqslant \mathbf{0}$

Since the dual of a primal linear program is itself a linear program the dual of a dual must also be a linear program, and in fact it turns out to be the primal itself. The proof in the general case is straightforward and is left to the Problem Set; however, an example is given in Exercise Set 3. One implication of this result is that the primal variables are the multipliers for the dual constraints—a fact we will use in Section 4.6.

We can summarise our present knowledge of the relationships between a primal and dual linear program as follows:

	Primal	Dual
Number of variables	n	m
Number of constraints	m	n
Variables	x	λ
Multipliers	λ	x
Objective function coefficients	c	b
Constraint constants	b	c

The examples of the next Exercise Set suggest further possible relationships between the primal and dual linear programs. Example (v) might suggest that a primal with an infinite optimal value generates a dual with no finite feasible points,[3] whereas example (iii) might suggest that a primal with redundant constraints has a dual with multiple solutions. The first proposition is always true, while the second is not, as we shall see in the Problem Set.

Exercise Set 3
For the primal linear programs:

$$\begin{aligned}
\text{maximise} \quad & c_1 x_1 + c_2 x_2 \\
\text{subject to} \quad & a_{11} x_1 + a_{12} x_2 \leqslant b_1 \\
& a_{21} x_1 + a_{22} x_2 \leqslant b_2 \\
& x_1 \geqslant 0,\ x_2 \geqslant 0,
\end{aligned}$$

with coefficients and constants defined in the accompanying table

(a) solve, using a contour map,
(b) find the dual and solve using a contour map,
(c) compare the optimal primal and dual values.

	c_1	c_2	b_1	b_2	a_{11}	a_{12}	a_{21}	a_{22}
(i)*	$-3/2$	-2	-1	1	-1	-2	-1	-1
(ii)	1	1	2	$1/2$	2	1	1	0
(iii)*	1	0	1	1	1	1	1	$1/2$
(iv)	1	1	2	3	1	2	2	2
(v)*	1	1	1	1	1	-1	-1	1
(vi)	1	1	-1	1	1	1	1	0

4.5 Convexity and Concavity

One important use of duality theory is in producing upper bounds to the optimal value of the primal problem for if we evaluate the dual objective function at any point in its feasible region then that value is never less than the value of the primal objective function at any point in the primal feasible region and hence never less than the primal optimal value. The proof is quite straightforward and relies upon the inequality:

$$h(\lambda) = \max_{x} \left(f(x) - \sum_i \lambda_i g_i(x) \right)$$

$$\geqslant f(x) - \sum_i \lambda_i g_i(x) \tag{4.9}$$

Since in the primal feasible region g_i is never positive and in the dual feasible region λ_i is never negative (unless g_i is zero) the second term in equation (4.9) is never negative, and hence:

$$h(\lambda) \geqslant f(x) - \sum_i \lambda_i g_i(x) \geqslant f(x)$$

for any feasible x and λ. In particular if x^* is an optimal point for the primal problem then

$$h(\lambda) \geqslant f(x^*) \equiv f^*$$

where f^* denotes the primal optimal value. By definition we also have the inequality $f(x^*) \geqslant f(x)$ for any feasible x and hence we have both upper and lower bounds on f^* given by:

$$h(\lambda) \geqslant f^* \geqslant f(x)$$

for any feasible x and λ.

To illustrate the use of this result let us consider the primal linear program:

maximise $x_1 + x_2 + x_3$
subject to
$x_1 + 2x_2 \leq 2$
$2x_1 + x_2 \leq 2$
$x_1 + x_3 \leq 1$
$x_1, x_2, x_3 \geq 0$

with dual:

minimise $2\lambda_1 + 2\lambda_2 + \lambda_3$
subject to
$\lambda_1 + 2\lambda_2 + \lambda_3 \geq 1$
$2\lambda_1 + \lambda_2 \geq 1$
$\lambda_3 \geq 1$
$\lambda_1, \lambda_2, \lambda_3 \geq 0.$

Being three-dimensional the primal cannot easily be solved by geometric techniques, and in fact we would need to use an algorithm of the type discussed in the next chapter. We can, however, gain some useful information concerning the primal optimal value by evaluating the primal and dual objective functions at various points of their respective feasible regions. For example:

(i) at $x = (1/2, 1/2, 1/2)$ the primal constraints are all satisfied and the primal objective function takes the value $3/2$ and consequently

$$f^* \geq 3/2$$

(ii) at $\lambda = (0, 1, 1)$ the dual constraints are all satisfied and the corresponding value of the dual objective function is 3 and hence

$$3 \geq f^* \geq 3/2.$$

We can find tighter bounds by noting that $\lambda = (1/3, 1/3, 1)$ also lies in the dual feasible region and yields the value $7/3$ for the dual objective function. So we have

$$7/3 \geq f^* \geq 3/2.$$

Such information could well be useful in giving some indication of, say, the profitability of a particular activity. Of course if the limits looked promising then a full solution would be undertaken.

Since no feasible dual objective function value is less than the primal optimal value, it follows that the optimal dual value is also not less than the primal optimal value. We might be tempted to go further and suggest a stronger result, that the optimal primal and dual values are necessarily equal. This is consistent with our experience in Section 4.2, where these optimal values equalled the value of the Lagrangean at a saddle point. Unfortunately this property does not always hold as the following example shows.

Example 8
The primal problem:

maximise x^2
subject to $0 \leq x \leq 1$

has optimal value 1 at $x = 1$ (Fig. 4.10a). Its Lagrangean is given by

$$L = x^2 - \lambda(x - 1), \quad (x \geq 0).$$

The maximum of this function with respect to x is always taken at ∞ with $L = \infty$. Hence

$$h(\lambda) = \infty \quad \text{for all } \lambda \geq 0$$

and the optimal value of the dual problem is ∞ (Fig. 4.10b). The optimal values of the primal and dual problems are therefore not equal. However, the Lagrangean does have a saddle point at $(x, \lambda) = (1, 2)$ as can be seen by sketching its contours (Fig. 4.10c). The

96 *Duality*

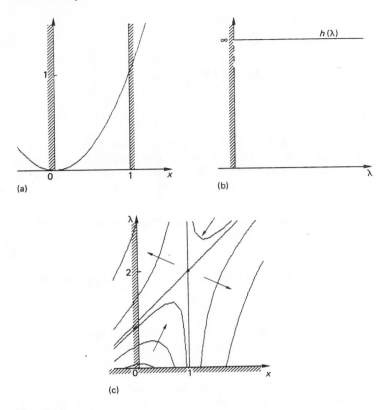

Fig. 4.10

crucial structural difference from the previous examples is in the positioning of the 'mountains' and 'valleys'. The saddle point in fact corresponds to the problem: minimise x^2 subject to $x \geqslant 1$.

We can avoid such breakdowns, if we impose certain extra conditions on the functions defining the optimisation problem. These extra conditions depend on the concepts of concave and convex functions. These concepts will be introduced in three stages according to the number of variables involved in the function. We will return to the primal–dual issue in the next section.

(a) Functions of One Variable[4]

A function $f(x)$ with a continuous second derivative $d^2 f/dx^2$ is said to be *concave* if that second derivative is nowhere *positive*. For a concave function, therefore, the derivative can never increase in value with x (Fig. 4.11), and hence any stationary point must be both a local and overall maximum. Examples of functions which are not concave are shown in Fig. 4.12. In the first case the derivative is increasing over the section CD of the graph and in the second it is increasing everywhere. This second case is an example of a *convex* function, defined as a function for which its second derivative is nowhere *negative*. Clearly if a function $f(x)$ is concave then the function $-f(x)$ is convex, and conversely.

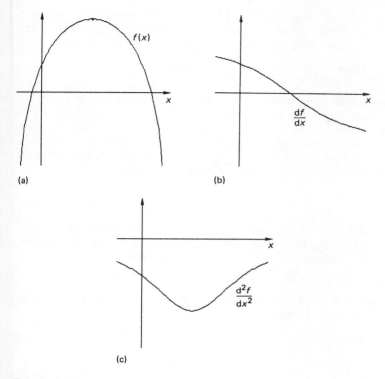

Fig. 4.11

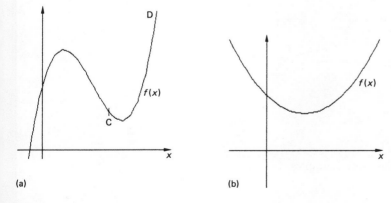

Fig. 4.12

Let us now apply this sign test on second derivatives to identify some concave and convex functions.

(i) The function $f(x) = a \exp(bx)$ is convex when $a \geqslant 0$ but concave when $a \leqslant 0$, since $d^2f/dx^2 = ab^2 \exp(bx)$.

(ii) The function $f(x) = (x^2 - a)^2$ has $d^2f/dx^2 = 12x^2 - 4a$, and hence is convex if a is negative. If, however, a is positive then f is neither concave nor convex, since there are points where the second derivative is positive and others where it is negative.

(b) Functions of Two Variables[5]

From a function of two variables we can generate functions of one variable by taking vertical planar cross-sections (Fig. 4.13), a technique we used to advantage in Chapter 1. If *all* these cross-sectional functions are concave then the function from which they are derived is said to be concave (similarly for convex functions). To make this precise let us choose two points $\mathbf{P}_1: (a_1, a_2)$ and $\mathbf{P}_2: (b_1, b_2)$ in the x_1, x_2 plane (Fig. 4.13) then the points $\mathbf{P}: (x_1, x_2)$ on the line segment $\mathbf{P}_1 \mathbf{P}_2$ can be parametrised by a variable u where

$$x_1 = a_1 + u(b_1 - a_1)$$
$$x_2 = a_2 + u(b_2 - a_2).$$

When $u = 0$ we are at point $\mathbf{P}_1$, and when $u = 1$ we are at $\mathbf{P}_2$. If we now take a cross-section of the surface $z = f(x_1, x_2)$ (Fig. 4.13b) with a vertical plane containing both $\mathbf{P}_1$ and $\mathbf{P}_2$ we obtain a curve of intersection whose equation is given by:

$$F(u) \equiv f(a_1 + u(b_1 - a_1), a_2 + u(b_2 - a_2)).$$

If $d^2 F/du^2 \leq 0$ for all choices of a_1, a_2, b_1, b_2 then $f(x_1, x_2)$ is said to be concave. On the other hand if $d^2 F/du^2 \geq 0$ then f is convex.

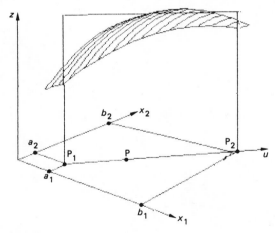

Fig. 4.13

Let us now consider some examples.

(i) For the function

$$f(x_1, x_2) = 2 \exp(3x_1 - 4x_2)$$

we have

$$F(u) = 2 \exp \{3(a_1 + u(b_1 - a_1)) - 4(a_2 + u(b_2 - a_2))\}$$

and hence

$$d^2 F/du^2 = \{3(b_1 - a_1) - 4(b_2 - a_2)\}^2 F(u).$$

This is always non-negative and so the function f is convex.

(ii) For the function:
$$f(x_1, x_2) = x_1 x_2$$
we have
$$F(u) = (a_1 + u(b_1 - a_1))(a_2 + u(b_2 - a_2))$$
with second derivative
$$d^2 F/du^2 = 2(b_1 - a_1)(b_2 - a_2).$$

The function f is therefore neither convex nor concave since this second derivative is sometimes positive (e.g. $b_1 = b_2 = 0$, and $a_1 = a_2$) and sometimes negative (e.g. $b_1 = b_2 = 0$ and $a_1 = -a_2$).

(c) Functions of n Variables

For a function f of an arbitrary number of variables $x_1, \ldots, x_n$ the procedure generalises straightforwardly. We define
$$F(u) = f(x_1, \ldots, x_n)$$
where $\quad x_i = a_i + u(b_i - a_i) \quad$ for $i = 1, \ldots, n$

and determine the sign of $d^2 F/du^2$. As an example, we can establish the result that any linear function is both convex and concave.

Let $\quad f = \sum_{i=1}^{n} c_i x_i$

then $\quad F(u) = \sum_i c_i (a_i + u(b_i - a_i))$

$$= \left(\sum_i c_i a_i \right) + u \left(\sum_i c_i (b_i - a_i) \right).$$

Hence $d^2 F/du^2 = 0$ everywhere, and the result follows.

An important property that simplifies the testing for concavity is that the sum of any number of concave functions is also concave (similarly for convex functions). This is easily established and will be left as an exercise for the reader.

For example
$$f(x) = (x^2 + 1)^2 + \exp(2x)$$
is convex since, as established earlier, the component functions are convex. This property will be useful in the following Exercise Set.

Exercise Set 4
Determine which of the following functions are concave, convex or neither.
1. (i)* $\quad x^{1/2}, \quad x = 0$ $\qquad$ (ii) $\quad x^4 + 1$
   (iii)* $\quad \ln x + x^{1/2}, \quad x > 0$ $\qquad$ (iv) $\quad x^4 + 1 - \ln x, \quad x > 0$
   (v)* $\quad \exp(x^3)$ $\qquad$ (vi) $\quad \ln(\ln x), \quad x > 1$.

2. (i)* $\quad -(x_1 - x_2)^2 \qquad$ (ii) $\quad x_1^2 + x_2^2$
   (iii)* $\quad -(x_1 - x_2)^2 + 2x_1 + 6x_2 - x_2^2 \qquad$ (iv) $\quad x_1^2 + x_2^2 + 2x_1 x_2$

(v)* $\exp(\ln x_1 + \ln x_2)/2$, $x_1 > 0$, $x_2 > 0$
(vi) $\ln(\exp x_1 + \exp x_2)$.

For notational convenience write $s_i = b_i - a_i$.

4.6 The Primal–Dual Structure

We are now in a position to state conditions that will guarantee the saddle point properties that were found to hold for the particular examples discussed in the first four sections of this chapter.

We will take the general primal problem to be in the form:

$$\text{maximise} \quad f(x)$$
$$x \in X$$
$$\text{subject to} \quad g_j(x) \leq 0 \quad \text{for } j = 1, \ldots, r.$$

By $x \in X$ we mean that only those points x which belong to the set X (defined by those constraints that are *not* put into the Lagrangean) are to be considered. In this book we shall only be interested in two cases:

(i) X denotes the whole space i.e. no restriction on x;

or (ii) X consists only of those points where none of the coordinates are negative,

since these occur most frequently in practice.

The dual problem is defined by

$$\min_{\lambda \geq 0} h(\lambda)$$

where

$$h(\lambda) = \max_{x \in X} L(x, \lambda)$$

and $\quad L(x, \lambda) = f(x) - \sum_{j=1}^{r} \lambda_j g_j(x)$.

If we impose the conditions that

(i) $f(x)$ is a concave function over the feasible region
(ii) all constraint functions $g_j(x)$ and those defining the set X are convex, and
(iii) there is at least one feasible point where none of the r constraints are binding,

then the following properties can be proved[6].

The Saddle Point Properties

1 If x^* is an optimal point for the primal and λ^* for the dual, then the Lagrangean has a saddle point at (x^*, λ^*).
2 The optimal values of the primal and dual problems are both equal to $L(x^*, \lambda^*)$.
3 The solutions x^*, λ^* satisfy the complementary slackness conditions (3.14) and Lagrange equations (3.13) when X denotes the whole space, and the Kuhn–Tucker conditions (3.24, 3.25) when X corresponds to the set of non-negative vectors.
4 Conversely, any solution of the complementary slackness conditions and Lagrangean equations yields optimal solutions to the primal and dual problems (similarly for the Kuhn-Tucker conditions).

There are three observations we should make about these conditions and properties before we put them to work.

(a) The rather strange condition (iii) is called the *constraint qualification*. Its function is to eliminate degeneracies that will force the saddle point to infinity.

To see how this comes about, let us analyse the problem:

maximise $\quad x+1$

subject to $\quad x^2 \leqslant \varepsilon^2 \quad$ (ε is a positive parameter)

with concave objective function and convex constraint function. We found the dual problem (Exercise Set 1) to be:

minimise $1/(4\lambda) + 1 + \lambda\varepsilon^2$

with solution $\lambda = 1/(2\varepsilon)$. The Lagrangean has a saddle point at $(\varepsilon, 1/(2\varepsilon))$ (Fig. 4.14a). As ε tends to zero the primal feasible region $(-\varepsilon \leqslant x \leqslant \varepsilon)$ collapses onto the single point $x = 0$ with the single constraint being binding and the saddle point of the Lagrangean moving to infinity (Fig. 4.14b).

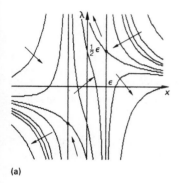
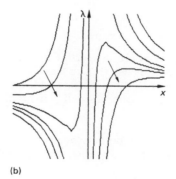

(a)   (b)

Fig. 4.14

(b) We met two types of saddle point in Section 4.2, unconstrained and boundary. A continuum of such points can also exist as the example:

maximise $\quad 1 - \varepsilon x$

subject to $\quad x \geqslant 0 \quad$ (where ε is a positive parameter)

discussed in Exercise Set 1 shows. Its Lagrangean is given by

$L = 1 + x(\lambda - \varepsilon)$

with a normal saddle point at $(0, \varepsilon)$ (Fig. 4.15a). As ε tends to zero the objective function of the primal tends to the constant value 1 with all non-negative x as optimal solutions. The saddle point shifts to the point $(0,0)$ on the boundary $\lambda = 0$. In the process all other points on the positive half of the x-axis become saddle points (corresponding to the multiplicity of solutions to the primal problem) in the sense that moving in the x direction does not increase the value of L, and moving in the λ direction does not decrease it (Fig. 4.15b).

One can in fact show that under the stated conditions the Lagrangean at (x^*, λ^*) satisfies the 'global' saddle point definition:

$L(x, \lambda^*) \leqslant L(x^*, \lambda^*) \leqslant L(x^*, \lambda)$

for *any* $x \in X$ and *any* $\lambda \geqslant 0$.

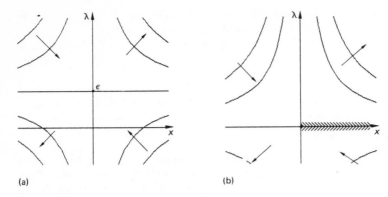

Fig. 4.15

(c) If any of the saddle point properties do not hold then it must be that one of the conditions is not satisfied. (E.g. in Example 8 the objective function is convex.) On the other hand, if one or more of the conditions does not hold we cannot conclude that the saddle point properties will not hold; counter examples will be given in the Problem Set.

One implication of the saddle point properties is that it may be worthwhile solving the primal problem by first solving the dual. As an example consider the linear program:

$$\begin{aligned} \text{maximise} \quad & 2x_1 - x_2 + 2x_3 \\ \text{subject to} \quad & 2x_1 + x_2 + x_3 \leqslant 2 \\ & x_1 + 2x_2 + 2x_3 \leqslant 2 \\ & x_1, x_2, x_3 \geqslant 0 \end{aligned}$$

with dual:

$$\begin{aligned} \text{minimise} \quad & 2\lambda_1 + 2\lambda_2 \\ \text{subject to} \quad & 2\lambda_1 + \lambda_2 \geqslant 2 \\ & \lambda_1 + 2\lambda_2 \geqslant -1 \\ & \lambda_1 + 2\lambda_2 \geqslant 2 \\ & \lambda_1, \lambda_2 \geqslant 0. \end{aligned}$$

Being two-dimensional the dual can be solved geometrically (Fig. 4.16). The minimum is taken at $(\lambda_1, \lambda_2) = (2/3, 2/3)$ with the first and third constraints being binding. Since the second dual constraint is not binding, its multiplier x_2 must be zero. Also, since λ_1 and λ_2 are not zero it must be that in the primal problem the corresponding constraints must be binding. Hence x_1 and x_3 satisfy the equations

$$\begin{aligned} 2x_1 + x_3 &= 2 \\ x_1 + 2x_3 &= 2 \end{aligned}$$

giving $x_1 = x_3 = 2/3$ as the solution to the primal.

In the next section we apply these various ideas to solve a particular class of optimisation problem, called the geometric program, by first solving the dual.

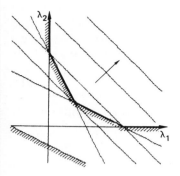

Fig. 4.16

4.7 Geometric Programming: An Application of Duality Theory

Another important class of optimisation problem is the *geometric program* originally used by Zener[7] and collaborators in optimal engineering design. Subsequent applications have spread into the area of Operational Research and include such topics as reliability theory[8] and regional planning[9]. It has even penetrated into the heart of economic theory as the example at the end of the next section will show.

To give some idea of the structural features of a geometric program we will model the following problem.

Another Chemical Tank Design Problem

A company wishes to install an open rectangular tank (Fig. 4.17) to store a chemical required in a new process that it is developing. The problem is to choose the dimensions of the tank so that it will hold a given volume V of liquid but at minimum capital outlay. If x_1, x_2, x_3 are the dimensions of the tank (Fig. 4.17a) and we ignore labour and incidental costs, then the outlay will be proportional to the total surface area of the tank, i.e. proportional to

$$x_1 x_3 + 2x_2 x_3 + 2x_1 x_2$$

where $x_1 x_2 x_3 = V$.

If we eliminate variable x_3 using the equality constraint[10], the design problem can be written as:

$$\text{minimise } bVx_2^{-1} + 2bVx_1^{-1} + 2bx_1 x_2 \qquad (4.10)$$
$$x \geq 0$$

where b is the unit cost of sheet steel.

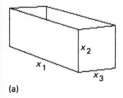
(a)

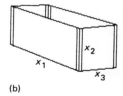
(b)

Fig. 4.17

On reflection the designer may wish to give extra strength to the tank by adding steel pillars to each vertical edge (Fig. 4.17b). This will add an extra cost $4dx_2^\alpha$ where $\alpha \geq 1$ and d is the unit cost of these strengthening girders. The purpose of the parameter α is to model the effect that the longer the girder the thicker it will need to be. The optimisation problem now becomes:

$$\underset{x \geq 0}{\text{minimise}} \ bVx_2^{-1} + 2bVx_1^{-1} + 2bx_1x_2 + 4dx_2^\alpha \tag{4.11}$$

These problems are examples of the simplest type of geometric program with form[11]:

$$\underset{x \geq 0}{\text{minimise}} \ \sum_{i=1}^{l} U_i \tag{4.12}$$

where each U term is a product of powers of the x variables multiplied by a positive coefficient. Hence in the case of n variables we can write U_i as:

$$U_i = C_i x_1^{a_{i1}} x_2^{a_{i2}} \ldots x_n^{a_{in}}, \quad C_i > 0,$$

or, using the product notation Π, as:

$$U_i = C_i \prod_{j=1}^{n} x_j^{a_{ij}}, \quad C_i > 0.$$

A sum of such terms is often called a *posynomial*. The exponents in the various U terms can be assembled for convenience into a matrix A with the (i,j) element equal to the exponent a_{ij}. For our first example,

$$\underset{x \geq 0}{\text{minimise}} \ bVx_1^0 x_2^{-1} + 2bVx_1^{-1}x_2^0 + 2bx_1^1 x_2^1$$

we have $C_1 = bV, C_2 = 2bV, C_3 = 2b$ and

$$A = \begin{bmatrix} 0 & -1 \\ -1 & 0 \\ 1 & 1 \end{bmatrix}.$$

The obvious way to solve the geometric program (4.12) is to treat it as a simple unconstrained optimisation problem in n variables, checking that the resulting solution also satisfies the non-negativity conditions $x \geq 0$. An alternative method that

(i) in most practical problems is computationally simpler,
(ii) highlights the essential structure of the program, and
(iii) is easy to generalise to geometric programs with posynomial constraints,

is based on duality theory. This might seem rather surprising since the problem (4.12) not only has no non-trivial constraints with which to generate dual variables but the objective function is seldom convex (e.g. consider $x_1 x_2$). However, we can put the primal problem (4.12) into a form suitable for the construction of a dual by applying the following two transformations.

(i) We rewrite each U term in an equivalent exponential form:

$$U_i = \exp(\ln C_i)\exp(a_{i1} \ln x_1) \ldots \exp(a_{in} \ln x_n)$$

$$= \exp\left(\sum_{j=i}^{n} a_{ij} \ln x_j + \ln C_i \right)$$

which can be simplified by defining new variables:

$$X_i = \ln x_i, \quad Y_i = \sum_{j=1}^{n} a_{ij} X_j + \ln C_i.$$

4.7 Geometric programming: an application of duality theory

Problem (4.12) then takes the form:

$$\text{minimise} \quad \sum_{i=1}^{l} \exp(Y_i) \quad (4.13)$$

$$\text{subject to} \quad Y_i = \sum_j a_{ij} X_j + \ln C_i \quad \text{for } i = 1, \ldots, l.$$

(Note that as a consequence of the transformations the new variables are not restricted to non-negative values.) The objective function is now convex, being a sum of linear exponentials and hence its negative (for the equivalent maximisation problem) is concave; the constraints are also convex being linear in the variables[12].

(ii) We now take the logarithm of the objective function to obtain the problem:

$$\text{minimise} \quad \ln\left(\sum_{i=1}^{l} \exp(Y_i)\right) \quad (4.14)$$

$$\text{subject to} \quad Y_i = \sum_{j=1}^{n} a_{ij} X_j + \ln C_i \quad \text{for } i = 1, \ldots, l.$$

That this problem is equivalent to problem (4.13) follows from the 'monotonicity' property:

$$\ln a < \ln b \quad \text{for } 0 < a < b$$

which indicates that the logarithmic function preserves the ordering of a set of numbers. Hence if X_i^*, Y_j^* yield the minimum objective value for (4.13) then they will also yield the minimum objective value for (4.14).

If we take the dual of (4.14) and then 'unwrap' the logarithm we obtain the following problem dual to (4.12)

$$\text{maximise} \quad \prod_{i=1}^{l} (C_i/\lambda_i)^{\lambda_i} \quad (4.15a)$$

$$\text{subject to} \quad \sum_i \lambda_i a_{ij} = 0 \quad \text{for } j = 1, \ldots, n \quad (4.15b)$$

$$\sum_i \lambda_i = 1 \quad (4.15c)$$

Details of the derivation are given in the Problem Set.

Equations (4.15b) are known as *orthogonality conditions* and can be calculated most simply in particular cases by multiplying matrix A on the *left* with a *row* vector of multipliers i.e. $\lambda A = 0$.

Equation (4.15c) is known as the *normality condition*. The primal optimal values are determined from the equations:

$$U_i = \lambda_i \left(\sum_j U_j\right) \quad \text{for } i = 1, \ldots, l \quad (4.16)$$

which are obtained when deriving the dual. The right hand sides are known from the optimal dual solution since $\sum_j U_j$ is the common optimal value of the primal and dual problems and λ_i are the optimal dual variables. The left hand sides are known functions of the primal variables and hence equations (4.16) define algebraic equations for those variables. It is clear from these that all λ_i must be non-negative.

106   *Duality*

A negative λ_i corresponds to the primal problem taking its minimum on the boundary—in deriving the dual (4.15) we assumed an internal solution. A sufficient condition for an internal solution and also for the saddle point properties to hold is that there is a solution to equations (4.15b, c) with all multipliers *positive*.

Example 9
For the problem (4.10) the dual is:

$$\text{maximise} \quad (bV/\lambda_1)^{\lambda_1} (2bV/\lambda_2)^{\lambda_2} (2b/\lambda_3)^{\lambda_3} \tag{4.17}$$

$$\text{subject to} \quad \begin{aligned} -\lambda_2 + \lambda_3 &= 0 \\ -\lambda_1 \quad\quad + \lambda_3 &= 0 \\ \lambda_1 + \lambda_2 + \lambda_3 &= 1. \end{aligned}$$

The equality constraints give us three equations for three multipliers and the unique solution is $\lambda_1 = \lambda_2 = \lambda_3 = 1/3$. There is no optimisation to perform now since there is only one feasible point. The optimal value of the dual and hence the primal is therefore given by (4.17) as

$$(3bV)^{1/3} (6bV)^{1/3} (6b)^{1/3} = 3b(2V)^{2/3} \tag{4.18}$$

The primal variables are determined from the conditions

$$U_1 = U_2 = U_3 = b(2V)^{2/3} \tag{4.19}$$

where $U_1 = bVx_2^{-1}, U_2 = 2bVx_1^{-1}, U_3 = 2bx_1x_2$. Substituting in equation (4.19) we find that $x_2 = 2^{-2/3} V^{1/3}$, $x_1 = x_3 = 2^{1/3} V^{1/3}$ are the optimal dimensions of the tank. If $V = 1$, for example, we have $x_2 = 0.63$, $x_1 = x_3 = 1.25$ and the total cost 4.78.

The dual and hence the primal problems (4.10) were trivial to solve because the dimension of the dual space was zero. In general the dimension can be seen from equations (4.15) to be:

(number of dual variables − number of constraints)

i.e.   (number of dual variables − number of primal variables − 1)

provided none of the orthogonality conditions are redundant. (We shall always assume this to be the case unless otherwise stated.) This dimensional statement can be rewritten solely in primal terms if we use the fact that the number of dual variables is equal to the number of U terms, i.e. the dimension equals

(number of U terms − number of primal variables − 1).

This dimension is often called the *degree of difficulty* of the problem. The Strengthened Tank Problem (4.11) has dimension 1, with its dual being given by:

$$\text{maximise} \quad (bV/\lambda_1)^{\lambda_1} (2bV/\lambda_2)^{\lambda_2} (2b/\lambda_3)^{\lambda_3} (4d/\lambda_4)^{\lambda_4}$$

$$\text{subject to} \quad \begin{aligned} -\lambda_2 + \lambda_3 \quad\quad &= 0 \\ -\lambda_1 \quad\quad + \lambda_3 + \lambda_4 &= 0 \\ \lambda_1 + \lambda_2 + \lambda_3 + \lambda_4 &= 1 \end{aligned} \tag{4.20}$$

when we take $\alpha = 1$.

We use the constraint equations to express $\lambda_1, \lambda_2, \lambda_3$ in terms of λ_4:

$$\lambda_2 = \lambda_3 = 1/3 - 2\lambda_4/3; \quad \lambda_1 = 1/3 + \lambda_4/3 \tag{4.21}$$

Since none of these multipliers may be negative for an internal solution we must have

$$0 \leqslant \lambda_4 \leqslant 1/2.$$

The dual problem (4.20) is therefore a constrained one-variable optimisation problem. Its objective function is rather complicated so it may be preferable to proceed numerically. A rough method for finding the maximum would be to plot the graph of the dual objective function over the range $0 \leqslant \lambda_4 \leqslant 1/2$ and visually identify the maximum point. More sophisticated numerical methods are available, the simplest of which, probably, is the method used for detecting the zeros of a function, in which smaller and smaller intervals are constructed containing the zero of interest. This method can be used to find the zeros of the derivative of the dual objective function. We shall, however, use an alternative method that exploits the primal–dual structure. We proceed as follows, taking for simplicity $b = d/2 = V = 1$.

(a) Arbitrarily choose $\lambda_4 = 1/4$, then $\lambda_1 = 5/12$, $\lambda_2 = \lambda_3 = 1/6$, and the dual objective function value is 7.84, so that[13]

minimum cost $\geqslant 7.84$.

(b) Calculate values for x_1, x_2 from equations (4.16) with $i = 1, 2$ using the dual variable and objective function values in (a) as approximations to their optimal values. We find $x_1 = 0.31$, $x_2 = 1.53$ with 7.96 as the value of the primal objective function. Hence

$7.96 \geqslant$ minimum cost $\geqslant 7.84$.

(c) Calculate a value for λ_4 from equation (4.16) with $i = 4$ using the values from (b) as approximations to the optimal values. In fact $\lambda_4 = 0.31$ and the dual objective function is 7.92 i.e.

$7.96 \geqslant$ minimum cost $\geqslant 7.92$.

This iterative procedure therefore gives reasonably tight bounds for the optimal cost after only three stages and in the process we have obtained approximations for both the primal and dual variables. Tighter bounds can be obtained by continuing the iterative process. For example the next step gives $x_1 = 0.29$, $x_2 = 1.96$ and an upper bound of 7.93. The power of this sort of iteration method becomes increasingly evident as the number of variables and U terms increases[7].

Before attempting to solve more geometric programs in the next Exercise Set let us summarise our solution method in Procedure Box 4.2.

PROCEDURE 4.2
To solve Geometric Program (4.12)

1 Write down the dual (4.15).
2 If the degree of difficulty is zero solve the linear orthogonality and normality equations (4.15b, c) for the, in general, unique dual feasible point. Determine the primal solution by solving equation (4.16). (Degeneracies can sometimes occur with a non-unique dual solution or negative dual variables.)
3 If the degree of difficulty is greater than zero use the orthogonality and normality equations to reduce the dual problem to a problem in d non-negative variables where d denotes the degree of difficulty. Solve this and hence the primal problem numerically.

Exercise Set 5

1 Find the primal and dual solutions to the following geometric programs:

(i)* $\quad$ minimise $\quad x_1 x_2^{-1} + 2x_2 + x_1^{-1/2}$
$\qquad x_1, x_2 \geq 0$

(ii) $\quad$ minimise $\quad x_1^{1/2} x_2^{-1} + 2x_1^{-1} + x_1^{-1/3} x_2^2$.
$\qquad x_1, x_2 \geq 0$

2 Find the one-dimensional duals of the following primal programs. Obtain appropriate solutions by plotting the dual objective function on graph paper.

(i)* $\quad$ minimise $\quad x_1 x_2^{-1} + 2x_2 + x_1^{-1/2} + x_1$.
$\qquad x_1, x_2 \geq 0$

(ii) $\quad$ minimise $\quad x_1^{1/2} x_2^{-1} + 2x_1^{-1} + x_1^{1/3} x_2^2 + x_2$.
$\qquad x_1, x_2 \geq 0$

3 Show that there are no solutions to the dual constraints with non-negative dual variables for the following problems:

(i)* $\quad$ minimise $\quad x_1 + x_2 + x_1 x_2$
$\qquad x_1, x_2 \geq 0$

(ii) $\quad$ minimise $\quad x_1 + x_1 x_2 + x_1 x_2^2$.
$\qquad x_1, x_2 \geq 0$

By drawing a contour map for the primal, or otherwise, check that the optimum occurs at the boundary.

*4.8 The Geometric Program with Constraints

The general geometric program has the form:

$$\text{minimise} \quad \sum_{i=1}^{l_0} U_i$$

$$\text{subject to} \quad \sum_{i=l_{j-1}+1}^{l_j} U_i \leq 1 \quad j = 1, \ldots, m, \tag{4.22}$$

possessing m inequality constraints, each a sum of terms of the form:

$$U_i = C_i \prod_{k=1}^{n} x_k^{a_{ik}} \quad \text{(with } C_i > 0\text{)}.$$

The dual problem is

$$\text{maximise} \quad \prod_{i=1}^{l_m} (C_i / \lambda_i)^{\lambda_i} \prod_{j=1}^{m} (\mu_j)^{\mu_j}$$

$$\text{subject to} \quad \sum_{i=1}^{l_m} \lambda_i a_{ik} = 0 \tag{4.23}$$

$$\sum_{i=1}^{l_0} \lambda_i = 1 \tag{4.24}$$

$$\mu_j = \sum_{i=l_{j-1}+1}^{l_j} \lambda_i, \quad \lambda_i \geq 0$$

for $i = 1, \ldots, l_m;\ j = 1, \ldots, m;\ k = 1, \ldots, n$.

The procedure for constructing the dual problem is therefore as follows.

(a) The objective function is the product of the following terms:
   (i) for the ith U function there is a term $(C_i/\lambda_i)^{\lambda_i}$,
   (ii) for the jth constraint there is a term $(\mu_j)^{\mu_j}$ where μ_j is the sum of the multipliers for the U terms in the jth constraint.
(b) Constraints:
   (i) the multipliers for the U terms in the objective function *only* sum to 1
   (ii) the multipliers for *all* U terms satisfy the orthogonality relationship (4.23).

The primal optimal values are derived from the following relationships:

$$\text{Objective Function: } U_i = \lambda_i \left(\sum_k U_k \right) \quad 1 \leqslant i \leqslant l_0 \quad (4.25)$$

where the sum is taken over only those U terms in the primal objective function;

$$j^{\text{th}} \text{ Constraint: } U_i = \lambda_i \left(\sum_k U_k \right) \bigg/ \mu_j, \quad l_{j-1}+1 \leqslant i \leqslant l_j \quad \text{if } \mu_j > 0$$

where the sum is taken over only those U terms in the jth constraint of the primal.

To ensure that the dual–primal structure is not invalidated by the primal taking its optimum on the boundary we have the sufficient condition that there exist a positive set of multipliers λ_i that satisfy the constraints (4.23, 4.24).

An Application: The Cobb–Douglas Minimum Cost Function

As an example of a geometric program with constraints let us consider a manufacturer who is operating a process where the output Q is related to the inputs of capital (equipment) x_1 and labour x_2 by the production function:

$$Q = f(x_1, x_2).$$

There are various choices for the mix of capital and labour to achieve a given output, and the optimal choice (from the manufacturer's point of view) is one that minimises total cost. If we take this production function to have the so-called Cobb–Douglas form[14]: $f = C x_1^{\alpha_1} x_2^{\alpha_2}$, with $\alpha_1 + \alpha_2 = 1$, then the optimisation problem can be written as:

minimise $\quad p_1 x_1 + p_2 x_2$

subject to $\quad C x_1^{\alpha_1} x_2^{\alpha_2} \geqslant Q$

where p_1 and p_2 are the prices of capital and labour respectively. (The inequality is to allow for the unlikely possibility of achieving more than the expected output Q at minimum cost.)

This problem is of the form (4.22) if we rearrange the inequality as

$$(Q/C) x_1^{-\alpha_1} x_2^{-\alpha_2} \leqslant 1.$$

In this case we have $U_1 = p_1 x_1$, $U_2 = p_2 x_2$, $U_3 = (Q/C) x_1^{-\alpha_1} x_2^{-\alpha_2}$, and

$$A = \begin{bmatrix} 1 & 0 \\ 0 & 1 \\ -\alpha_1 & -\alpha_2 \end{bmatrix}.$$

110 *Duality*

The dual problem is therefore

maximise $(p_1/\lambda_1)^{\lambda_1}(p_2/\lambda_2)^{\lambda_2}(Q/(C\lambda_3))^{\lambda_3}\mu_1^{\mu_1}$

subject to
$$\lambda_1 - \alpha_1\lambda_3 = 0$$
$$\lambda_2 - \alpha_2\lambda_3 = 0$$
$$\lambda_1 + \lambda_2 = 1$$
$$\lambda_3 = \mu_1.$$

The constraint equations give, with $\alpha_1 + \alpha_2 = 1$, the unique solution

$$\lambda_1 = \alpha_1, \quad \lambda_2 = \alpha_2, \quad \lambda_3 = \mu_1 = 1,$$

and therefore the minimum cost is given by:

$$(Q/C)(p_1/\alpha_1)^{\alpha_1}(p_2/\alpha_2)^{\alpha_2}.$$

From equation (4.25) we find the optimal choices of the inputs to be

$$x_1 = (Q/C)(\alpha_1/\alpha_2)^{\alpha_2}(p_2/p_1)^{\alpha_2}, \quad x_2 = (Q/C)(\alpha_2/\alpha_1)^{\alpha_1}(p_1/p_2)^{\alpha_1}.$$

The analysis generalises straightforwardly to the case of n inputs to the production process.

Problem Set 4

1 Find and solve the duals of the following primal problems (omitting the non-negativity constraints from the Lagrangean in parts (i), (ii), (v) and (vi)).

(i)*   maximise $-x$, subject to $x^2 \geq 1$, $x \geq 0$.
(ii)   maximise $-x^2$, subject to $x^2 \geq 1$, $x \geq 0$.
(iii)*   maximise x_2, subject to $x_1 + x_2^2 \leq 1$, $x_1 + 1 \geq 0$.
(iv)   maximise x_1, subject to $x_1^2 + x_2^2 \leq 1$, $x_2 \geq 0$.
(v)*   maximise $x_1 + x_2$, subject to $2x_1 + x_2 \leq 2$, $x_1 \leq 1$, $x_1 \geq 0$, $x_2 \geq 0$.
(vi)   minimise $x_1 + x_2$, subject to $x_1 + x_2 \geq 0$, $x_1 \geq 1$, $x_1 \geq 0$, $x_2 \geq 0$.

In each case check whether the optimal primal and dual values are equal and whether the conditions for the saddle point properties to hold are satisfied.

2 Solve the following geometric programs by first solving the dual.

(i)*   minimise $x_1^{1/2} + 2x_1^{-1}x_2^{-1} + 3x_1^{-1/3}x_2^2$.
       $x_1, x_2 \geq 0$
(ii)   minimise $x_1^{1/2}x_2^{-1} + 2x_1^{-1}x_2^{-1} + x_1^{-1/3}x_2^2$.
       $x_1, x_2 \geq 0$
(iii)*   minimise $x_1 + x_2$, subject to $x_1^{-1}x_2^{-1} \leq 1$.
       $x_1, x_2 \geq 0$
(iv)   minimise $x_1^{-1}x_2^{-1}$, subject to $x_1^{1/3}x_2 + x_1 x_2^{1/3} \leq 1$.
       $x_1, x_2 \geq 0$
(v)*   minimise $x_1^{1/3}x_2^{2/3}$, subject to $x_1^{-1} \leq 1$, $x_2^{-1} \leq 1$.
       $x_1, x_2 \geq 0$
(vi)   minimise $3x_1^{-1/2}x_2^{-1}$, subject to $x_1^{1/2}x_2 \leq 1$, $2x_1 x_2^{1/2} \leq 1$.
       $x_1, x_2 \geq 0$
(vii)*   minimise $x_1^{1/2}x_2^{1/2} + x_1 x_2 + x_1^{-1}x_2^{-1}$.
       $x_1, x_2 \geq 0$
(viii)   minimise $x_1 x_2^{-1} + x_1 x_2 + x_1 x_2^{-1}$.
       $x_1, x_2 \geq 0$

3 (a) Prove that the dual of the dual of a linear program is the original linear program. (*Hint*: multiply the inequalities by -1 and convert the minimisation problem into a maximisation problem.)
(b) By first performing the minimisation with respect to $\lambda \geq 0$ in the saddle point problem derive the primal problem (4.3).
(c)* Find the dual of the geometric program (4.14). (Since the constraints are equalities there will be no sign constraints on the multipliers.)

4 Show that the optimal value for the program:

$$\text{minimise} \sum_{i=1}^{n} p_i x_i$$

$$\text{subject to } C \prod_{i=1}^{n} x_i^{\alpha_i} \geq Q \quad \text{with} \sum_{i=1}^{n} \alpha_i = 1$$

is given by:

$$(Q/C) \prod_{i=1}^{n} (p_i/\alpha_i)^{\alpha_i}.$$

5 (a)* For functions of one variable $f(x)$ we can use the sign of the second derivative at a stationary point to distinguish between local maxima and minima. Indeed, if this second derivative is continuous and has negative sign at the stationary point then the function is concave in an interval about the stationary point, and hence the point must be a local maximum. Similarly, if the sign is positive it must be a local minimum. Check this property for the stationary points of the functions discussed in Exercise Set 1 of Chapter 1.
(b) For functions of two or more variables we can also derive conditions in terms of the second order partial derivatives to distinguish between local maxima and minima.
(i) Show, using the chain rule (Section J of the Appendix), that if

$$F(u) = f(a_1 + us_1, a_2 + us_2)$$

then

$$\frac{d^2 F}{du^2} = \frac{\partial^2 f}{\partial x_1^2} s_1^2 + 2 \frac{\partial^2 f}{\partial x_1 \partial x_2} s_1 s_2 + \frac{\partial^2 f}{\partial x_2^2} s_2^2 \tag{4.26}$$

where $\partial^2 f/\partial x_i \partial x_j$ denotes the function obtained from f by differentiating partially first with respect to x_j and then with respect to x_i. (Write $\partial^2 f/\partial x_i^2 = \partial^2 f/\partial x_i \partial x_i$, and use the fact that $\partial^2 f/\partial x_i \partial x_j = \partial^2 f/\partial x_j \partial x_i$, i.e. the order of differentiation is unimportant when the derivatives are continuous.)
(ii) Show that $d^2 F/du^2$ is negative in all directions when

$$\frac{\partial^2 f}{\partial x_1^2} < 0 \quad \text{and} \quad \left(\frac{\partial^2 f}{\partial x_1^2}\right)\left(\frac{\partial^2 f}{\partial x_2^2}\right) < \frac{\partial^2 f}{\partial x_1 \partial x_2}.$$

(iii)* If the conditions of part (ii) hold at a stationary point then this point is clearly a local maximum. Verify this for the functions listed in Exercise Set 2 of Chapter 1.
(iv) Derive the equivalent conditions for a local minimum and test against the functions in Exercise Set 2 of Chapter 1.

6 In this question, we shall be assuming that you have experience in inverses and eigenvalues of matrices.

(a) The general quadratic program has the form

$$\text{maximise} \sum_{i=1}^{n} \left(\sum_{j=1}^{n} q_{ij} x_i x_j + q_i x_i \right) \quad \text{with} \quad q_{ij} = q_{ji}; \text{ not all } q_{ij} \text{ zero}$$

$$\text{subject to} \sum_{i=1}^{n} a_{ji} x_i \leqslant b_j \quad \text{for } j = 1, \ldots, m$$

with quadratic objective function and linear constraints. In matrix form it can be written as

maximise $x^T Q x + q^T x$
subject to $A x \leqslant b$.

Prove that its dual is also quadratic and is given by:

$$\underset{\lambda \geqslant 0}{\text{minimise}} \; -(1/4)(\lambda^T A - q^T) Q^{-1} (A^T \lambda - q) + \lambda^T b$$

(provided Q has an inverse and is such that the primal objective function is concave).

(b) Use this result to find the duals of the following concave quadratic programs:

(i)* maximise $\quad -(x_1^2 + x_2^2) + x_1 - x_2$
subject to $\quad x_1 + x_2 \leqslant 1, \quad x_2 \leqslant 1$.

(ii) maximise $\quad -(x_1^2 + x_1 x_2 + x_2^2) + 3x_1 + x_2$
subject to $\quad x_1 + 2x_2 \leqslant 2, \quad 2x_1 \leqslant -1$.

References

1. For a consistent development of the theory we shall always start with a maximisation problem. For this electricity problem this will require a little juggling with the signs. For simplicity we ignore non-negativity constraints on the outputs in the expectation that none of them will be binding.
2. When $\lambda = 1$ it is also taken at all other allowed values of x_1.
3. If there are no feasible points for the dual linear program then we know from the observation made following (4.8) that the dual optimal value must be infinite, again equalling that of the primal.
4. The definitions we shall use are not the most general since we assume that functions have continuous second derivatives. They are, however, the simplest in this context. We also assume that the functions are defined for all x or for x in a range $a < x < b$. Extension to the boundary points is by continuity.
5. We assume all second order partial derivatives of f to be continuous and that the functions are defined either for all (x_1, x_2) or for the interior of a set of points with the property that the line segment joining any two points in the set lies entirely in the set. Such a set is known as a *convex set*. Extension to the boundary is by continuity.
6. The proof is somewhat complicated and technical so we shall not develop it here. If you are interested, you may wish to refer to M. D. Intriligator, *Mathematical Optimisation and Economic Theory*, Prentice-Hall (1971). The purpose of the convexity condition on the constraint functions is to guarantee that the feasible region is a convex set, essential for the proof of the results. Note that the solutions (x^*, λ^*) are supposed finite.
7. See R. J. Duffin, E. L. Peterson and C. Zener, *Geometric Programming: Theory and Applications*, Wiley (1967) and W. I. Zangwill, *Nonlinear Programming—A Unified Approach*, Prentice-Hall (1969).
8. F. A. Tillman et al., *Optimisation of Systems Reliability*, Marcel Dekker (1980).
9. P. Nijkamp, *Planning of Industrial Complexes by Means of Geometric Programming*, Rotterdam University Press (1972).
10. Although we have not recommended this procedure in the earlier chapters we use it here to produce a simple example of a geometric program.

11 In many texts the boundary, with any x variable zero, is excluded from discussion. Indeed the derivation of the dual assumes that the optimum is taken at an internal point. But in some cases the optimum is taken at the boundary as we shall see in Exercise Set 5.
12 It does not, however, satisfy the constraint qualification since the equality constraints are always binding. In fact we can relax the constraint qualification if the equalities are linear. See M. Bazaraa and C. M. Shetty, *Nonlinear Programming*, Wiley (1979) for details.
13 The dual feasible value is not *greater* than the primal optimal value because the primal problem is posed as a minimisation rather than a maximisation problem.
14 See, for example, the book by A. Sen, *Growth Economics*, Penguin (1970).

5
Linear Programming

5.1 Introduction

On several occasions in the last two chapters we have come across examples of linear programs, that is optimisation problems in which both the objective function and the constraints are linear. Although simple in form these programs are important in practice since a large number of problems—many within the realms of Operational Research—are naturally linear, the transportation problem discussed in Chapter 3 being a good example. Others include the following.

(i) *The Diet Problem.*
This is of particular use in hospitals and other institutions where a balanced diet has to be produced economically. Each potential type of food is listed together with its constituents (vitamins, calories, fibre, etc.) and its current price. The dietician specifies daily limits on the intake of each constituent—these provide the constraints. The objective function represents the total cost of the food plus preparation.

(ii) *The Blending Problem.*
A contract to supply fuel to a customer will usually specify limits on its characteristics (such as calorific value and sulphur content). To fulfil a contract the producer will blend several grades of fuels to meet the specification at minimum cost. The objective function in this case is therefore the total cost of supplying the fuel, and the constraints are the limits in the specification.

(iii) *The Assignment Problem.*
A manufacturer wishes to find the best assignment of men to jobs. From past experience (or aptitude tests) he has estimates of the efficiencies of each man for each job. If each man is assumed to spend a proportion of his time with each job, the problem becomes one of maximising the total efficiency subject to the men being fully occupied and all the jobs being filled. It turns out that the optimal solution requires each man to spend all his time with one job, a result that can be used as an argument against polygamy!

As a specific example of a linear program, let us consider the problem of a microcomputer manufacturer who is having difficulty getting enough silicon 'chips' with which to build his computers. He is retailing two types of computer—the 'Standard' which requires 2 memory chips and one processor chip and earns a profit of £50 and the 'Super' which requires 8 memory chips and one processor chip and earns a profit of £60. With his workforce and equipment he is able to produce the equivalent of 60 Standards per day—three Supers taking as long as five Standards. The problem is how many Standards, x_1, and how many Supers, x_2, he should produce with his present stock of 232 memory and 50 processor chips in order to maximise his profit P, where

$$P = 50x_1 + 60x_2.$$

The maximisation is to be performed under the constraints implicit in the above discussion.

(i) If T is the time taken to produce one Standard then

$$x_1 T + x_2 5T/3 \leqslant 60 T$$

which can be simplified to

$$3x_1 + 5x_2 \leqslant 180.$$

(ii) The processor and memory chip requirements for the two types of computer imply that

$$x_1 + x_2 \leqslant 50$$
$$2x_1 + 8x_2 \leqslant 232.$$

(iii) Finally we have the non-negativity conditions

$$x_1 \geqslant 0, \quad x_2 \geqslant 0.$$

The coefficients in this problem have been chosen so that at the optimum both x_1 and x_2 are integers, as in fact they should be. If this had not been the case we should either have to use the non-integer solution as an approximation to the actual solution, or use a more sophisticated technique to obtain the optimal integer solution.

In summary the problem faced by the computer manufacturer can be written as:

maximise $P = 50x_1 + 60x_2$
subject to $3x_1 + 5x_2 \leqslant 180$
$x_1 + x_2 \leqslant 50$ (5.1)
$2x_1 + 8x_2 \leqslant 232$
$x_1, x_2 \geqslant 0.$

We shall use this example extensively in the development of an algorithm for solving a linear program. This algorithm—known as the Simplex Algorithm—is based on the property that the optimum, in general, occurs at a vertex of the feasible region. This might suggest that all we need to do is locate all the vertices, eliminate those not in the feasible region and then determine at which of the remaining vertices the objective function takes its optimal value. However, a few simple calculations should convince you that this is seldom a practical strategy. If the number of variables is 2 and the number of constraints 3 there are just 10 vertices, but if we double both the number of variables and the number of inequality constraints then the number of vertices rises to 210. With problems of a practical size, the number of vertices can become astronomical. The Simplex Algorithm provides a far more efficient search procedure. It achieves the optimum by moving along a path on the boundary of the feasible region, at each stage choosing the next vertex on the path by finding a direction in which the Lagrangean is increasing in value. Hence most of the vertices are not even considered because of their evident non-optimality.

The Simplex Algorithm is just one of several such algorithms, but it is the most amenable to calculation by hand and so it is the one we will concentrate on in this chapter. For some particular types of linear program there are even more efficient algorithms that exploit their special structure: this is the case for both the transportation and assignment problems. We shall not have the space to discuss them here but details can be found in numerous books on Operational Research techniques.[1]

In previous chapters, the intention has been to present simple examples to illustrate the theory rather than become too involved in the algebra and long arithmetic calculations that practical problems usually imply. This chapter is exceptional in this respect: the Simplex Algorithm is a practical method for solving a linear program and hence to understand it will require a little more algebra and arithmetic.

116 Linear programming

Our strategy for deriving the Simplex Algorithm is based on the Kuhn–Tucker conditions discussed in Chapter 3. The geometric idea of tracing a path on the boundary of the feasible region is translated into a systematic procedure for solving the sets of linear inequalities defined by the Kuhn–Tucker conditions. This solution procedure is considerably simplified by the introduction of a new set of variables called *slack variables*, which transform all the non-trivial inequality constraints of the linear program into equality constraints. We can then use the special techniques developed for the solution of linear simultaneous equations.

5.2 Slack Variables

Let us start our analysis with the mathematical formulation of the linear program involving r non-negative variables $x_1, x_2, \ldots, x_r$ and m inequality constraints:

$$\text{maximise} \sum_{i=1}^{r} c_i x_i \tag{5.2a}$$

$$\text{subject to} \sum_{i=1}^{r} a_{ji} x_i \leqslant b_j \qquad j = 1, \ldots, m \tag{5.2b}$$

$$x_i \geqslant 0 \qquad i = 1, \ldots, r \tag{5.2c}$$

problem (5.1) clearly being a particular case.

We now convert the inequality constraints (5.2b) into a set of equalities by defining a new non-negative variable for each constraint as the difference between the two sides of that constraint. Precisely, for the jth constraint we define the variable x_{r+j} as

$$x_{r+j} = b_j - \sum_{i=1}^{r} a_{ji} x_i$$

or, more symmetrically, as

$$\sum_{i=1}^{r} a_{ji} x_i + x_{r+j} = b_j.$$

From the direction of the inequality we can see that x_{r+j} must be non-negative as required. Each slack variable provides a measure of how far the constraint is from being binding—when it is zero the constraint is binding and as it increases from zero the distance from the corresponding boundary progressively increases.

With the addition of slack variables problem (5.2) becomes

$$\text{maximise} \sum_{i=1}^{r} c_i x_i$$

$$\text{subject to} \sum_{i=1}^{r} a_{ji} x_i + x_{r+j} = b_j \qquad j = 1, \ldots, m \tag{5.3}$$

$$x_i \geqslant 0 \qquad i = 1, \ldots, r + m.$$

To gain some intuitive understanding of the role of these new variables in the development of the theory let us look again at problem (5.1). Here three slack variables x_3, x_4, x_5 are required, defined by the equations:

$$\begin{aligned} 3x_1 + 5x_2 + x_3 &= 180 \\ x_1 + x_2 \phantom{{}+x_3} + x_4 &= 50 \\ 2x_1 + 8x_2 \phantom{{}+x_3+x_4} + x_5 &= 232 \\ x_1, \ldots, x_5 &\geqslant 0. \end{aligned}$$

The feasible region (Fig. 5.1a) is bounded by five lines corresponding to the five constraints of (5.1) being treated as equalities. For example the boundary of $3x_1 + 5x_2 \leq 180$ is given by $3x_1 + 5x_2 = 180$, and so can be equivalently defined by $x_3 = 0$. Similarly, each boundary line (including the x_1 and x_2 axes) can now be specified by setting *one* of $x_1, \ldots, x_5$ zero (Fig. 5.1b). The vertices, being the intersections of two boundary lines, are therefore specified by setting *two* of the variables zero.

These properties generalise immediately to higher dimensions. For $r = 3$ each constraint defines a plane specified by one variable being zero (Fig. 5.1c). The edges are specified by two variables being zero, and the vertices by three being zero. In the general case (5.3) the vertices will be specified by a minimum of r variables being zero. The remaining m variables are known as *basis variables*, often referred to collectively as *the basis*. For example, the vertex A_1 in problem (5.1) is defined by the non-basis variables x_2 and x_4 being zero. The basic variables are x_1, x_3, and x_5, taking the values 50, 30 and 132 respectively.

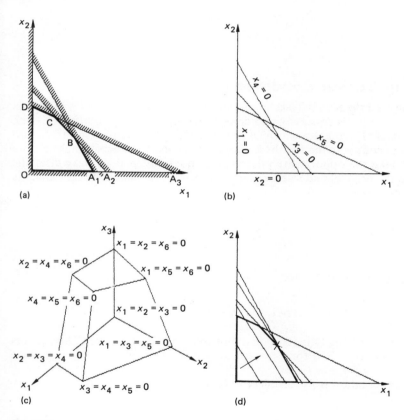

Fig. 5.1

Exercise Set 1

1 (i) Plot the following feasible regions and obtain the coordinates of each vertex.
(ii) Add slack variables to the constraints and associate with each vertex found in (i) its defining non-basis variables.

(a) $2x_1 + x_2 \leq 10$
$x_2 \leq 8$
$-x_1 + x_2 \leq 5$
$2x_1 - x_2 \leq 5$
$x_1, x_2 \geq 0$

(b)* $x_1 - 2x_2 \leq -1$
$x_1 + x_2 \leq 2$
$x_1 + 3x_2 \geq 3$
$x_1, x_2 \geq 0$.

2 With the feasible region defined as in 1(b), find points to minimise each of the following objective functions
   (i) by graphical methods, and
   (ii) by evaluating each objective function at every vertex.

(a)* $x_1 + 2x_2$
(b) $2x_1 + x_2$
(c)* $x_1 + 4x_2$
(d) $x_1 - x_2$
(e)* $-x_1 - x_2$.

5.3 The Kuhn–Tucker Conditions

In Chapter 3 we derived the Kuhn–Tucker conditions for problem (5.2) and we noted in Chapter 4 that because of its concavity and convexity properties any solution of these conditions is optimal. Having transformed the inequality constraints of (5.2) into equalities we have, of course, to modify the Kuhn–Tucker conditions accordingly. Before doing this we will take the opportunity of rewriting equations (5.3) to make clear the symmetry between the original and slack variables. Precisely, the Kuhn–Tucker analysis will be based on the program[2]:

$$\text{maximise } \sum_{i=1}^{n} c_i x_i$$

$$\text{subject to } \sum_{i=1}^{n} a_{ji} x_i = b_j \qquad j = 1, \ldots, m \qquad (5.4)$$

$$x_i \geq 0 \qquad i = 1, \ldots, n$$

where the constraints are linearly independent and consistent with $n > m$. (This form is identical to (5.3) when all $c_i = 0$ for $i > r$, all $a_{ji} = 0$ for $i > r$ except $i = r+j$; when $a_{jr+j} = 1$ for all j; and $n = r+m$.)

The Kuhn–Tucker conditions for problem (5.4) are given by:

$$\partial L/\partial x_i = c_i - \sum_{j=1}^{m} \lambda_j a_j \leq 0 \qquad (5.5)$$

$$x_i \geq 0 \qquad (5.6)$$

$$x_i \partial L/\partial x_i = 0 \qquad (5.7)$$

for $i = 1, \ldots, n$ and

$$\partial L/\partial \lambda_j = b_j - \sum_{i=1}^{n} a_{ji} x_i = 0 \qquad (5.8)$$

for $j = 1, \ldots, m$

where $L = \sum_{i=1}^{n} c_i x_i - \sum_{j=1}^{m} \lambda_j \sum_{i=1}^{n} (a_{ji} x_i - b_j)$

$= c_1 x_1 + c_2 x_2 + \ldots + c_n x_n - \lambda_1 (a_{11} x_1 + a_{12} x_2 + \ldots + a_{1n} x_n - b_1)$
$\quad - \lambda_2 (a_{21} x_1 + a_{22} x_2 + \ldots + a_{2n} x_n - b_2) - \ldots$
$\quad - \lambda_m (a_{m1} x_1 + a_{m2} x_2 + \ldots + a_{mn} x_n - b_m).$ (5.9)

Note that since the constraints are equalities the complementary slackness conditions corresponding to equations (5.8) are automatically satisfied and so $\lambda_1 \ldots \lambda_m$ are not required a priori to be non-negative. Note also that the complementary slackness conditions (5.7) ensure that at any vertex the Lagrangean and objective functions are numerically equal.

Conditions (5.6) and (5.8) of course correspond to the feasible region of problem (5.4) and so just confirm that any solution must be feasible. Consequently, the conditions (5.5) and (5.7) must contain the extra information necessary to pinpoint the position of the optimum. For each $x_i > 0$, we see from (5.7) that the λs must satisfy a linear equation ($\partial L/\partial x_i = 0$), and since there are only m λs, they cannot, in general, be made to satisfy more than m such equations. It follows that at most m of the xs may be positive, or equivalently, at least $n - m$ must be zero. This corresponds in general to the condition for a vertex. (It can be shown that the optimum of a linear program with a finite feasible region is *always* attained at a vertex—maybe elsewhere also.)

Suppose that at the optimum $x_i > 0$ (i.e. x_i is a basis variable) then from (5.7) we must have $\partial L/\partial x_i = 0$: but $\partial L/\partial x_i$ is just the coefficient of x_i in L. Consequently, at the optimum the λs have to be chosen so that in L the coefficients of the basis variables vanish, and the remaining coefficients (of the non-basis variables) must then be non-positive according to (5.5).

We now consider the systematic search procedure referred to in the introduction. In this algorithm we start with any convenient vertex of the feasible region and then choose the λs in L in such a way as to make the coefficients of the corresponding basis variables zero. Such a vertex satisfies all the Kuhn–Tucker conditions except, perhaps, (5.5). If (5.5) is also satisfied then we are at an optimum, otherwise we must move from the present vertex to a neighbouring one at which the Lagrangean has a higher value. Neighbouring vertices are such that they have all but one of their non-basis variables in common (or equivalently have all but one of their basis variables in common). This corresponds to the normal use of the word in two or three dimensions. The choice of which variable should enter the basis is made by considering the sign of $\partial L/\partial x_i$ amongst the non-basis variables. A positive sign indicates that the corresponding condition (5.5) is violated, and also that with these values for λ, L will increase in that direction. Consequently, any such variable when brought into the basis will result in an increased value for L. Having decided which variable is to enter the basis, we must now decide which variable must leave, in order to keep the total number of basis variables unchanged. Which it will be is determined by the constraints (5.6) and (5.8) as we shall see in the next section. In this way, the Lagrangean (and hence the objective function) is increased at each step, leading to the optimum in relatively few steps[3].

In summary, the steps of the proposed algorithm are as follows.

1 Start at a vertex of the feasible region (i.e. a feasible vertex): this defines a basis.
2 Choose the λs in L so as to make the coefficients of the basis variables zero.
3 (a) If the coefficients of the xs in L are all non-positive then an optimum has been found.
   (b) If not, then choose any x_i whose coefficient is positive as a new basis variable. The variable that must leave the basis is determined by the constraints.

A new feasible vertex (and basis) has now been defined.
Repeat from Step 2.

5.4 Development of the Algorithm

Let us follow the algorithm proposed in the last section to solve problem (5.1). This problem with slack variables included is:

maximise $\quad P = 50x_1 + 60x_2$

subject to

$$3x_1 + 5x_2 + x_3 \qquad\qquad = 180 \quad (5.9a)$$
$$x_1 + x_2 + x_4 \qquad = 50 \quad (5.9b)$$
$$2x_1 + 8x_2 + x_5 = 232 \quad (5.9c)$$
$$x_1, x_2, \ldots, x_5 \geqslant 0$$

with $n = 5$ and $m = 3$. At a vertex $n - m = 2$ of the variables must be zero and the remaining three must form the basis. The steps of the algorithm are as follows.

1. We must find a feasible vertex (any will do). Luckily the constraint equations (5.9) are in such a form as to suggest an 'obvious' one; $x_1 = x_2 = 0$ with $x_3 = 180$, $x_4 = 50$, and $x_5 = 232$, the value of the objective function, P, being zero. Our starting vertex is therefore $\mathbf{x}_0 = (0, 0; 180, 50, 232)$, which corresponds to the origin in the two-dimensional space of Fig. 5.1a. For clarity the original variables x_1 and x_2 are shown separated from the slack variables by a semicolon.

2. The Lagrangean is:

$$L = 50x_1 + 60x_2 - \lambda_1(3x_1 + 5x_2 + x_3 - 180) - \lambda_2(x_1 + x_2 + x_4 - 50)$$
$$- \lambda_3(2x_1 + 8x_2 + x_5 - 232) \quad (5.10)$$

and we choose the λs to make the coefficients of the basis variables x_3, x_4, x_5 zero. Hence $\lambda_1 = \lambda_2 = \lambda_3 = 0$ (corresponding to none of the constraints in the Lagrangean being binding), giving

$$L = 50x_1 + 60x_2.$$

3. The coefficients in L of both x_1 and x_2 are positive so $\mathbf{x}_0$ does not correspond to an optimum. Either x_1 or x_2 could now be chosen to become a basis variable—for the sake of argument let us choose x_1. This means that we have in fact chosen to move towards the next vertex in the direction of x_1 increasing, i.e. along the x_1 axis towards A_1, A_2, and A_3 (Fig. 5.1a). As there can be only three basis variables we must now choose one of x_3, x_4 and x_5 to leave the basis and become zero—but which? Let us try each in turn, remembering that x_2 is still zero.

    Setting $x_3 = 0$ in equation (5.9a) gives $x_1 = 180/3 = 60$, and consequently $x_4 = -10$, $x_5 = 112$ from the other equations.

    Setting $x_4 = 0$ in equation (5.9b) gives $x_1 = 50$, and consequently $x_3 = 30$, $x_5 = 132$.

    Setting $x_5 = 0$ in equation (5.9c) gives $x_1 = 232/2 = 116$, and consequently $x_3 = -168$, $x_4 = -66$.

    These points correspond to A_2, A_1 and A_3 respectively in Fig. 5.1a, only A_1 being in the feasible region since the other points have at least one element negative. That this is the only point from amongst the three that is feasible can be deduced from the values of x_1 alone. Only the nearest vertex to the origin along the x_1 axis could be feasible (since the others are on the 'wrong side' of a boundary line defining the nearest vertex) and this corresponds to the point with the lowest x_1 value. Amongst the three values, 60, 50, 116, obtained for x_1 we must therefore choose $x_1 = 50$, which resulted from setting x_4 zero, so x_4 leaves the basis. The new feasible vertex is:

    $$\mathbf{x}_1 = (50, 0; 30, 0, 132), \text{ with } P = 2500.$$

2 We now apply the algorithm again, starting from x_1. Referring to equation (5.10) we must now choose the λs to make the coefficients of x_1, x_3, and x_5 zero. Setting each zero gives:
$$50 - 3\lambda_1 - \lambda_2 - 2\lambda_3 = 0; \quad 0 - \lambda_1 = 0; \quad 0 - \lambda_3 = 0,$$
i.e. $\lambda_1 = 0, \quad \lambda_2 = 50, \quad \lambda_3 = 0$, resulting in:
$$L = 10x_2 - 50x_4 + 2500.$$

3 Now the coefficient of x_2 is positive so we need to make x_2 a basis variable, and one of x_1, x_3, x_5 zero. (We are moving along AB in Fig. 5.1a.) As before, we try each possibility in turn, remembering that $x_4 = 0$.

    Setting $x_1 = 0$ in equation (5.9b) gives $x_2 = 50$.
    Setting $x_3 = 0$ in equation (5.9a) gives $3x_1 + 5x_2 = 180$ and
                  in equation (5.9b) gives $x_1 + x_2 = 50$.
Solving these equations simultaneously results in $x_2 = 15$.
    Setting $x_5 = 0$ in equation (5.9c) gives $2x_1 + 8x_2 = 232$ and
                  in equation (5.9b) gives $x_1 + x_2 = 50$, resulting in $x_2 = 132/6 = 22$.

Arguing as before, we choose the smallest of these values for x_2, namely $x_2 = 15$, which followed from setting $x_3 = 0$. The other coordinates for this point are found by substituting these values back into the constraints to give the new feasible vertex

$$x_2 = (35, 15; 0, 0, 42) \text{ with } P = 2650$$

corresponding to the point B in Fig. 5.1a.

2 Applying the algorithm again, starting with x_2 and referring to equation (5.10), we choose the λs to make the coefficients of x_1, x_2 and x_5 zero, i.e.
$$50 - 3\lambda_1 - \lambda_2 - 2\lambda_3 = 0$$
$$60 - 5\lambda_1 - \lambda_2 - 8\lambda_3 = 0$$
and $\quad\quad 0 - \lambda_3 = 0.$
Solving these simultaneously gives:
$$\lambda_1 = 5, \quad \lambda_2 = 35, \quad \lambda_3 = 0$$
and
$$L = 5x_3 - 35x_4 + 2650.$$

3 Neither of these coefficients is positive, hence x_2 must correspond to an optimum of the problem, as can be confirmed by graphical techniques (Fig. 5.1d).

5.5 Streamlining the Algorithm

Although this procedure provides us with a workable algorithm for solving a linear program, it was very noticeable towards the end of our short example that the steps were becoming longer and more involved. In order to obtain the values of the λs a set of simultaneous equations had to be solved, and yet another set to find the corresponding xs—in contrast to the initial steps at which all these values were obvious. The reason for this was that initially the constraints were in a very convenient form: the basis variables appeared in one and only one equation (each with unit coefficient) and consequently the constant terms (necessarily non-negative for a feasible solution) gave the corresponding values for these basis variables. Since this form for the equations is so useful we shall refer to it as the *standard form* with respect to a particular basis. If we now *arrange* for our equations to be in standard form with respect to the current basis at each iteration of the algorithm—by a method known as 'pivoting'—we should be able to benefit from simpler algebra. The idea behind pivoting is quite straightforward: the unique equation (known as the pivot equation) in which both the variable to leave the basis and the one to enter the basis (the pivot variable) occur is divided by the coefficient of the new basis variable and is then used to substitute out the new basis variable from each of the other constraints. The equations will then be in standard form with respect to the new basis. These new equations will, of course, be equivalent to the previous set in that a solution of the one will always be a solution of the other.

Let us look at this pivoting operation at the second iteration of the algorithm after we have decided that x_1 should replace x_4 as a basis variable. At this stage the constraint equations are (5.9), the pivot variable is x_1 and the pivot equation is (5.9b):

$$x_1 + x_2 + x_4 = 50 \tag{5.9b}$$

since it contains both x_1 and x_4. This equation is ready for the substitution process since the coefficient of x_1 happens to be unity. Now we must ensure that x_1 appears in this pivot equation only, so we must use it to eliminate x_1 from the other constraints. For example, to remove x_1 from equation (5.9a) we subtract 3 times (5.9b), and to remove x_1 from equation (5.9c) we subtract 2 times (5.9b). The new set of equations is therefore:

$$2x_2 + x_3 - 3x_4 = 30 \tag{5.11a}$$
$$x_1 + x_2 \phantom{{}+{}} + x_4 = 50 \tag{5.11b}$$
$$6x_2 \phantom{{}+{}} - 2x_4 + x_5 = 132 \tag{5.11c}$$

which is in standard form with respect to the new basis variables x_1, x_3 and x_5, whose values can be read off as 30, 50 and 132 respectively with x_2 and x_4 zero.

In our algorithm we found that the determination of the λs also became complicated as the iterations progressed: but even their evaluation can be simplified. This is achieved by the introduction of 'local' multipliers which are just linear combinations of the λs. For example the original Lagrangean (5.10) expressed in terms of the constraints (5.9)

$$L = 50x_1 + 60x_2 - \lambda_1(3x_1 + 5x_2 + x_3 - 180) - \lambda_2(x_1 + x_2 + x_4 - 50) \\ - \lambda_3(2x_1 + 8x_2 + x_5 - 232)$$

can be written in terms of the new constraints (5.11) as

$$L = 50x_1 + 60x_2 - \mu_1(2x_2 + x_3 - 3x_4 - 30) - \mu_2(x_1 + x_2 + x_4 - 50) \\ - \mu_3(6x_2 - 2x_4 + x_5 - 132) \tag{5.12}$$

where

$$\mu_1 = \lambda_1, \quad \mu_2 = 3\lambda_1 + \lambda_2 + 2\lambda_3 \quad \text{and} \quad \mu_3 = \lambda_3 \tag{5.13}$$

are our new 'local' multipliers.

The values of the μs are now obtained quite straightforwardly, since in order to have the basis variable coefficients zero we must have

$$50 - \mu_2 = 0, \quad 0 - \mu_1 = 0, \quad 0 - \mu_3 = 0.$$

As we see the μs are the coefficients of the corresponding basis variables in P. For example, since x_1 is defined in the second constraint (5.11b) which has multiplier μ_2, we have $\mu_2 = 50$, the coefficient of x_1 in $P = 50x_1 + 60x_2$. This result is quite general and follows from the fact that the constraints are in standard form. What is more, the λs will be found to be the negatives of the coefficients of the slack variables in L after the μs have been given their numerical values. This follows from the fact that the λs were these coefficients in equation (5.10), and we showed this to be equivalent to equation (5.12). As $L = 10x_2 - 50x_4 + 2500$ we must have $\lambda_1 = 0$, $\lambda_2 = 50$ and $\lambda_3 = 0$ (the coefficients of the slack variables x_3, x_4 and x_5) confirming the values found earlier.

5.6 The Simplex Tabular Algorithm

We will now rework problem (5.1) incorporating these modifications. At the same time we will give a condensed 'tabular' method which eliminates the unnecessary repetition of variable names and 'mechanises' all the decision processes. It is this tabular method that is usually referred to as the Simplex Tabular Algorithm.

For convenience we will restate the problem here, together with the first iteration which was performed in the last section.

$$\begin{aligned}
\text{Maximise} \quad & P = 50x_1 + 60x_2 \\
\text{subject to} \quad & 3x_1 + 5x_2 + x_3 = 180 \\
& x_1 + x_2 + x_4 = 50 \\
& 2x_1 + 8x_2 + x_5 = 232 \\
& x_1, x_2, \ldots, x_5 \geq 0.
\end{aligned}$$

1 These constraints are in standard form with respect to the basis variables x_3, x_4, x_5, and result in the 'obvious' initial solution $x_0 = (0, 0; 180, 50, 232)$.

2 $L = 50x_1 + 60x_2 - \lambda_1(3x_1 + 5x_2 + x_3 - 180) - \lambda_2(x_1 + x_2 + x_4 - 50)$
$- \lambda_3(2x_1 + 8x_2 + x_5 - 232)$.

For the coefficients of the basis variables to be zero we need $\lambda_1 = \lambda_2 = \lambda_3 = 0$. In this case we have the trivial result that $L = 50x_1 + 60x_2$. We can tabulate this information more concisely as in Table 5.1, where the coefficients of the objective function and the constraints are shown in the centre section of the table and the corresponding constants in the RHS (Right-Hand-Side) column. The second column (on the left) contains the names of the basis variables opposite the equation in which each appears, so that we can read off their values as 180, 50 and 232 from the RHS column. The values of the 'local' multipliers are shown in the μ column. In this first tableau the μs are of course just the λs and are found, as explained in the last section, as the coefficients in P of the corresponding basis variables. Each coefficient in the L row can be calculated mechanically by the rule:

coefficient of x_i in L = coefficient of x_i in $P -$

$$\sum_j \mu_j \times (\text{coefficient of } x_i \text{ in the } j^{th} \text{ constraint}) \tag{5.14}$$

This rule also applies to finding minus the constant in L, if we remember that the constant in P is (usually) zero. In terms of the tableau, each coefficient of the L row is evaluated by multiplying each element in its column by the corresponding element in the μ column, adding their products and subtracting the sum from the P coefficient. For example the coefficient of x_1 in L is obtained as $50 - (0 \times (3) + 0 \times (1) + 0 \times (2)) = 50$, since the μ values are $(0, 0, 0)$ and the x_1 coefficients are $(3, 1, 2)$.

Table 5.1 The Initial Tableau

Basis		P	50	60	0	0	0	0	
μ	x		x_1	x_2	x_3	x_4	x_5	RHS	Ratio
0	x_3		3	5	1	0	0	180	180/3 (1a)
0	x_4		☐1	1	0	1	0	50	ⓐ50/1 (1b)
0	x_5		2	8	0	0	1	232	232/2 (1c)
	L		ⓐ50	60	0	0	0	0	

124 *Linear programming*

The purpose of the Ratio column and the significance of the circles and square will be described shortly.

3 In order to decide which variable should enter the basis, we look for a variable with a positive coefficient in L: here there are two, and we shall arbitrarily choose x_1 which then becomes the pivot variable—shown circled in the tableau.

The variable to leave the basis is chosen by inspecting the possible values for x_1 after setting each of the basis variables zero in turn. From the first constraint (setting $x_3 = 0$) we get $x_1 = 180/3 = 60$, from the second ($x_4 = 0$), $x_1 = 50/1 = 50$ and from the third ($x_5 = 0$), $x_1 = 232/2 = 116$. These values for x_1 are obtained by dividing the elements of the RHS column by the corresponding coefficients of the pivot variable and are shown in the Ratio column.

We choose the smallest of these values, $x_1 = 50$ (shown circled), since it corresponds to the nearest vertex in the direction of L increasing. The second constraint therefore becomes our pivot equation. For our constraints to be in standard form with respect to our new basis x_3, x_1, x_5 the coefficient of x_1 must be unity in the pivot equation (shown boxed), and zero in the others. Performing the same pivoting operation as before we obtain:

$$2x_2 + x_3 - 3x_4 = 30$$
$$x_1 + x_2 \phantom{{}-{}} + x_4 \phantom{{}+x_5} = 50$$
$$6x_2 \phantom{{}+x_3} - 2x_4 + x_5 = 132,$$

or, in terms of the tableau rows shown in Tables 5.1 and 5.2:

$$(2b) = (1b); \quad (2a) = (1a) - 3(2b); \quad (2c) = (1c) - 2(2b)$$

where, for example, the second expression means that each element of row (2a) is formed by subtracting 3 times the corresponding element in row (2b) from the corresponding element in row (1a), i.e. (0 2 1 −3 0 30) = (3 5 1 0 0 180) − 3(1 1 0 1 0 50). In forming the second tableau, we have replaced x_4 by x_1 in the basis column, the corresponding μ being 50, the coefficient of x_1 in P. Using equation (5.14) the coefficients in L are obtained as:

x_1: $50 - (0(0) + 50(1) + 0(0)) = 0$ since $(0, 50, 0)$ and $(0, 1, 0)$ are the elements in the μ and x_1 columns respectively).

Similarly:

x_2: $60 - (0(2) + 50(1) + 0(6)) = 10$
x_3: $0 - (0(1) + 50(0) + 0(0)) = 0$
x_4: $0 - (0(-3) + 50(1) + 0(-2)) = -50$
x_5: $0 - (0(0) + 50(0) + 0(1)) = 0,$

and minus the constant as:

$$0 - (0(30) + 50(50) + 0(132)) = -2500.$$

Table 5.2 The Second Tableau

μ	x	x_1	x_2	x_3	x_4	x_5	RHS	Ratio	
0	x_3	0	⬜2	1	−3	0	30	(30/2)	(2a)
50	x_1	1	1	0	1	0	50	50/1	(2b)
0	x_5	0	6	0	−2	1	132	132/6	(2c)
	L	0	(10)	0	−50	0	−2500		

The new feasible vertex is (50, 0; 30, 0, 132) with $P = 2500$ (since at a vertex $L = P$, and $-L = -2500$ from the tableau). The corresponding λs are 0, 50, 0 respectively, the negatives of the slack variable coefficients in the L row. We can now dispense with the full equations and work in terms of the tableaux alone.

2 The only positive L coefficient is that of x_2, which becomes the new pivot variable. The Ratio column is next evaluated and the smallest feasible value for x_2 chosen and circled, indicating the new pivot equation. We now pivot on the boxed element which this time is not unity, and so we must divide (2a) by 2 in order to obtain (3a) (Table 5.3). The other rows of this new tableau are obtained by forming: (3b) = (2b) − (3a) and (3c) = (2c) − 6(3a), and the new basis variable x_2 replaces x_3 in the basis column. The x_1, x_2 and x_5 coefficients of L must be zero, the others being evaluated using equation (5.14):

x_3: $0 - (60(1/2) + 50(-1/2) + 0(-3)) = -5$
x_4: $0 - (60(-1/2) + 50(5/2) + 0(7)) = -35$
constant: $0 - (60(15) + 50(35) + 0(42)) = -2650$.

In summary, the new feasible vertex is (35, 15; 0, 0, 42) with $P = 2650$ and the Lagrange multipliers $(\lambda_1, \lambda_2, \lambda_3) = (5, 35, 0)$.

As none of the coefficients in the L row are positive the optimal solution has been obtained, confirming the value obtained earlier.

Table 5.3 The Third Tableau

μ	x	x_1	x_2	x_3	x_4	x_5	RHS	
60	x_2	0	1	1/2	−3/2	0	15	(3a)
50	x_1	1	0	−1/2	5/2	0	35	(3b)
0	x_5	0	0	−3	7	1	42	(3c)
	L	0	0	−5	−35	0	−2650	

The steps performed in the Simplex Tabular Algorithm are summarised in Procedure Box 5.1. (Note that in certain circumstances the algorithm may break down—these will be discussed later.)

PROCEDURE 5.1
The Simplex Tabular Algorithm
(To solve Problem 5.4)

1. Find a feasible vertex. (If the constraints are in the form (5.2) and $b_j \geq 0$ for all j, then the origin is always such a point with the slack variables forming the basis.)
2. If the equations are not in standard form with respect to the current basis perform the necessary pivoting operation.
3. Obtain the local multipliers (the coefficients of the corresponding basis variables in P), and evaluate the coefficients and constant in L using equation (5.14).
4. If none of the coefficients in L are positive an optimal solution has been found.

    Otherwise choose as the new pivot variable an x_i such that its coefficient in L is positive. This variable enters the basis. Evaluate the Ratio column: the smallest non-negative entry defines[4] the pivot equation and hence the variable to leave the basis. The current basis variables are now known.
    Repeat from step 2.

We will apply this procedure to another example, this time one in three variables for which a graphical solution would be more difficult. Here the tableaux have been joined together, demonstrating the compactness of the procedure.

Example 1

Maximise $5x_1 + 4x_2 + 6x_3$
subject to
$$x_1 - x_2 + x_3 \leq 20$$
$$3x_1 + 2x_2 + 4x_3 \leq 42$$
$$3x_1 + 2x_2 \leq 30$$
$$x_1, x_2, x_3 \geq 0.$$

We must first put the problem into the form of (5.4) by adding slack variables x_4, x_5 and x_6, giving:

maximise $5x_1 + 4x_2 + 6x_3$
subject to
$$x_1 - x_2 + x_3 + x_4 = 20$$
$$3x_1 + 2x_2 + 4x_3 + x_5 = 42$$
$$3x_1 + 2x_2 + x_6 = 30$$
$$x_1, \ldots, x_6 \geq 0.$$

Here again we have an obvious initial feasible vertex defined by $x_1 = x_2 = x_3 = 0$, giving $x_4 = 20$, $x_5 = 42$ and $x_6 = 30$. Fortunately the constraints are already in standard form with respect to this basis. A complete solution is given in Table 5.4, the optimal vertex being (0, 15, 3; 32, 0, 0) with 78 as the value of the objective function. Note that in the second tableau the negative element in the Ratio column is not considered as it must correspond to a non-feasible vertex.

Table 5.4 A Solution to Example 1

| Basis | | P | 5 | 4 | 6 | 0 | 0 | 0 | | |
μ	x		x_1	x_2	x_3	x_4	x_5	x_6	RHS	Ratio
0	x_4		1	−1	1	1	0	0	20	20
0	x_5		3	2	④	0	1	0	42	㉑/2
0	x_6		3	2	0	0	0	1	30	∞
	L		5	4	⑥	0	0	0	0	
0	x_4		1/4	−3/2	0	1	−1/4	0	19/2	−19/3
6	x_3		3/4	1/2	1	0	1/4	0	21/2	21
0	x_6		3	②	0	0	0	1	30	⑮
	L		1/2	①	0	0	−3/2	0	−63	
0	x_4		5/2	0	0	1	−1/4	3/4	32	
6	x_3		0	0	1	0	1/4	−1/4	3	
4	x_2		3/2	1	0	0	0	1/2	15	
	L		−1	0	0	0	−3/2	−1/2	−78	

In each of the tableaux we had a choice as to which variable should be brought into the basis and in each case we took the one with the largest positive coefficient in L. We *could* have chosen any with a positive coefficient, but it has been found that *on average* the one with the largest element requires the fewest tableaux.

Procedure 5.1, as stated, assumes that we know a suitable starting point and that no degeneracies are present: we will discuss these provisos in the next section.

Exercise Set 2

1 (a)* Maximise $P = 11x_1 + 10x_2$
subject to
$$x_1 \leqslant 80$$
$$3x_1 + 2x_2 \leqslant 320$$
$$x_1 + 3x_2 \leqslant 240$$
$$x_1 + 2x_2 \leqslant 180$$
$$x_1, x_2 \geqslant 0.$$

(i) Obtain a graphical solution to the above problem.
(ii) Obtain a solution by the Simplex Tabular Algorithm, noting on the graph drawn in (i) the positions of the vertices visited.
(b) Repeat part (a) for the problem:

maximise $P = 3x_1 - 2x_2$
subject to
$$x_1 + x_2 \leqslant 10$$
$$x_2 \leqslant 8$$
$$-x_1 + x_2 \leqslant 5$$
$$x_1 - x_2 \leqslant 5$$
$$x_1, x_2 \geqslant 0.$$

2 Use the Simplex Tabular Algorithm to solve the following linear programs.
(i) Maximise $P = 8x_1 + 16x_2 + 7x_3$
subject to
$$3x_1 + 4x_2 + x_3 \leqslant 24$$
$$x_1 + 4x_2 + 3x_3 \leqslant 48$$
$$x_1, x_2, x_3 \geqslant 0.$$

(ii)* Maximise $P = 3x_1 + 2x_2 + 4x_3$
subject to
$$3x_1 + x_2 + 4x_3 \leqslant 60$$
$$x_1 + 2x_2 + 3x_3 \leqslant 30$$
$$2x_1 + 2x_2 + 3x_3 \leqslant 60$$
$$x_1, x_2, x_3 \geqslant 0.$$

5.7 Applying the Simplex Tabular Algorithm

In this section we consider methods that can be employed to enable the Simplex Tabular Algorithm to be used on a wider variety of problems than those already considered.

Setting up the problem

1 *Minimisation* The problem (5.4) is assumed to be one of maximisation. However, as usual, a minimisation problem can be solved by maximising the negative of the objective function. Alternatively, appropriate changes can be made to the algorithm.

2 *The initial vertex* The simplex method requires that we start with a feasible vertex, and that the equations are in standard form with respect to the corresponding basis. This implies, as mentioned earlier, that the constraint constants (RHS) must be non-negative. In the problems so far considered such a vertex was obvious, with the basis being formed by the slack variables. In many problems, such as those in which the constraints are either equalities or in which the coefficients of the slack variables have a negative sign, we have more difficulty in finding an initial vertex. For example, consider the following constants:

$$x_1 - x_2 \geqslant 3$$
$$2x_1 + x_2 = 4$$
$$5x_1 - 3x_2 \leqslant -2$$
$$-x_1 + x_2 \geqslant -1.$$

If we add slack variables x_3, x_4 and x_5 (all non-negative), to make the inequality constraints into equalities we obtain:

$$x_1 - x_2 - x_3 \qquad\qquad = 3$$
$$2x_1 + x_2 \qquad\qquad\qquad = 4$$
$$-5x_1 + 3x_2 \quad - x_4 \qquad = 2$$
$$x_1 - x_2 \qquad\quad + x_5 = 1$$

where we have rearranged the last two constraints to make the constants positive. Note the signs of x_3 and x_4! These constraints are not in standard form with respect to any basis, so there is no obvious solution. Now, we *could* in this case solve the set of equations by setting two variables zero at a time until a feasible vertex appeared, but this would not be sensible in a large problem.

A more useful approach requires the introduction of yet more non-negative variables—*artificial variables*—to 'prime' the problem. We add as many of these variables as necessary to put the equations into standard form. In this example we would need to add just three artificial variables x_6, x_7, x_8, as x_5 is already of the correct sign, so that the constraints become:

$$x_1 - x_2 - x_3 \qquad + x_6 \qquad\qquad\qquad = 3$$
$$2x_1 + x_2 \qquad\qquad\qquad + x_7 \qquad\qquad = 4$$
$$-5x_1 + 3x_2 \quad - x_4 \qquad\qquad\quad + x_8 = 2$$
$$x_1 - x_2 \qquad\quad + x_5 \qquad\qquad\qquad = 1$$
$$x_1, \ldots, x_8 \geqslant 0.$$

These equations are now in standard form with respect to the feasible vertex (0, 0; 0, 0, 1; 3, 4, 2) of the extended problem, and so we can proceed with the simplex algorithm. However, these artificial variables have nothing to do with the real problem and so it is necessary for them to be zero in the optimal solution—but there is no guarantee that this in fact will happen. In order to ensure that it does, we use a 'penalty' technique to force them out of the final basis. One such method requires us to incorporate each of the artificial variables into the objective function with a very large penalty.

For example, if our objective function had been

$$\text{maximise} \quad P = 10x_1 + 5x_2,$$

we would replace it by

$$\text{maximise} \quad P = 10x_1 + 5x_2 - Mx_6 - Mx_7 - Mx_8$$

where M is some very large (usually unspecified) number—larger than any other in the problem. The expectation is that the artificial variables will be forced out of the basis and will subsequently never return.

If it is found to be impossible to eliminate an artificial variable from the basis, then we must conclude that there is no feasible solution to the original problem.

3 *Unconstrained Variables* In our working we have assumed that all the variables in the problem are non-negative. If one of them, x say, is not restricted in this way, then we introduce two new non-negative variables y_1, y_2 and replace x by $y_1 - y_2$. As the difference between two non-negative variables can be either positive or negative this 'trick' enables unconstrained variables to be handled within the framework of the Simplex Tabular Algorithm. As the coefficients of y_1 and y_2 must always be of opposite sign, it follows that they cannot *both* be in the same basis, for if they were, their columns would not be of opposite sign.

The Ratio Problem

1 *Degeneracy* It is sometimes the case that one or more of the basis variables at a particular feasible vertex also turn out to be zero. This feature is called degeneracy, and corresponds in two dimensions, for example, to three (or more) boundary lines passing through the vertex being considered. It follows that the Ratio corresponding to that constraint must be zero. The question is, should we discard it or use it to define the pivot row? A standard method of arriving at the correct conclusion is to replace the offending RHS by ε, a small (undefined) positive number, keeping it in the working only while there would be degeneracy without it or until an optimal solution has been attained. Graphically, all that is happening (in two dimensions) is that the three lines no longer meet in a single point but in three very close points (Figs 5.2a, b). If there are several degeneracies it is *possible* for 'cycling' to occur, that is, for a set of tableaux to be repeated indefinitely. The easiest way of ensuring that this does not happen (when there are degeneracies present) is to choose the pivot variable *at random* from amongst those with a positive coefficient in L. In *practical* problems, however, cycling seems never to occur!

2 *Infinite Solutions* If all the finite Ratios are negative then we must in general conclude that there is no neighbouring vertex in the direction of L increasing, and so it is possible to move along one boundary without limit—that is we must have an infinite solution. This situation generally corresponds to a problem that has been badly modelled.

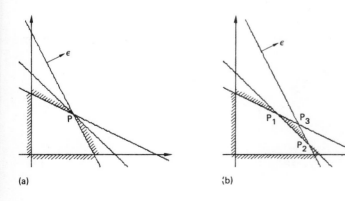

Fig. 5.2

Independence

Throughout the working we have assumed that the constraint equations are linearly independent (that is that none are redundant in that they can be obtained from the others). If we start with inequalities linear independence follows automatically from the definition of the slack variables. With equality constraints, however, there is no guarantee that they will be independent. Where possible, dependent constraints should be omitted from the analysis. If, by chance, a linearly dependent constraint should 'slip through the net', at some stage in the solution a complete row of zeros might emerge. To continue the solution process it is then only necessary to delete this whole line from that and subsequent tableaux.

Alternative Solutions

It should be pointed out that the optimum is often attained at more than one vertex. The presence of such alternatives is indicated in an optimal tableau by a zero coefficient in L of a non-basis variable. This variable can therefore be brought into the basis without changing the value of L (or the objective function), and hence provides an alternative vertex with the same value of the objective function.

If two or more vertices correspond to the optimum then there are an *infinite* number of feasible points which give the same value for the objective function. For example, any point on the line segment joining any pair of optimal vertices is also optimal, since the Lagrangean is not increasing in moving from one optimal vertex to the next.

The following example includes many of the above difficulties.

Example 2

$$\begin{aligned}
\text{Minimise} \quad & P = -2x - 2y \\
\text{subject to} \quad & x - 6y \geq -43 \\
& x + y \leq 13 \\
& x + y \geq 6 \\
& x \leq 6 \\
& y \geq 0, \quad x \text{ otherwise unrestricted.}
\end{aligned}$$

A plot of the feasible region is shown in Fig. 5.3.

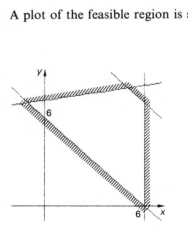

Fig. 5.3

5.7 Applying the simplex tabular algorithm

First we transform the problem into one of maximisation by considering the objective function $Q = 2x + 2y$.

Next we rewrite the first constraint to make the constant term positive:

$$-x + 6y \leqslant 43.$$

As x is not restricted to be non-negative we substitute:

$$x = x_1 - x_2, \quad \text{with } x_1 \geqslant 0, \quad x_2 \geqslant 0.$$

If we rename y to be x_3 the problem becomes:

maximise $\quad 2x_1 - 2x_2 + 2x_3$
subject to
$$-x_1 + x_2 + 6x_3 \leqslant 43$$
$$x_1 - x_2 + x_3 \leqslant 13$$
$$x_1 - x_2 + x_3 \geqslant 6$$
$$x_1 - x_2 \leqslant 6$$
$$x_1, x_2, x_3 \geqslant 0.$$

In order to make the constraints into equalities we now add slack variables x_4, x_5, x_6, x_7:

$$-x_1 + x_2 + 6x_3 + x_4 = 43$$
$$x_1 - x_2 + x_3 + x_5 = 13$$
$$x_1 - x_2 + x_3 - x_6 = 6$$
$$x_1 - x_2 + x_7 = 6.$$

These equations are *almost* in standard form (with respect to x_4, x_5, x_7), but because of the sign of x_6 we must add an artificial variable x_8, so that the complete problem becomes:

maximise $\quad 2x_1 - 2x_2 + 2x_3 \qquad\qquad\qquad - Mx_8$
subject to
$$-x_1 + x_2 + 6x_3 + x_4 = 43$$
$$x_1 - x_2 + x_3 + x_5 = 13$$
$$x_1 - x_2 + x_3 - x_6 + x_8 = 6$$
$$x_1 - x_2 + x_7 = 6$$
$$x_1, \ldots, x_8 \geqslant 0, \quad M \text{ very large}.$$

A complete solution is shown in Table 5.5, but the following notes may help you to understand all the steps in the calculations.

Notes (the numbers refer to those given in Table 5.5)

1. The problem is in standard form with respect to the basis variables x_4, x_5, x_8, x_7.
2. There is a choice of pivot equation: it is more convenient to choose the third since it forces x_8 from the basis.
3. As x_8 is now zero, a feasible vertex solution to the *original* problem has now been found: x_8 is no longer required and so we need not continue calculating its coefficients.
4. Since the RHS element of the last constraint is zero, there is degeneracy. Temporarily replace 0 by $\varepsilon(>0)$.
5. All the coefficients in L are non-negative: so setting ε zero, the vertex $\mathbf{x}_1 = (x_1, x_2, x_3; x_4, x_5, x_6, x_7; x_8) = (6, 0, 7; 7, 0, 7, 0; 0)$ with $Q = 26$ is a solution to the maximisation problem and $P = -26$ to the original problem. The zeros corresponding to the basis variables are underlined—the remaining zeros corresponding to x_2 and x_7 indicate that alternative solutions may exist. In the last tableau we have brought x_7 into the basis leading to an alternative optimal vertex $\mathbf{x}_2 = (5, 0, 8; 0, 0, 7, 1; 0)$ with $Q = 26$ as before. Since $\mathbf{x}_1, \mathbf{x}_2$ are both optimal vertices, all the points on the line segment joining

132 *Linear programming*

Table 5.5 Solution to Example 2

Basis		Q	2	−2	2	0	0	0	0	−M	0		Notes
μ	x		x_1	x_2	x_3	x_4	x_5	x_6	x_7	x_8	RHS	Ratio	
0	x_4		−1	1	6	1	0	0	0	0	43	<0	1
0	x_5		1	−1	1	0	1	0	0	0	13	13	
−M	x_8		☐1	−1	1	0	0	−1	0	1	6	⑥	2
0	x_7		1	−1	0	0	0	0	1	0	6	6	
	L		(2+M)	−2−M	2+M	0	0	−M	0	0	6M		
0	x_4		0	0	7	1	0	−1	0		49	<0	3
0	x_5		0	0	0	0	1	1	0		7	7	
2	x_1		1	−1	1	0	0	−1	0		6	<0	
0	x_7		0	0	−1	0	0	☐1	1		∅ε	ⓔ	4
	L		0	0	0	0	0	2	0		−12		
0	x_4		0	0	6	1	0	0	1		49+ε	(49+ε)/6	
0	x_5		0	0	☐1	0	1	0	−1		7−ε	ⓘ7−ε	
2	x_1		1	−1	0	0	0	0	1		6+ε	∞	
0	x_6		0	0	−1	0	0	1	1		ε	<0	
	L		0	0	②	0	0	0	−2		−12−2ε		
0	x_4		0	0	0	1	−6	0	☐7		7+7ε	ⓘ1+ε	
2	x_3		0	0	1	0	1	0	−1		7−ε	<0	
2	x_1		1	−1	0	0	0	0	1		6+ε	6+ε	
0	x_6		0	0	0	0	1	1	0		7	∞	
	L		<u>0</u>	0	<u>0</u>	<u>0</u>	−2	<u>0</u>	⓪		−26		5
0	x_7		0	0	0	1/7	−6/7	0	1		1		6
2	x_3		0	0	1	1/7	1/7	0	0		8		
2	x_1		1	−1	0	−1/7	6/7	0	0		5		
0	x_6		0	0	0	0	1	1	0		7		
	L		0	0	0	0	−2	0	0		−26		

these two points are also optimal. In particular $x_3 = (5.5, 0, 7.5; 3.5, 0, 7, 0.5; 0)$ also gives $Q = 26$, though it is not a vertex[5]. An attempt to bring x_2 into the basis in place of x_1 leads to the only finite ratio being negative. This implies that although the objective function remains finite (at 26), x_1 and x_2 may take infinite values provided they satisfy $x_1 - x_2 = 6$. All these points map onto the single point ($x = 6, y = 7$) in our original (2-dimensional) problem (Fig. 5.3). This corresponds to the vertex x_1 already found.

6 Since the degeneracy has been 'broken', the εs need no longer be used—they can be set zero.

Exercise Set 3
Use Procedure 5.1 to solve the following problems.

(i)* Maximise $x_1 + x_2 + x_3$
subject to $2x_1 - 3x_2 + 2x_3 \leqslant 2$
$-3x_1 + 2x_2 + 2x_3 \leqslant 2$
$2x_1 + 2x_2 - 3x_3 \leqslant 2$
$x_1, x_2, x_3 \geqslant 0.$

(ii) Minimise $-x_1 + 2x_2 - 2x_3$
subject to $-x_1 + x_2 \leqslant 10$
$x_1 - x_2 \leqslant 1$
$x_1 + 2x_3 \leqslant 8$
$x_1, x_2, x_3 \geqslant 0.$

(iii)* Minimise $60x_1 + 30x_2 + 60x_3$
subject to $3x_1 + x_2 + 2x_3 \geqslant 3$
$x_1 + 2x_2 + 2x_3 \geqslant 2$
$4x_1 + 3x_2 + 3x_3 \geqslant 4$
$x_1, x_2, x_3 \geqslant 0.$

(iv) Minimise $3x_1 + 2x_2 + 3x_3$
subject to $3x_1 + 4x_2 + 4x_3 \geqslant 2$
$4x_1 + 2x_2 + 3x_3 \geqslant 3$
$7x_1 + 6x_2 + 6x_3 \geqslant 4$
$x_1, x_2, x_3 \geqslant 0.$

(v)* Maximise $-3x_1 + 4x_2 - x_3$
subject to $x_2 + x_3 \leqslant 4$
$2x_1 + 3x_2 - x_3 \leqslant 8$
$-x_1 + 2x_2 + 4x_3 \leqslant 11$
$x_2, x_3 \geqslant 0, \quad x_1 \text{ unconstrained}.$

(vi) Maximise $4x_1 - 2x_2 - 3x_3$
subject to $x_1 - x_2 + x_3 \leqslant 8$
$3x_1 + x_2 \leqslant 4$
$20x_1 + 8x_2 - x_3 \leqslant 22$
$x_1, x_2 \geqslant 0, \quad x_3 \text{ unconstrained}.$

5.8 Sensitivity Analysis

In many instances the coefficients in either the objective function or the constraints will only be estimated, and we might like to know how sensitive the solution is to changes in them.

1 *Sensitivity to changes in the objective function*
Since the objective function appears just once in the tableaux (the top line), it is very easy to adjust the coefficients in this row and note the effect on the L row of the final tableau. For

example if we required to find what range of values for the coefficient of x_1 in the computer manufacturer problem (5.1) still gives rise to the point (35, 15; 0, 0, 42) as optimum, the P row and final tableau would be as in Table 5.6, where k is the coefficient of x_1.

Table 5.6 Change in the Objective Function

Basis μ	x	P x_1	k	60 x_2	0 x_3	0 x_4	0 x_5	0 RHS
60	x_2	0		1	1/2	−3/2	0	15
k	x_1	1		0	−1/2	5/2	0	35
0	x_5	0		0	−3	7	1	42
	L	0		0	−30 + k/2	90 − 5k/2	0	−900 − 35k

Note that k is now the local multiplier for the second constraint and so the calculation of L must take this into account. For the optimum vertex to remain unaltered we must have

$$-30 + k/2 \leqslant 0 \text{ and } 90 - 5k/2 \leqslant 0, \quad \text{i.e. } k \leqslant 60 \text{ and } k \geqslant 36.$$

That is, the coefficient of x_1 can take any value in the range 36 to 60 (inclusive) and vertex (35, 15; 0, 0, 42) will still remain optimal. Note that the *value* of the objective function depends on the actual value used for k.

2 Sensitivity to changes in the constraint coefficients

The effect of changes in a coefficient (or RHS) of a constraint is obtained by noting the effect of changes to the initial basis variables. The method is best explained through an example. Suppose we wish to discover by how much the coefficient of x_1 in the second constraint of Example 1 of Section 5.6 can be changed without affecting the optimal point.

We write out the first and last tableaux (Table 5.7), adding an extra column corresponding to the *change* in the x_1 coefficient (from 3 to k).

Table 5.7 Change in a Constraint Coefficient

Basis μ	x	P x_1	5 x_2	4 x_3	6 x_4	0 x_5	0 x_6	0 RHS	Change
0	x_4	1	−1	1	1	0	0	20	0
0	x_5	3	2	4	0	1	0	42	$k − 3$
0	x_6	3	2	0	0	0	1	30	0
	L	5	4	6	0	0	0	0	
0	x_4	5/2	0	0	1	−1/4	3/4	32	$−(k−3)/4$
6	x_3	0	0	1	0	1/4	−1/4	3	$(k−3)/4$
4	x_2	3/2	1	0	0	0	1/2	15	0
	L	−1	0	0	0	−3/2	−1/2	−78	

This is $(k − 3)$ times the x_5 column in the first tableau. As we move from tableau to tableau the 'change' column is always $(k − 3)$ times the x_5 column, consequently, the x_1 column in the final tableau is changed

$$\text{from } \begin{pmatrix} 5/2 \\ 0 \\ 3/2 \end{pmatrix} \text{ to } \begin{pmatrix} 5/2 - (k-3)/4 \\ 0 + (k-3)/4 \\ 3/2 + 0 \end{pmatrix}.$$

The corresponding coefficient in L then becomes:

$$5 - 0(5/2 - (k-3)/4) - 6((k-3)/4) - 4(3/2) = (7 - 3k)/2.$$

Hence the vertex remains optimal for $(7 - 3k)/2 \leq 0$, i.e. $k \geq 7/3$. Note that in this case, provided $k \geq 7/3$, the optimal value of P does not change.

This technique can only be employed with guaranteed success for coefficients corresponding to variables not in the basis in the final tableau[6].

5.9 Applications

A Least Absolute Deviation Regression

In Chapter 1 we discussed the problem of finding a line of 'best fit' to a set of data. The technique used (known as the method of least squares) involved minimising the sum of squares of the errors of the points from the line. The disadvantage of this method is that it puts a disproportionate weight on those observations far from the 'true' line. This means that any misreadings (which tend to be in this category) will automatically introduce a large bias in the position of the line. To overcome this disadvantage, instead of using the least squares criterion we often minimise the sum of absolute errors, the resulting line being known as the least absolute deviation (LAD) regression line.

For the sake of definiteness let us consider once again the simple numerical data of Chapter 1, where the data points were $(x, y) = (1, 1), (2, 2), (3, 4)$, and the desired line is of the form $y = ux + v$.

We now wish to find constants u and v to minimise the sum of absolute errors, $\sum_{i=1}^{3} |e_i|$, where

$$e_1 = 1 - 1u - v$$
$$e_2 = 2 - 2u - v$$
and $$e_3 = 4 - 3u - v.$$

Obviously we cannot use the calculus directly to solve this problem because of the non-differentiable objective function. However we can turn it into a linear program.

We first note that there is no reason why e_i, u and v cannot take positive or negative values, so using the standard trick discussed in Section 5.7 we substitute $e_i = f_i - g_i$ ($i = 1, 2, 3$); $u = u_1 - u_2$; $v = v_1 - v_2$, where each of the new variables is required to be non-negative. Our next problem is to rewrite the objective function in a linear form. We use the observation made in Section 5.7 that f_i and g_i cannot both be basis variables at the same vertex. Hence at the optimum

(and any vertex) $\quad |f_i - g_i| = f_i \quad$ if $g_i = 0$
$ = g_i \quad$ if $f_i = 0$,

or, more compactly $|f_i - g_i| = f_i + g_i$.

The problem can now be written:

$$\text{minimise} \quad \sum_{i=1}^{3} (f_i + g_i)$$

subject to
$$f_1 - g_1 + 1(u_1 - u_2) + (v_1 - v_2) = 1$$
$$f_2 - g_2 + 2(u_1 - u_2) + (v_1 - v_2) = 2$$
$$f_3 - g_3 + 3(u_1 - u_2) + (v_1 - v_2) = 4$$
$$f_1, f_2, f_3, g_1, g_2, g_3, u_1, u_2, v_1, v_2 \geq 0$$

which is a linear program with equality constraints. The solution (using artificial variables) gives the LAD regression line as:

$$y = 3x/2 - 1/2.$$

B Quadratic Programming

In Quadratic Programming we are concerned with maximising a quadratic function of non-negative variables subject to linear constraints. As we shall see the Kuhn–Tucker conditions in this case reduce to a (larger) set of linear constraints defining a feasible region in the extended space of variables and multipliers, together with a set of complementary slackness conditions. Provided the objective function is concave[7] it follows from the Kuhn–Tucker conditions that the optimum occurs at any vertex of this feasible region which also satisfies the complementary slackness conditions. To find such a vertex, is to some extent similar to the problem of finding an initial feasible vertex to a linear program, and we shall adapt the Simplex Tabular Algorithm to accomplish this in the context of the following example.

$$\begin{aligned}
\text{Maximise} \quad & -x_1^2 + 2x_1 x_2 - 2x_2^2 + 2x_1 + 6x_2 \\
\text{subject to} \quad & x_1 + x_2 \leq 2 \\
& -x_1 + 2x_2 \leq 2 \\
& 2x_1 + x_2 \leq 3 \\
& x_1, x_2 \geq 0.
\end{aligned}$$

It was shown (Exercise Set 4 of Chapter 4) that the objective function is concave and so the following Kuhn–Tucker conditions define an overall optimum to the problem:

$$\begin{aligned}
\partial L/\partial x_1 &= -2x_1 + 2x_2 + 2 - \lambda_1 + \lambda_2 - 2\lambda_3 \leq 0; & x_1 &\geq 0; & x_1 \partial L/\partial x_1 &= 0 \\
\partial L/\partial x_2 &= 2x_1 - 4x_2 + 6 - \lambda_1 - 2\lambda_2 - \lambda_3 \leq 0; & x_2 &\geq 0; & x_2 \partial L/\partial x_2 &= 0 \\
\partial L/\partial \lambda_1 &= -(x_1 + x_2 - 2) \geq 0; & \lambda_1 &\geq 0; & \lambda_1 \partial L/\partial \lambda_1 &= 0 \\
\partial L/\partial \lambda_2 &= -(-x_1 + 2x_2 - 2) \geq 0; & \lambda_2 &\geq 0; & \lambda_2 \partial L/\partial \lambda_2 &= 0 \\
\partial L/\partial \lambda_3 &= -(2x_1 + x_2 - 3) \geq 0; & \lambda_3 &\geq 0; & \lambda_3 \partial L/\partial \lambda_3 &= 0.
\end{aligned}$$

If we now add slack variables $s_1, \ldots, s_5 \geq 0$, these conditions become:

$$\left. \begin{aligned}
2x_1 - 2x_2 + \lambda_1 - \lambda_2 + 2\lambda_3 - s_1 &= 2 \\
-2x_1 + 4x_2 + \lambda_1 + 2\lambda_2 + \lambda_3 \phantom{{}-s_1} - s_2 &= 6 \\
x_1 + x_2 \phantom{{}+\lambda_1+2\lambda_2+\lambda_3-s_1-s_2} + s_3 &= 2 \\
-x_1 + 2x_2 \phantom{{}+\lambda_1+2\lambda_2+\lambda_3-s_1-s_2+s_3} + s_4 &= 2 \\
2x_1 + x_2 \phantom{{}+\lambda_1+2\lambda_2+\lambda_3-s_1-s_2+s_3+s_4} + s_5 &= 3
\end{aligned} \right\} (5.15) \quad \left. \begin{aligned} x_1 s_1 &= 0 \\ x_2 s_2 &= 0 \\ \lambda_1 s_3 &= 0 \\ \lambda_2 s_4 &= 0 \\ \lambda_3 s_5 &= 0 \end{aligned} \right\} (5.16)$$

$$x_1, \ldots, s_5 \geq 0$$

where the complementary slackness conditions (5.16) have been written in terms of the slack variables. These conditions imply that at least 5 of the variables should be zero. Since (5.15) consists of 5 linear constraints in 10 variables then in the absence of degeneracy, this implies that the solution of the problem corresponds to a vertex. Another interpretation of conditions (5.16) is that they restrict which variables may be in the basis together. For example as $x_1 s_1 = 0$, we cannot have them both as (non-degenerate) basis variables (i.e. they cannot both be positive) at the same vertex. Our problem then is to find a feasible vertex to (5.15) which also satisfies (5.16). The equations (5.15) are not even in standard form with respect to any basis, and so we have a problem akin to the initialisation problem of Linear Programming. We must introduce artificial variables s_6 and s_7 into the first and second equations of (5.15). However, in order to ensure that they do not appear in the final solution we have to set up a penalty objective function of the form:

minimise $s_6 + s_7$.

Note, there is no need to include Ms here as there are no other terms in the objective function: the minimum value is zero whether they are included or not.

We can now write our problem as:

maximise $\quad -s_6 - s_7$

subject to

$$\begin{aligned}
2x_1 - 2x_2 + \lambda_1 - \lambda_2 + 2\lambda_3 - s_1 + s_6 &= 2 & x_1 s_1 &= 0 \\
-2x_1 + 4x_2 + \lambda_1 + 2\lambda_2 + \lambda_3 - s_2 + s_7 &= 6 & x_2 s_2 &= 0 \\
x_1 + x_2 + s_3 &= 2 & \lambda_1 s_3 &= 0 \\
-x_1 + 2x_2 + s_4 &= 2 & \lambda_2 s_4 &= 0 \\
2x_1 + x_2 + s_5 &= 3 & \lambda_3 s_5 &= 0
\end{aligned}$$

$$x_1, \ldots, s_7 \geq 0$$

so that the usual Simplex Technique can be used, but with the added conditions concerning which variables may not be in the basis at the same time. A complete solution is shown in Table 5.8, where we have included in parentheses each variable that must *not* enter the basis with a current basis variable. For example in the first tableau s_5 is in the basis, so since $\lambda_3 s_5 = 0$, λ_3 may not enter the basis at the next step. The final tableau does not include any artificial variables and all the Kuhn–Tucker conditions are satisfied so $x_1 = 4/5$ and $x_2 = 6/5$ must correspond to an optimum of the original problem. The value of the objective function is found on substitution to be 5.76.

That this is the correct solution can be checked by referring to the contour map of the original problem (Fig. 5.4). The algorithm takes us along the path A → B → C → D from the origin, A, to the constrained maximum at D. If we had started by making x_1 a basic variable we would have moved along the alternative path A → E → F → G → D. EF is a section of the line $\partial f/\partial x_1 = 0$ where f is the objective function. If we travelled along this line we would eventually reach the unconstrained stationary point at (5, 4). We are prevented from doing so by the third constraint along which we move to reach the constrained maximum at D.

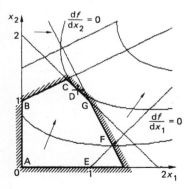

Fig. 5.4

138 *Linear programming*

Table 5.8 An Example of a Quadratic Program

μ	Basis x	P	x_1	x_2	λ_1	λ_3	λ_2	s_1	s_2	s_3	s_4	s_5	s_6	s_7	RHS	Ratio
			0	0	0	0	0	0	0	0	0	0	−1	−1	0	
−1	s_6		2	−2	1	−1	2	−1	0	0	0	0	1	0	2	<0
−1	s_7		−2	4	1	2	1	0	−1	0	0	0	0	1	6	3/2
0	$s_3(\lambda_1)$		1	−1	0	0	0	0	0	1	0	0	0	0	2	2
0	$s_4(\lambda_2)$		−1	[2]	0	0	0	0	0	0	1	0	0	0	2	①
0	$s_5(\lambda_3)$		2	1	0	1	0	0	0	0	0	1	0	0	3	3
	L		0	②	2	1	3	−1	−1	0	0	0	0	0	8	
−1	s_6		1	0	1	−1	2	−1	0	0	1	0	1	0	4	<0
−1	s_7		0	0	1	[2]	1	0	−1	0	−2	0	0	1	2	①
0	$s_3(\lambda_1)$		3/2	0	0	0	0	0	0	1	−1/2	0	0	0	1	∞
0	$x_2(s_2)$		−1/2	1	0	0	0	0	0	0	1/2	0	0	0	1	∞
0	$s_5(\lambda_3)$		5/2	0	0	0	0	0	0	0	−1/2	1	0	0	2	∞
	L		1	0	2	①	3	−1	−1	0	−1	0	0	0	6	
−1	s_6		1	0	3/2	0	5/2	−1	−1/2	0	0	0	1	0	5	5
0	$\lambda_2(s_4)$		0	0	1/2	1	1/2	0	−1/2	0	−1	0	0	1	1	∞
0	$s_3(\lambda_1)$		[3/2]	0	0	0	0	0	0	1	−1/2	0	0	0	1	(2/3)
0	$x_2(s_2)$		−1/2	1	0	0	0	0	0	0	1/2	0	0	0	1	<0
0	$s_5(\lambda_3)$		5/2	0	0	0	0	0	0	0	−1/2	1	0	0	2	4/5
	L		①	0	3/2	0	5/2	−1	−1/2	0	0	0	0	0	5	

−1	s_6	0	0	3/2	0	5/2	−1	−1/2	−2/3	1/3	0	1	13/3	26/9
0	$\lambda_2(s_4)$	0	0	⟨1/2⟩	1	1/2	0	−1/2	0	−1	0	0	1	②
0	$x_1(s_1)$	1	0	0	0	0	0	0	2/3	−1/3	0	0	2/3	∞
0	$x_2(s_2)$	0	1	0	0	0	0	0	1/3	1/3	0	0	4/3	∞
0	$s_5(\lambda_3)$	0	0	0	0	0	0	0	−5/3	1/3	1	0	1/3	∞
	L	0	0	③/2	0	5/2	−1	−1/2	−2/3	1/3	0	0	13/3	
−1	s_6	0	0	0	−3	1	−1	1	−2/3	10/3	0	1	4/3	②/5
0	$\lambda_1(s_3)$	0	0	1	2	1	0	−1	0	−2	0	0	2	<0
0	$x_1(s_1)$	1	0	0	0	0	0	0	2/3	−1/3	0	0	2/3	<0
0	$x_2(s_2)$	0	1	0	0	0	0	0	1/3	1/3	0	0	4/3	4
0	$s_5(\lambda_3)$	0	0	0	0	0	0	0	−5/3	1/3	1	0	1/3	1
	L	0	0	0	−3	1	−1	1	−2/3	⟨10/3⟩	0	0	4/3	
0	$s_4(\lambda_2)$	0	0	0	−0.9	0.3	−0.3	0.3	−0.2	1	0	0	0.4	
0	$\lambda_1(s_3)$	0	0	1	0.2	1.6	−0.6	−0.4	−0.4	0	0	0	2.8	
0	$x_1(s_1)$	1	0	0	−0.6	0.1	−0.1	0.1	0.6	0	0	0	0.8	
0	$x_2(s_2)$	0	1	0	0.6	−0.1	0.1	−0.1	0.4	0	0	0	1.2	
0	$s_5(\lambda_3)$	0	0	0	0.3	−0.1	0.1	−0.1	−1.6	0	1	0	0.2	
	L	0	0	0	0	0	0	0	0	0	0	−1	0	

140 Linear programming

Problem Set 5

1 Obtain the optimal solutions of the following linear programs. Include in each answer not only the coordinates of the point at which the optimum is attained, but also the values of the Lagrange multipliers.

(i)*   maximise $\quad -9x_1 - x_2 + x_3 - 3x_4$
$$\text{subject to} \quad \begin{aligned} -3x_1 \quad\quad + 4x_3 - x_4 &\leq 3 \\ 2x_1 + 5x_2 + x_3 + x_4 &\geq 1 \\ x_1 + x_2 \quad\quad - x_4 &\leq 2 \\ x_1, x_2, x_3, x_4 &\geq 0. \end{aligned}$$

(ii)   minimise $\quad 3w_1 - w_2 + 2w_3$
$$\text{subject to} \quad \begin{aligned} -3w_1 - 2w_2 + w_3 &\geq -9 \\ 5w_2 - w_3 &\leq 1 \\ 4w_1 - w_2 &\geq 1 \\ w_1 + w_2 + w_3 &\leq 3 \\ w_1, w_2, w_3 &\geq 0. \end{aligned}$$

2 Show that the two problems of Question 1 are duals. Using the arguments of Section 4.6 obtain the solution of (ii) directly from that of (i).

3* Using the fact that $x = (8, 6, 3, 0)$ is a feasible vertex for the following problem, obtain a solution using the Simplex Technique.

$$\begin{aligned} \text{Minimise} \quad & x_1 + 2x_2 + x_3 \\ \text{subject to} \quad & x_1 - x_2 + x_3 + 7x_4 = 5 \\ & x_2 + x_3 + x_4 = 9 \\ & x_3 + 2x_4 = 3 \\ & x_1, x_2, x_3, x_4 \geq 0. \end{aligned}$$

[*Hint*: put the equations in standard form with respect to the basis variables x_1, x_2, x_3.]

4* A Trim-Loss Problem.
A manufacturer of polythene bags buys rolls of polythene sheeting 2.3 metres wide. From these he cuts narrower rolls of widths 1.2, 0.8 and 0.45 metres as required for the bags he produces. On a particular day he has orders which would require 30 of the 1.2, 40 of the 0.8, and 50 of the 0.45 metre rolls. How best should he cut the rolls so as to minimise his waste sheeting?
[*Hint*: (i) List the 5 possible knife settings, and calculate the amount of waste produced by each.
(ii) Denoting the number of rolls produced at each of these settings by $x_1, \ldots, x_5$, obtain the linear program to minimise waste, assuming over-production of useful widths can be used on subsequent days.]

5 Obtain the dual of Question 2(ii) (Exercise Set 2), and *deduce* its solution from that of the primal.
Check your answer with that of Question 1(iii) (Exercise Set 3).

6 Consider again the LAD regression problem of Section 5.9. Suppose now that we wish to make the extra condition that the predicted value of y (i.e. $ux + v$) should never be less than the actual value as given by the data. Formulate this as a linear program—but do not attempt to solve it.

7 (i)* Solve the following linear program:

maximise $\quad 3x_1 + 2x_2 + x_3$
subject to $\quad x_1 + x_2 + x_3 \leq 3$
$\qquad\qquad 2x_1 \quad\;\; + x_3 \geq 2$
$\qquad\quad -x_1 + x_2 + x_3 \geq 0$
$\qquad\qquad x_1, x_2, x_3 \geq 0.$

(ii) If the objective function were to be changed to

$$kx_1 + 2x_2 + x_3$$

for what values of k would the position of the optimum found in (i) remain unchanged?

(iii)* If to the problem in (i) a new variable y is added so that the problem becomes:

maximise $\quad 3x_1 + 2x_2 + x_3 + 3y$
subject to $\quad x_1 + x_2 + x_3 + y \leq 3$
$\qquad\qquad 2x_1 \quad\;\; + x_3 - y \geq 2$
$\qquad\quad -x_1 + x_2 + x_3 + y \geq 0$
$\qquad\qquad x_1, x_2, x_3, y \geq 0,$

find a solution to this modified problem using the solution found in (i).
[*Hint*: write the extra column of coefficients as a combination of the slack variable columns in the first tableau. In the final tableau the new variable column will be the same combination of the final slack variable columns.]

8 Obtain the dual of the problem of 7(i) and find its solution. If the dual variables are denoted by w_1, w_2 and w_3 what will be the effect of adding the extra constraint

$$w_1 + w_2 - w_3 \geq 3?$$

[*Hint*: what is the dual of this modified problem?]

References

1. The interested reader might consult one of the following: H. M. Wagner, *Principles of Operations Research*, Prentice-Hall 2nd Ed. (1975), H. A. Taha, *Operations Research, an Introduction*, Collier-Macmillan 3rd Ed. (1982), F. S. Hillier and G. J. Lieberman, *Introduction to Operations Research*, Holden-Day 3rd Ed. (1980).
2. Problem (5.4) is considered in its full generality in Section 5.7.
3. In the case of a degenerate feasible vertex, the value of the objective function may not change in moving from one vertex to the next.
4. If we took a negative entry to correspond to our pivot row then the corresponding value for the new basis variable would automatically be negative, giving an infeasible solution.
5. The general formula for such a point is $ax_1 + (1-a)x_2$, where a is a constant satisfying $0 \leq a \leq 1$. In our example we took $a = 1/2$.
6. For a more general discussion see M. S. Bazaraa and J. J. Jarvis, *Linear Programming and Network Flows*, Wiley (1977).
7. For those who are familiar with quadratic forms, this is equivalent to the quadratic coefficient matrix being negative definite.

6
Optimal Control Theory: The Basics

6.1 Introduction

Optimal Control Theory addresses the problem of how best to operate a dynamic system to optimise some given criterion of performance. It was originally developed from the need for greater precision and control in industrial engineering systems, a development that was greatly boosted by the complex design problems that arose in military and space programs that were undertaken, particularly in the 1960s. More recently the theory has been used by Economists to solve planning problems at both the micro- and macro-economic levels. This is the area of interest to us in this chapter.

Optimal Control Theory is a particular example of the optimisation theory discussed in previous chapters and hence can be solved by standard Lagrangean techniques. The reason for discussing it further in this and the next chapter is that the Lagrange equations have a special form—they are either difference or differential equations depending on whether the system changes at discrete points in time or continuously—and hence the many algebraic and graphical techniques that are available for solving such equations can be used to solve the optimal control problem.

Before getting into the mathematics let us first discuss an example that will identify the essential features of an optimal control problem.

The Pioneer Problem

Imagine a group of pioneers who have set up a rural community in their new homeland. After each harvest they have to decide how much corn to set aside for consumption over the next year and how much to use as seedcorn for the next year's harvest. This is not an easy question to answer since the decision in one year will affect the decisions in all subsequent years. For example, if more is consumed in a particular year then all subsequent harvests will be smaller unless there are compensating reductions in consumption later.

The linkage between consumption decisions and subsequent harvests can be made clear by introducing the following notation. Let x_k be the size of the harvest in the kth year and u_k the amount set aside from that harvest for consumption during the next year. The amount left over for planting is therefore $(x_k - u_k)$. If the size of a harvest is proportional to the amount planted then the harvest in the $(k+1)$th year is given by:

$$x_{k+1} = c(x_k - u_k) \qquad \text{for } k = 0, 1, 2, \ldots \tag{6.1}$$

where c is the so-called growth factor (assumed constant from year to year). From equation (6.1) we can calculate the harvests in each successive year, given the initial stock of seedcorn, x_0, and any set of consumption choices that the community may wish to make.

To determine the *optimal* set of consumption choices the community must first decide on its criterion of optimality. If it comes to the conclusion that it wishes to maximise total consumption over, say, n years then it will have to:

$$\text{maximise} \sum_{k=0}^{n} u_k$$

subject to the constraints (6.1) and the inequalities:

$$x_k \geq 0, \quad u_k \geq \underline{u} \quad \text{for each } k \tag{6.2}$$

where $\underline{u}$ is the lowest level of consumption that the community is willing to tolerate. To complete the specification of the problem the community must also decide on the amount of corn to leave at the end of the planning period for future generations, since presumably it is their wish that the community should survive well beyond n years.

As we shall see in Section 6.3 the optimal solution of this problem is to remain at subsistence level for n years and then to consume as much as possible in the final year. Starving for n years and then having too much to eat in the final year does not appear to be a very sensible policy. What the community should maximise is not consumption but 'satisfaction' from consumption. The distinction is illustrated by the graph in Fig. 6.1 which shows that the increase in satisfaction due to a given increase in consumption is much greater at low consumption than at high consumption. (An extra crumb when you are starving is 'worth' more than an extra loaf when you are well fed.) The technical term for our intuitive concept of satisfaction is Utility, a term we will use where necessary in later examples. To model the saturation effect in Utility we will employ the simple functional form u_k^α where α takes some fixed value in the interval $0 < \alpha < 1$. With this choice the community's optimisation problem becomes:

$$\text{maximise} \sum_{k=0}^{n} u_k^\alpha$$

$$\begin{aligned}
\text{subject to} \quad & x_{k+1} = c(x_k - u_k) && \text{for } k = 0, 1, \ldots, n \\
& x_0 = a, \, x_{n+1} = b \\
& x_k \geq 0, \, u_k \geq \underline{u} && \text{for } k = 0, 1, \ldots, n
\end{aligned} \tag{6.3}$$

where a and b are given constants. The variables x_k specifying the size of the harvests are referred to as the *state variables* and the variables u_k quantifying the consumption decisions as the *control variables* of the problem.

The Pioneer Problem is an example of a discrete optimal control problem whose

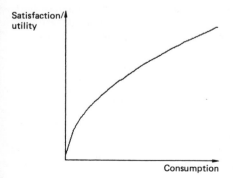

Fig. 6.1

general form is given by:

$$\text{maximise} \sum_{k=0}^{n} f_k(x_k, u_k) \tag{6.4a}$$

$$\text{subject to} \quad \begin{array}{l} x_{k+1} - x_k = g_k(x_k, u_k) \\ x_0 = a, x_{n+1} = b \end{array} \quad \text{for } k = 0, 1, \ldots, n \tag{6.4b}$$

together with inequality constraints on x_k and u_k where appropriate. In the Pioneer Problem we have the identification:

$$f_k = u_k^a, \quad g_k = (c-1)x_k - cu_k.$$

A discrete optimal control problem has therefore the following three distinguishing features.

(i) The variables can be divided into two sets, the *state* variables, x_k, and the *control* variables, u_k.
(ii) The problem is subject to a set of equality constraints (6.4b) that relate the difference in the state variable, $x_{k+1} - x_k$, to the current values x_k, u_k, of the state and control variables. This set of equations is collectively called the *dynamic equation* of the control problem.
(iii) The objective function (6.4a) is expressed as a sum of $(n+1)$ terms, the kth depending only on k and the kth values of the state and control variables. Such an objective function is said to be *separable*.

In the next section we derive from the Lagrangean formalism Hamilton's method for solving problem (6.4). In this method the optimum is found by solving a pair of difference equations, called the *Hamilton equations*, using the special techniques that have been developed for this type of equation. In Section 6.3 we will apply this method to obtain a full solution of the Pioneer Problem. The remainder of the chapter will be devoted to continuous optimal control problems which describe those systems that change continuously in time. This would be the case for example if the crop in the Pioneer Problem were timber rather than corn since timber is or can be harvested continuously. A modern industrial society could perhaps also be described by a continuous harvesting model with the crop being consumption goods and the growth mechanism the capital investment process. The Ramsey model is such a model and we will use the continuous version of Hamilton's method to solve it in Section 6.7.

It should be borne in mind that the Hamilton equations, like those of Lagrange, are necessary conditions for a finite optimum, i.e. not all solutions are necessarily optimal. In most of our examples, however, there will be a unique solution which will correspond to a finite optimum. This can be verified by concavity/convexity arguments, a point elaborated on in Section J of the Appendix. If the solution is not unique then as a last resort we will identify the optimal solution by evaluating the objective function for each solution.

6.2 Optimal Control Theory: The Discrete Case

Let us consider the general problem (6.4) ignoring for the moment any inequalities that may apply to the state and control variables. The Lagrangean is given by:

$$L = \sum_{k=0}^{n} f_k(x_k, u_k) - \sum_{k=0}^{n} \lambda_k (x_{k+1} - x_k - g_k(x_k, u_k))$$

where λ_k is the multiplier for the kth of the equations (6.4b). To derive the Lagrange equations:

$$\frac{\partial L}{\partial x_k} = 0, \quad \frac{\partial L}{\partial u_k} = 0 \tag{6.5}$$

for the stationary points we first note that the variable x_k appears in both the kth and $(k-1)^{\text{th}}$ terms of the second sum in L but only in the kth term of the first sum. This can be seen more easily by explicitly writing out the terms in L:

$$\begin{aligned} L = & f_0(x_0, u_0) - \lambda_0(x_1 - x_0 - g_0(x_0, u_0)) \\ & + f_1(x_1, u_1) - \lambda_1(x_2 - x_1 - g_1(x_1, u_1)) \\ & \cdots \cdots \cdots \cdots \cdots \\ & + f_{k-1}(x_{k-1}, u_{k-1}) - \lambda_{k-1}(x_k - x_{k-1} - g_{k-1}(x_{k-1}, u_{k-1})) \\ & + f_k(x_k, u_k) - \lambda_k(x_{k+1} - x_k - g_k(x_k, u_k)) \\ & \cdots \cdots \cdots \cdots \cdots \\ & + f_n(x_n, u_n) - \lambda_n(x_{n+1} - x_n - g_n(x_n, u_n)) \end{aligned} \tag{6.6}$$

Hence:

$$\frac{\partial L}{\partial x_1} = \frac{\partial f_1}{\partial x_1} + \lambda_1 \frac{\partial g_1}{\partial x_1} - \lambda_0 + \lambda_1 = 0$$

$$\frac{\partial L}{\partial x_2} = \frac{\partial f_2}{\partial x_2} + \lambda_2 \frac{\partial g_2}{\partial x_2} - \lambda_1 + \lambda_2 = 0$$

and generally:

$$\frac{\partial L}{\partial x_k} = \frac{\partial f_k}{\partial x_k} + \lambda_k \frac{\partial g_k}{\partial x_k} - \lambda_{k-1} + \lambda_k = 0 \quad \text{for } k = 1, \ldots, n \tag{6.7}$$

The second Lagrange equation in (6.5) can be obtained from equation (6.6) more straightforwardly since u_k only appears in the kth term of both sums. In fact:

$$\frac{\partial L}{\partial u_k} = \frac{\partial f_k}{\partial u_k} + \lambda_k \frac{\partial g_k}{\partial u_k} = 0 \tag{6.8}$$

These equations, (6.7) and (6.8), can be specified more compactly if we define the function H_k as:

$$H_k = f_k + \lambda_k g_k$$

for then they can be written in the form:

$$0 = \frac{\partial H_k}{\partial u_k} \quad \text{for } k = 0, 1, \ldots, n \tag{6.9}$$

$$\lambda_{k-1} - \lambda_k = \frac{\partial H_k}{\partial x_k} \quad \text{for } k = 1, 2, \ldots, n \tag{6.10}$$

Together with the dynamic equation:

$$x_{k+1} - x_k = \frac{\partial H_k}{\partial \lambda_k} \quad \text{for } k = 0, 1, \ldots, n \tag{6.11}$$

written in symmetric form using $\partial H_k/\partial \lambda_k = g_k$, equations (6.9), (6.10) provide three equations to determine the three sets of unknowns: x_k, u_k and λ_k. The 'master' function H_k is called the *Hamiltonian* for the problem after a 19th century Irish mathematician,

146 Optimal control theory: the basics

Hamilton, who made many significant contributions to Optimisation Theory. The first equation (6.9) reflects the fact that at a finite optimum each H_k is maximised with respect to its own control variable u_k, a fact commonly referred to as the *maximum principle*. This equation determines the stationary points amongst which the maximum is to be found.

The strategy we adopt to solve the set of equations (6.9–6.11) is to first use (6.9) to express u_k in terms of y_k and λ_k and then to substitute for u_k in equations (6.10), (6.11) which, as a consequence, become equations for x_k and λ_k only. These equations are examples of what are known as difference equations. Their distinguishing feature is that the difference in the values of the variables as k is changed by 1 depends only on the values of those variables at k. These particular difference equations are the Hamilton equations for the problem. Techniques for solving them are discussed in the context of the following examples[1].

Example 1
The Hamiltonian for the problem:

$$\text{maximise} \quad -\sum_{k=0}^{3}(x_k + u_k^2)$$

$$\text{subject to} \quad x_{k+1} - x_k = u_k$$

$$x_0 = x_4 = 0$$

is given by:

$$H_k = -(x_k + u_k^2) + \lambda_k u_k$$

and hence:

$$0 = \frac{\partial H_k}{\partial u_k} = -2u_k + \lambda_k \quad \text{for } k = 0, 1, 2, 3 \tag{6.12}$$

$$\lambda_{k-1} - \lambda_k = \frac{\partial H_k}{\partial x_k} = -1 \quad \text{for } k = 1, 2, 3 \tag{6.13}$$

$$x_{k+1} - x_k = \frac{\partial H_k}{\partial \lambda_k} = u_k \quad \text{for } k = 0, 1, 2, 3 \tag{6.14}$$

From (6.12) we deduce that $u_k = \lambda_k/2$ at which point H_k takes its maximum value for each x_k and λ_k. Substitution in the other two equations, (6.13) and (6.14), for u_k gives, on rearrangement:

$$\lambda_k = \lambda_{k-1} + 1 \quad \text{for } k = 1, 2, 3 \tag{6.15}$$

$$x_{k+1} = x_k + \tfrac{1}{2}\lambda_k \quad \text{for } k = 0, 1, 2, 3 \tag{6.16}$$

If we write out equation (6.15) explicitly for each value of k we obtain the set of equations:

$$k = 1: \quad \lambda_1 = \lambda_0 + 1$$
$$k = 2: \quad \lambda_2 = \lambda_1 + 1$$
$$k = 3: \quad \lambda_3 = \lambda_2 + 1$$

which can be solved recursively by substituting for λ_1 from the first of these equations into the second and for λ_2 from the second into the third. Precisely we have:

$$k = 1: \quad \lambda_1 = \lambda_0 + 1$$
$$k = 2: \quad \lambda_2 = (\lambda_0 + 1) + 1 = \lambda_0 + 2 \tag{6.17}$$
$$k = 3: \quad \lambda_3 = (\lambda_0 + 2) + 1 = \lambda_0 + 3$$

The second equation (6.16), written out for each k:

$k = 0$: $x_1 = x_0 + \frac{1}{2}\lambda_0$
$k = 1$: $x_2 = x_1 + \frac{1}{2}\lambda_1$
$k = 2$: $x_3 = x_2 + \frac{1}{2}\lambda_2$
$k = 3$: $x_4 = x_3 + \frac{1}{2}\lambda_3$

can also be solved recursively using the results (6.17), so that

$k = 0$: $x_1 = 0 + \frac{1}{2}\lambda_0$ (since we are told that $x_0 = 0$)
$k = 1$: $x_2 = \frac{1}{2}\lambda_0 + \frac{1}{2}(\lambda_0 + 1) = \lambda_0 + \frac{1}{2}$
$k = 2$: $x_3 = (\lambda_0 + \frac{1}{2}) + \frac{1}{2}(\lambda_0 + 2) = \frac{3}{2}\lambda_0 + \frac{3}{2}$
$k = 3$: $x_4 = (\frac{3}{2}\lambda_0 + \frac{3}{2}) + \frac{1}{2}(\lambda_0 + 3) = 2\lambda_0 + 3$ (6.18)

Since we are also told that $x_4 = 0$ we can deduce from (6.18) that $\lambda_0 = -\frac{3}{2}$. Knowing λ_0 we can substitute back to determine explicitly the sequences of state and control variables:

$\mathbf{x} = (x_0, x_1, x_2, x_3, x_4) = (0, -\frac{3}{4}, -1, -\frac{3}{4}, 0)$
$\mathbf{u} = (u_0, u_1, u_2, u_3) = (-\frac{3}{4}, -\frac{1}{4}, \frac{1}{4}, \frac{3}{4})$.

These sequences can be represented by the graphs of Fig. 6.2 where the values x_k and u_k are plotted against k. For emphasis the points in the sequences (k, x_k) and (k, u_k) are joined by line segments (although the intermediate points on the segments have no meaning).

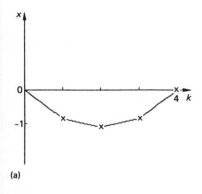
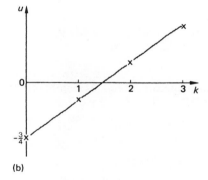

(a) (b)

Fig. 6.2

Example 2
For the problem[2]

$$\text{maximise} \quad -\tfrac{1}{2}\sum_{k=0}^{3}(x_k^2 + u_k^2)$$
$$\text{subject to} \quad x_{k+1} = x_k - u_k + 1 \quad (\text{i.e. } g_k = 1 - u_k)$$
$$x_0 = 3, \quad x_4 = 0$$

we have the Hamiltonian:

$$H_k = -\tfrac{1}{2}(x_k^2 + u_k^2) + \lambda_k(1 - u_k)$$

148 *Optimal control theory: the basics*

and the equation system:

$$0 = \frac{\partial H_k}{\partial u_k} = -u_k - \lambda_k, \quad \text{giving } u_k = -\lambda_k$$

$$\lambda_{k-1} - \lambda_k = \frac{\partial H_k}{\partial x_k} = -x_k$$

$$x_{k+1} - x_k = \frac{\partial H_k}{\partial \lambda_k} = 1 - u_k.$$

Eliminating u_k we obtain the difference equations:

$$\lambda_k = \lambda_{k-1} + x_k \tag{6.19}$$

$$x_{k+1} = x_k + 1 + \lambda_k \tag{6.20}$$

In Example 1 the λ_k equation was *uncoupled* in that it did not involve the other sequence x_k and hence could be solved on its own. The x_k equation could then be solved using the solution for λ_k. This is not the case in this example and so we need another line of attack. The method we adopt is to eliminate the λ_k sequence between the two equations to obtain a difference equation involving just the x_k sequence. To carry through the elimination we first rewrite equation (6.20) as:

$$\lambda_k = x_{k+1} - x_k - 1 \tag{6.21}$$

and, replacing k by $k-1$, we then have:

$$\lambda_{k-1} = x_k - x_{k-1} - 1.$$

Substituting these relations in equation (6.19) we obtain after rearrangement the second order difference equation for x_k:

$$x_{k+1} = 3x_k - x_{k-1} \tag{6.22}$$

This is second order because it relates members of the sequence over two intervals. Writing out equation (6.22) for $k = 1, 2, 3$ we find that:

$k = 1$: $\quad x_2 = 3x_1 - x_0$
$k = 2$: $\quad x_3 = 3x_2 - x_1$
$k = 3$: $\quad x_4 = 3x_3 - x_2.$

These equations can be solved recursively:

$k = 1$: $\quad x_2 = 3x_1 - 3$
$k = 2$: $\quad x_3 = 8x_1 - 9$
$k = 3$: $\quad x_4 = 21x_1 - 24.$

But $x_4 = 0$ so $x_1 = 8/7$ and, substituting back,

$$x_2 = 3/7, \quad x_3 = 1/7.$$

The multipliers and hence the control variables can most easily be calculated from equation (6.21). In fact:

$$\lambda = (\lambda_0, \lambda_1, \lambda_2, \lambda_3) = (-20/7, -12/7, -9/7, -8/7)$$

with $u_k = -\lambda_k$. The graphs of the state and control variables are shown in Fig. 6.3.

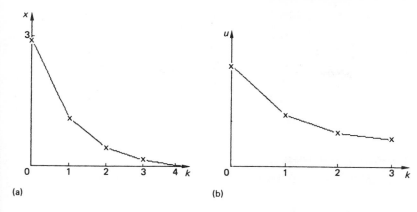

(a)
(b)

Fig. 6.3

Before asking you to solve some similar examples we first summarise our experience with discrete optimal control problems in Procedure Box 6.1.

PROCEDURE 6.1
To solve the Discrete Optimal Control Problem (6.4)
(without bounds on the state and control variables)

1 **Construct the Hamiltonian:**

$$H_k = f_k + \lambda_k g_k.$$

2 **Derive the equations:**

$$0 = \frac{\partial H_k}{\partial u_k} \quad \text{(Maximum Principle)} \tag{6.9}$$

$$\lambda_{k-1} - \lambda_k = \frac{\partial H_k}{\partial x_k} \tag{6.10}$$

(Hamilton Equations)

$$x_{k+1} - x_k = \frac{\partial H_k}{\partial \lambda_k} \tag{6.11}$$

3 **Solve these equations in the following fashion.**
   (i)   Eliminate u_k using the first equation.
   (ii) If one of the resulting difference equations is uncoupled then solve in sequence. If not then eliminate λ_k to obtain a second order difference equation for x_k.
   (iii) Determine any unknowns by imposing the given end-point values for x_0, x_{n+1}.

Exercise Set 1
Solve the discrete optimal control problem (6.4) for the following choices of the functions f, g and parameters n, x_0, x_{n+1}.

150 *Optimal control theory: the basics*

	(i)*	(ii)	(iii)*	(iv)	(v)*	(vi)
f	$-(x_k + \tfrac{1}{2}u_k^2)$	$-(x_k^2 + u_k^2)$	$-u_k^2$	$-(1+u_k^2)^{1/2}$	$u_k^{1/2}$	$u_k^{1/2}$
g	u_k	$1 - 2u_k$	u_k	u_k	$x_k - 2u_k$	$2x_k - 3u_k$
n	2	2	3	4	2	2
x_0	4	6	0	0	1	1
x_{n+1}	1	-1	4	5	4	21

*6.3 The Pioneer Problem

We are now in a position to solve the following generalised form of the Pioneer Problem:

$$\text{maximise} \quad \sum_{k=0}^{n} \rho^k u_k^\alpha \quad \text{with } 0 < \alpha < 1$$

$$\text{subject to} \quad x_{k+1} = c(x_k - u_k) \quad \text{with } g_k = (c-1)x_k - cu_k \tag{6.23}$$

$$x_0 = a, \quad x_{n+1} = b,$$

in which we have included a discount factor ρ^k where $0 < \rho < 1$. As k tends to infinity ρ^k decreases geometrically to zero to model the psychological experience that events further into the future become less and less important. The factor ρ^k can also be used to incorporate population growth into the model. Suppose the population is increasing in size each year by a factor ρ_1 where $\rho_1 > 1$ then the population in the kth year is given by $A\rho_1^k$ where A is the initial size of the community. In this situation the sensible objective is to maximise total (discounted) utility per head of the population: $\sum_{k=0}^{k} \rho^k (u_k/A\rho_1^k)^\alpha$, i.e.

$$\text{maximise} \quad A^{-\alpha} \sum_{k=0}^{n} \rho_2^k u_k^\alpha$$

where $\rho_2 = \rho/\rho_1^\alpha$. This is essentially the same objective function as in (6.23) but with a trivial multiplying factor $A^{-\alpha}$ and an adjusted discount factor satisfying $0 < \rho_2 < 1$.

In specifying (6.23) we have, for simplicity, ignored inequality constraints on the state and control variables in the expectation that none of them will be binding, but we will have to check this after the solution has been found. Even with this simplification the analysis will still be somewhat more involved than that experienced in the last section since n, the length of the planning period, and the various constants in the model are left unspecified. The payoff will be in the generality of the results. Let us now solve problem (6.23) by directly applying Procedure 6.1.

1. $H_k = \rho^k u_k^\alpha + \lambda_k((c-1)x_k - cu_k)$

2. $\dfrac{\partial H_k}{\partial u_k} = \alpha \rho^k u_k^{\alpha-1} - c\lambda_k = 0$

    $\lambda_{k-1} - \lambda_k = \dfrac{\partial H_k}{\partial x_k} = (c-1)\lambda_k$

    $x_{k+1} - x_k = \dfrac{\partial H_k}{\partial \lambda_k} = (c-1)x_k - cu_k$

3. (i) The first equation gives:

    $$u_k = (\lambda_k c/(\alpha \rho^k))^{-\beta} \quad \text{where } \beta = 1/(1-\alpha) \tag{6.24}$$

(ii) The second equation becomes
$$\lambda_k = \frac{1}{c}\lambda_{k-1}$$
so each time we increase the index by 1 we multiply by $1/c$. Hence:
$$\lambda_k = \left(\frac{1}{c}\right)^k \lambda_0 \tag{6.25}$$

The third equation, on substituting for λ_k and $\dot{u}_k$, gives:
$$x_{k+1} = cx_k - c\theta a^k$$
where for algebraic convenience we have defined $a = (\rho c)^\beta$ and $\theta = (\lambda_0 c/\alpha)^{-\beta}$. To obtain a general expression for x_k let us write out the first few cases:

$$x_1 = cx_0 - c\theta \quad \text{with } a^0 = 1$$

$$x_2 = cx_1 - c\theta a = c^2 x_0 - c^2 \theta \left(1 + \frac{a}{c}\right)$$

$$x_3 = cx_2 - c\theta a^2 = c^3 x_0 - c^3 \theta \left(1 + \left(\frac{a}{c}\right) + \left(\frac{a}{c}\right)^2\right)$$

$$x_4 = cx_3 - c\theta a^3 = c^4 x_0 - c^4 \theta \left(1 + \left(\frac{a}{c}\right) + \left(\frac{a}{c}\right)^2 + \left(\frac{a}{c}\right)^3\right).$$

The pattern is now clear with the kth term being given by[3]:

$$x_k = c^k x_0 - c^k \theta \left(1 + \left(\frac{a}{c}\right) + \ldots + \left(\frac{a}{c}\right)^{k-1}\right)$$

$$= c^k x_0 - c^k \theta \left\{\frac{\left(\frac{a}{c}\right)^k - 1}{\left(\frac{a}{c}\right) - 1}\right\} \quad \text{when } a \neq c \tag{6.26}$$

using the formula for summing a geometric series.

(iii) If there is no consumption at all then the final harvest will be $x_{n+1} = c^{n+1} x_0$ as can be seen by solving the dynamic equation with $u_k = 0$ for all k. If there is consumption then the final harvest will be smaller, i.e. $x_{n+1} < c^{n+1} x_0$, or, equivalently:
$$x_{n+1} = \gamma c^{n+1} x_0, \quad 0 < \gamma < 1$$
where γ provides a convenient parametrisation of the final state.

If we impose this end-point condition on equation (6.26) with $k = n+1$ we find the unknown parameter θ to be:

$$\theta = x_0 (1 - \gamma) \left\{\frac{\left(\frac{a}{c}\right) - 1}{\left(\frac{a}{c}\right)^{n+1} - 1}\right\} \tag{6.27}$$

In summary, we conclude from equations (6.24) and (6.25) that the optimal consumption is given by:

$$u_k = \theta a^k \tag{6.28}$$

where $a = (\rho c)^\beta$ and θ is defined by equation (6.27). The optimal harvests are calculated from equation (6.26).

If ρc and hence a is greater than 1 then from equation (6.28) we see that consumption grows geometrically by a factor a each year (Fig. 6.4a). If, however, ρc is less than 1 then consumption decays geometrically (Fig. 6.4b). Since $c > 1$ the product ρc can only be less than 1 when the future is heavily discounted and hence the decreasing consumption does not then particularly cause concern.

Also of interest is what happens when the utility exponent α increases to 1, at which point utility equals consumption. In the case that $\rho c > 1$ we note that both the parameters $\beta = 1/(1 - \alpha)$ and $a = (\rho c)^\beta$ tend to plus infinity and that as a result u_k becomes proportional to a^{k-n}, according to equations (6.27) and (6.28). Since for $k < n$ the exponent is negative and for $k = n$ zero all but the last of the control variables in the limit become zero (Fig. 6.4c). This result supports the comments made in the introduction that maximising total *consumption* is not an appropriate objective to choose.

Two further points are worth making. Firstly it is clear from the solution that the inequality bounds $u_k \geq \underline{u}$, $x_k \geq 0$ can always be satisfied by our optimal solution if $\underline{u} = 0$, but that there could be situations (Figs 6.4a, b) where some of the values of u fall below a positive $\underline{u}$. In this case Procedure 6.1 would no longer be appropriate and would

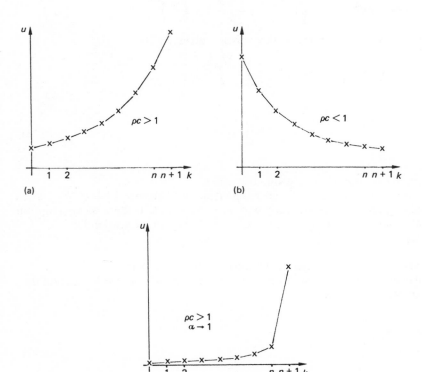

Fig. 6.4

have to be generalised. This is done for continuous control problems in the next chapter. Secondly we note that if we set the final state variable too high (e.g. $x_{n+1} > c^{n+1} x_0$) then there is no solution to the problem because the corn cannot grow fast enough! The possibility of there being no solution to an optimal control problem must always be borne in mind.

6.4 Continuous Optimal Control Theory

So far we have limited discussion to control problems where the system changes at only discrete points in time. We now generalise the theory to handle systems which change continuously by taking the limit as the parameter h, which measures the time between discrete system changes, tends to zero. This is equivalent to taking the limit as the number $n + 1$ of system changes over a planning period of length T tends to infinity, since $h = T/(n+1)$. In the limit the polygonal graphs (Fig. 6.5a) representing the sequences of state and control variables tend to continuous graphs (Fig. 6.6a) defining continuous (and usually differentiable) functions $x(t)$, $u(t)$ of the variable t which takes over the rôle of the index k in specifying time. Similarly the sequence $f_k(x_k, u_k)$ evaluated for these sets of state and control variables tends in the limit to a function $f(x(t), u(t), t)$ (Figs 6.5b, 6.6b).

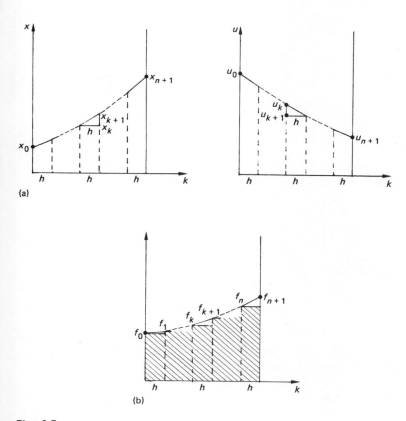

Fig. 6.5

154 *Optimal control theory: the basics*

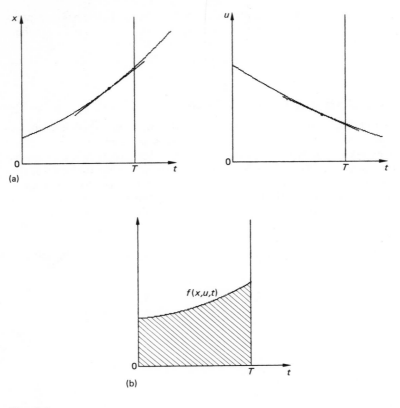

Fig. 6.6

In summary we have:

$$
\begin{array}{c}
\textit{Discrete} \\
\left.\begin{array}{c} k \\ x_k \\ u_k \end{array}\right\} \\
\left.\begin{array}{c} f_k(x_k, u_k) \\ g_k(x_k, u_k) \end{array}\right\}
\end{array}
\quad \xrightarrow{h \to 0} \quad
\begin{array}{c}
\textit{Continuous} \\
\left\{\begin{array}{c} t \\ x(t) \\ u(t) \end{array}\right. \\
\left\{\begin{array}{c} f(x(t), u(t), t) \\ g(x(t), u(t), t) \end{array}\right.
\end{array}
$$

and similarly:

$$\lambda_k \longrightarrow \lambda(t).$$

To obtain the limit of the discrete control problem (6.4) we first multiply the functions f_k and g_k by parameter h (which initially has value 1), to obtain:

$$\text{maximise} \quad \sum_{k=0}^{n} f_k(x_k, u_k) h \tag{6.29a}$$

$$\text{subject to} \quad x_{k+1} - x_k = g_k(x_k, u_k) h \tag{6.29b}$$

In the limit the sum in the objective function becomes an integral and the difference on the left hand side of the dynamic equation becomes a derivative, as we shall now argue.

(i) The sum in (6.29a) represents the total area of the rectangles shaded in Fig. 6.5b, approximating the area 'under' the polygonal graph and hence 'under' the continuous limit of this graph. The error in this approximation will in general reduce to zero as h tends to zero and hence the sum will tend to the area 'under' the graph of $f(x, u, t)$ (Fig. 6.6b), i.e. the integral:

$$\int_0^T f(x(t), u(t), t) \, dt.$$

(ii) On dividing by h the left hand side of the dynamic equation (6.29b) approximates the gradient of the tangent to the graph of $x = x(t)$ and in the limit equals its derivative (Figs 6.5a, 6.6a). For notational convenience this derivative will be denoted by $\dot{x}$, i.e.

$$\dot{x} = \frac{dx}{dt}.$$

In summary we shall take the basic continuous optimal control problem to have the form:

$$\text{maximise} \quad \int_0^T f(x, u, t) \, dt \tag{6.30a}$$

$$\text{subject to} \quad \begin{aligned} \dot{x} &= g(x, u, t) \\ x(0) &= a, \ x(T) = b \end{aligned} \tag{6.30b}$$

where for brevity we have written $x(t)$ as x and $u(t)$ as u. The integral in (6.30a) is usually referred to, for mathematical reasons, as the *objective functional* for the problem. To solve problem (6.30), i.e. to find functions x and u that satisfy the dynamic equation (6.30b) and maximise the objective functional (6.30a), we have to derive the continuous analogues of the equation system (6.9–6.11) by taking the limit as h tends to zero[4]. To do this we first construct the 'continuous' Hamiltonian H from the f and g functions in equations (6.30) as:

$$H = f + \lambda g$$

where the multiplier λ is now a function of t. (We will refer to it is as the multiplier function.) In terms of H the equation system (6.9–6.11) of Procedure 6.1 has the limit:

$$0 = \frac{\partial H_k}{\partial u_k} \tag{6.31}$$

$$-\left(\frac{\lambda_k - \lambda_{k-1}}{h}\right) = \frac{\partial H_k}{\partial x_k} \quad \xrightarrow{h \to 0} \quad -\dot{\lambda} = \frac{\partial H}{\partial x} \tag{6.32}$$

$$\left(\frac{x_{k+1} - x_k}{h}\right) = \frac{\partial H_k}{\partial \lambda_k} \qquad \dot{x} = \frac{\partial H}{\partial \lambda} \tag{6.33}$$

$$0 = \frac{\partial H}{\partial u}$$

using the rule that a difference becomes a derivative in the limit. To solve the equation system (6.31–6.33) we follow the same strategy as in the discrete case by solving (6.31) for u in terms of x and λ and substituting in (6.32), (6.33) to obtain, in this case, two *differential equations* for x and λ. To avoid getting too deeply involved in the theory of these equations we will restrict attention to those optimisation problems that generate differential equations that can, after appropriate manipulations, be solved analytically by integration. The necessary techniques and definitions are reviewed in Section G of the Appendix.

To gain experience with the equation system (6.31–6.33) let us look at two examples.

Example 3
The Hamiltonian for the problem:

$$\text{maximise} \quad \int_0^1 u^{1/2}\, dt$$

$$\text{subject to} \quad \begin{array}{l} \dot{x} = -u \\ x(0) = 1, \quad x(1) = 0 \end{array}$$

(with $f = u^{1/2}$ and $g = -u$), is given by:

$$H = u^{1/2} - \lambda u$$

and the equation system by:

$$\frac{\partial H}{\partial u} = \tfrac{1}{2} u^{-1/2} - \lambda = 0, \quad \text{i.e. } u = 1/(4\lambda^2)$$

$$-\dot{\lambda} = \frac{\partial H}{\partial x} = 0$$

$$\dot{x} = \frac{\partial H}{\partial \lambda} = -u.$$

Eliminating u we obtain the differential equations:

$$\dot{\lambda} = 0; \quad \dot{x} = -1(4\lambda^2).$$

The first of these equations is easy to solve since it states that λ is a function whose derivative is zero. So λ is a constant, A, say. The second equation can now be solved directly by integration:

$$x = -\int \frac{1}{4A^2}\, dt + B = -\frac{t}{4A^2} + B$$

where B is a constant of integration. A and B can be determined by imposing the end-point conditions:

$$1 = x(0) = B; \quad 0 = x(1) = -\frac{1}{4A^2} + B.$$

We deduce that $B = 1$ and $A^2 = 1/4$ and hence that the optimal state and control variables for this problem are given by (Fig. 6.7a):

$$x = 1 - t \quad \text{and} \quad u = 1.$$

Example 4
For the problem:

$$\text{maximise} \quad -\tfrac{1}{2} \int_0^1 (x^2 + u^2)\, dt$$

$$\text{subject to} \quad \begin{array}{l} \dot{x} = u \\ x(0) = 0, \quad x(1) = 1 \end{array}$$

we derive from its Hamiltonian:

$$H = -\tfrac{1}{2}(x^2 + u^2) + \lambda u \tag{6.34}$$

6.4 *Continuous optimal control theory* 157

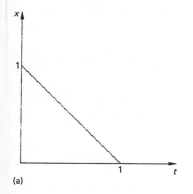

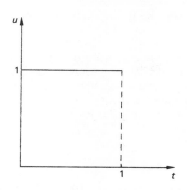

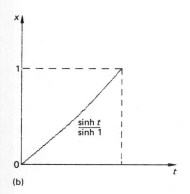

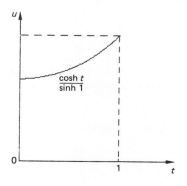

Fig. 6.7

the equations:

$$0 = \frac{\partial H}{\partial u} = -u + \lambda \quad \text{i.e. } u = \lambda$$

$$-\dot\lambda = \frac{\partial H}{\partial x} = -x$$

$$\dot x = \frac{\partial H}{\partial \lambda} = u.$$

After the elimination of u the Hamilton equations are found to be

$$\dot\lambda = x; \quad \dot x = \lambda \tag{6.35}$$

Unlike Example 3 the λ equation is not uncoupled since the right hand side involves the other variable x and hence we cannot solve the differential equations sequentially.

There are several ways of proceeding. In analogy with the discrete case we can eliminate one of the variables and obtain a single differential equation for the other variable. Precisely, if we differentiate the second of the equations (6.35) and then substitute for λ

from the first we obtain the following second order differential equation for x:
$$\ddot{x} - x = 0$$
where $\ddot{x}$ denotes the second derivative d^2x/dt^2 of x with respect to t. There is a special technique for solving such an equation, involving the determination of the so-called complementary and particular solutions. The details can be found in most elementary books on differential equation theory[5]. An alternative method is to equate the ratios of the two sides of the equations in (6.35) to obtain a single first order equation:

$$\frac{d\lambda}{dx} = \frac{x}{\lambda} \tag{6.36}$$

Here we have used the identity $(d\lambda/dt)/(dx/dt) = (d\lambda/dx)$ in which λ is considered a function of x with t eliminated between them. Equation (6.36) is a separable equation and can be solved by multiplying through by λ and integrating:

$$\int \lambda \, d\lambda = \int x \, dx + A$$

i.e. $\quad \tfrac{1}{2}\lambda^2 = \tfrac{1}{2}x^2 + A \tag{6.37}$

where A is a constant of integration. If we substitute for λ from (6.37) in the second of the equations (6.35) we obtain a single first order differential equation for x:

$$\dot{x} = \lambda = \pm (x^2 + 2A)^{1/2}.$$

This is again a separable equation and can be solved by integration:

$$\pm \int \frac{dx}{(x^2 + 2A)^{1/2}} = \int dt + B = t + B \tag{6.38}$$

To evaluate the left hand integral we use the hyperbolic functions, sinh and cosh, discussed in Section H of the Appendix. If you are not familiar with these functions you may wish to refer to the Appendix at this point.

The nature of the left hand integral in equation (6.38) depends on the sign of A. Let us suppose that A is positive then we can evaluate the integral by substituting $x = (2A)^{1/2} \sinh y$ to obtain:

$$\int \frac{dx}{(x^2 + 2A)^{1/2}} = \int \frac{(2A)^{1/2} \cosh y}{(2A)^{1/2} \cosh y} dy = y = \sinh^{-1} \frac{x}{(2A)^{1/2}}.$$

Substituting back in equation (6.38) we find, on rearrangement, that

$$x = A_1 \sinh(t + B) \tag{6.39}$$

where $A_1 = \pm (2A)^{1/2}$ and:

$$u = A_1 \cosh(t + B)$$

since $u = \dot{x}$. The constants A_1 and B are found by imposing the end-point conditions on (6.39):

$$x(0) = A_1 \sinh B = 0, \quad x(1) = A_1 \sinh(1 + B) = 1.$$

We conclude that $B = 0$ and $A_1 = 1/\sinh 1$. The optimal state and control variables are therefore:

$$x = \sinh t/\sinh 1 \quad \text{and} \quad u = \cosh t/\sinh 1$$

with the graphs shown in Fig. 6.7b. (If we had not been able to satisfy the end-point conditions we would have to go back to equation (6.38) and try the substitution $x = (-2A)^{1/2} \cosh y$ with A negative.)

This solution procedure only works if the Hamilton equations do not explicitly depend on t (only implicitly through the intermediary functions x and λ), for then we can eliminate t by simple division to obtain a differential equation (e.g. equation (6.36)) relating x and λ only. This will clearly be the case when the Hamiltonian itself does not depend explicitly on t—the problem is then called an *autonomous* control problem and has the special property that the Hamiltonian is constant in time, a result established in Section J of the Appendix. For the autonomous problem of Example 4, for instance, we have that:

$$H = -\tfrac{1}{2}(x^2 + u^2) + \lambda u = \tfrac{1}{2}\lambda^2 - \tfrac{1}{2}x^2 = \text{a constant}$$

using $u = \lambda$. This is just the result (6.37) obtained by integrating the differential equation (6.36).

Finally, let us summarise our discussion in Procedure Box 6.2.

PROCEDURE 6.2
To solve the Continuous Optimal Control Problem (6.30)

1 **Construct the Hamiltonian:**

$$H = f + \lambda g$$

2 **Derive the equations:**

$$0 = \frac{\partial H}{\partial u} \quad \text{(Maximum Principle)}$$

$$\left. \begin{array}{l} -\dot{\lambda} = \dfrac{\partial H}{\partial x} \\[6pt] \dot{x} = \dfrac{\partial H}{\partial \lambda} \end{array} \right\} \quad \text{(Hamilton Equations)}$$

3 **To solve these equations:**
(i) eliminate u using the first equation;
(ii) use the flowchart of Fig. 6.8 to solve the resulting differential equations for x and λ;
(iii) determine the constants of integration by imposing the given end-point values $x(0)$, $x(T)$.

Exercise Set 2
Solve problem (6.30) when $T = 1$ and functions f, g and parameters a, b are given by:

	(i)*	(ii)	(iii)*	(iv)	(v)*	(vi)
f	$-(x+\tfrac{1}{2}u^2)$	$-\tfrac{1}{2}(x+u)^2$	$-(1+u^2)^{1/2}$	$u^{1/2}$	$-\tfrac{1}{2}(x^2+u^2)$	$-\tfrac{1}{2}x^2 - \tfrac{1}{2}(u-x)^2$
g	u	$u-x$	u	$2x-u$	$x+u$	$2u$
a	2	1	0	1	0	0
b	1	1	1	2	1	1

160 *Optimal control theory: the basics*

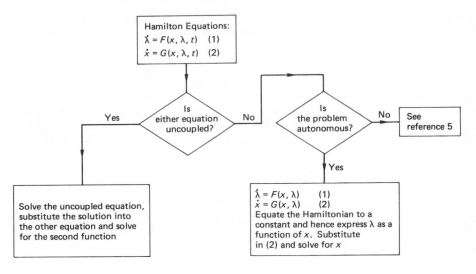

Fig. 6.8

In several of these problems a calculator will be useful when determining the constants of integration. To solve the last two problems use the fact that the Hamiltonian is constant.

6.5 An Exhaustible Resource Problem[6]

A company with an effective monopoly over the phosphate deposits located in a certain region wishes to determine how much phosphate to sell at each point in time in order to maximise its total profit over the period to the exhaustion of the deposits.

To model this problem let $x(t)$ denote the amount of phosphate left at time t and $u(t)$ the rate at which the phosphate is being produced (and sold). By definition it must be that:

$$\dot{x} = -u.$$

If R_0 is the initial size of the deposits and T is the time of exhaustion then x is subject to the end-point conditions:

$$x(0) = R_0, \quad x(T) = 0.$$

Assuming negligible costs the total profit made by the company is given by:

$$\int_0^T e^{-rt} u(t) p(t) \, dt \qquad (6.40)$$

where $p(t)$ is the market price for phosphate and e^{-rt} is the discount factor that models the economic principle that identical incomes received at different points in time are not to be valued equally. This is because income $I(0)$ received at time zero will be worth $I(0)e^{rt}$ in t years time if it is invested at a (continuous) rate of interest r and hence income $I(t)$ received at time t would be equivalent to income $I(t)e^{-rt}$ received at time zero. The total income in (6.40) is therefore the 'sum' of the equivalent 'component' incomes if they were all received at time zero, i.e. the *present value* of all these incomes. The price $p(t)$ and the rate $u(t)$ at which the phosphate is sold are related by a demand function $p = D(u)$ with the

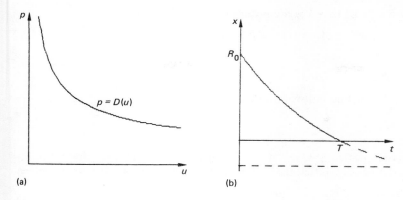

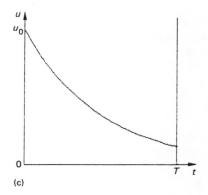

Fig. 6.9

general shape shown in Fig. 6.9a. In terms of D the optimisation problem can be written as:

$$\text{maximise} \quad \int_0^T e^{-rt} uD(u) \, dt \tag{6.40a}$$

$$\text{subject to} \quad \begin{array}{l} \dot{x} = -u \\ x(0) = R_0, \, x(T) = 0 \end{array} \tag{6.40b}$$

If, for definiteness, we suppose that $D(u) = \beta u^{-\alpha}$ with $0 < \alpha < 1$ and $\beta > 0$ (the so-called *iso-elastic* demand function) then the Hamiltonian, given by:

$$H = \beta e^{-rt} u^{1-\alpha} - \lambda u$$

generates the equations:

$$0 = \frac{\partial H}{\partial u} = \beta(1-\alpha) e^{-rt} u^{-\alpha} - \lambda$$

$$-\dot{\lambda} = \frac{\partial H}{\partial x} = 0$$

$$\dot{x} = \frac{\partial H}{\partial \lambda} = -u$$

The second equation tells us that λ is a constant, say A_0, the first that:
$$u = A_1 e^{-rt/\alpha} \qquad \text{where } A_1 = ((1-\alpha)\beta/A_0)^{1/\alpha}$$
and the third that:
$$\dot{x} = -A_1 e^{-rt/\alpha}.$$
Directly integrating this last equation we obtain:
$$x = \frac{\alpha}{r} A_1 e^{-rt/\alpha} + A_2$$
where the constants of integration A_1 and A_2 are found as usual by imposing the end-point conditions. The final conclusion is that the optimal state and control variables are given by:
$$x = R_0 \left\{ \frac{e^{-rt/\alpha} - e^{-rt/\alpha}}{1 - e^{-rt/\alpha}} \right\}; \qquad u = \frac{r}{\alpha} R_0 \frac{e^{-rt/\alpha}}{1 - e^{-rt/\alpha}} \tag{6.41}$$
with the graphs shown in Figs 6.9b, c.

The optimal depletion path clearly depends on the chosen exhaustion time T. The company may have definite views on what T should be. If not then we could suggest that it chooses T to maximise its optimal profit obtained by substituting (6.41) in the objective functional (6.40a):
$$\text{Optimal Profit} = \beta \left(\frac{\alpha}{r R_0} \right)^\alpha R_0 (1 - e^{-rt/\alpha})^\alpha.$$
This profit progressively increases with T (Fig. 6.10a) reaching its maximum at infinity. Hence the optimal choice for T is infinite with x and u decaying exponentially:
$$x = R_0 e^{-rt/\alpha}; \qquad u = \frac{r}{\alpha} R_0 e^{-rt/\alpha}.$$
From the first of these relations we note that a higher discount rate r leads to faster exploitation of the resources (Fig. 6.10b), corresponding to a higher discounting of the future.

It is not always the case that the optimal exhaustion period T is infinite. In fact, if we choose Hotelling's demand function[7]:
$$D(u) = \beta(1 - e^{-bu})u^{-1}, \qquad \beta > 0$$
where b denotes the price at which consumers cease buying phosphate (Fig. 6.10c) the optimal policy is to exhaust the resource in a finite time as we shall show in the next chapter.

6.6 Phase Space

From the examples in Section 6.4 it is clear that it can often be a little difficult to obtain analytic solutions to the Hamilton equations. In some cases we might have to be content with a solution in numerical form. Even if an analytic solution can be obtained it may be so complicated that it is not easy to unravel its main characteristics. In this situation it is sometimes possible to replace or augment the quantitative analysis with a graphical procedure that will determine the main qualitative properties without requiring the differential equations to be solved directly.

To set up the graphical procedure we first note that each solution from the infinity of

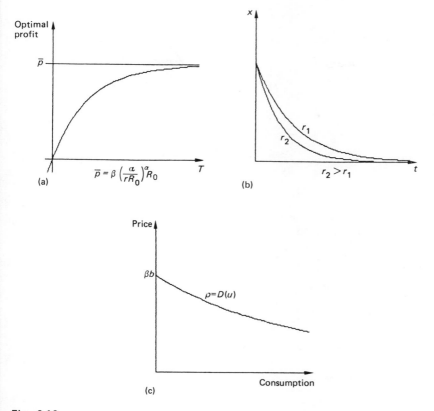

Fig. 6.10

solutions to the Hamilton equations:

$$\frac{d\lambda}{dt} = -\frac{\partial H}{\partial x} \equiv F(x, \lambda, t) \tag{6.42}$$

$$\frac{dx}{dt} = \frac{\partial H}{\partial \lambda} \equiv G(x, \lambda, t) \tag{6.43}$$

can be represented by a curve in *phase space*, the space of the variables x and λ. For each t the values $x(t)$, $\lambda(t)$ define the coordinates of a point in this space and the set of such points for different values of t defines the curve (Fig. 6.11a). As t increases we move in a particular direction on this curve, a direction indicated by superimposing an arrow pointing in the direction of movement (Fig. 6.11a). With this directional information added we will refer to the curve as a *trajectory* of the system.

If we superimpose trajectories for different solutions of the Hamilton equations on the same diagram we will get a rather complicated pattern unless the functions F and G do not explicitly depend on t, i.e.

$$F = F(x, \lambda), \quad G = G(x, \lambda).$$

164 *Optimal control theory: the basics*

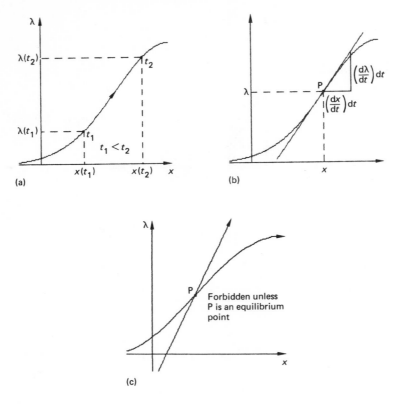

Fig. 6.11

In this case the gradient of the tangent to a trajectory at a point P (Fig. 6.11b):

$$\frac{d\lambda}{dx} = \frac{F(x, \lambda)}{G(x, \lambda)} \tag{6.44}$$

is uniquely determined by the coordinates of P unless both F and G are zero[8]. If this is not the case then we conclude that two trajectories can only pass through P if they touch, since the tangent at P is unique. The touching of trajectories, however, is such a rare occurrence that we shall ignore the possibility and work with the rule that at most one trajectory passes through any point in phase space where F and G are not both zero (Fig. 6.11c). At the exceptional points where F and G are both zero the one trajectory rule can break down as examples will later show. These exceptional points are called *equilibrium points* because once at such a point we will always stay there with $\dot{x}$ and $\dot{\lambda}$ both being zero.

We can use equation (6.44) to determine the tangents at a set of points in phase space which can then be pieced together to form smooth trajectories for the problem. This process is illustrated in Fig. 6.12, with the arrow directions determined by referring back to equations (6.42), (6.43) and determining the signs of $\dot{x}$ and $\dot{\lambda}$. This procedure can be made quite accurate by locating the tangents at a sufficiently large number of points. We will not, however, require such detail and will instead use a construction method that only uses information about the signs of the functions F and G, not their magnitudes. This simplified method is called the *Phase Space Method* and involves in general the following steps.

6.6 *Phase space* 165

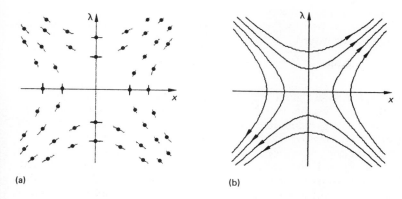

Fig. 6.12

(i) Sketch the curve:

$$F(x, \lambda) = 0$$

in the phase space plane. This curve is called the *λ-isocline* and has the property that at each point P belonging to it, the tangent to the trajectory at P is necessarily horizontal (see equation (6.44)). Register this fact by superimposing horizontal tangents on the isocline (Fig. 6.13a).[9] At all points in any region defined by this isocline, F and hence $\dot{\lambda}$ is either always positive or always negative. If positive then λ is increasing with t, i.e. we

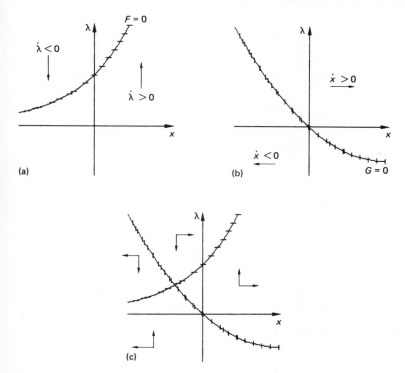

Fig. 6.13

166 *Optimal control theory: the basics*

are moving up rather than down on any trajectory in this region. We indicate this fact by superimposing an arrow pointing up (Fig. 6.13a). If the sign of $\dot{\lambda}$ is negative then the arrow direction is downward.

(ii) Sketch the curve:
$$G(x, \lambda) = 0$$

This is the *x-isocline*. It defines the set of points in phase space where the tangents are vertical (Fig. 6.13b). Register this fact by superimposing vertical tangents on the isocline. At all points in any region defined by this isocline, G and hence $\dot{x}$ is either always positive or always negative. If positive then x is increasing with t, i.e. we are moving to the right on any trajectory in this region. We indicate this fact with an arrow pointing to the right. If the sign is negative then the arrow points to the left.

(iii) Superimpose the information obtained in steps (i), (ii) to obtain a phase space divided in general into 4 regions by the isocline curves. In each of these regions there is a pair of arrows, one indicating movement to the left or right and the other movement up or down (Fig. 6.13c).

(iv) With this arrow and isocline information we are now in a position to sketch in the system trajectories. The best way is to start at a point on one of the isoclines and draw in the trajectory through that point by obeying the arrow instructions. For example if we start at point P on the isocline $G = 0$ in Fig. 6.14a we move up and to the right following the arrow instructions in region I, the tangent to the trajectory being initially vertical. For the other part of the trajectory in region II we approach point P from below and from the right.

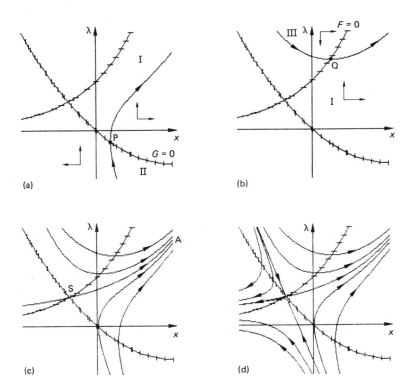

Fig. 6.14

Similarly starting at point Q on the other isocline and following the arrows in regions I and III we obtain the trajectory sketched in Fig. 6.14b.

The two sets of trajectories crossing the isoclines into region I (Fig. 6.14c) are not allowed to cross or touch each other and hence there must be a special trajectory SA emanating from the equilibrium point at S that divides them. This property also holds in the other three regions as shown in the complete trajectory plot of Fig. 6.14d.

To illustrate this construction technique and to discuss how to interpret the phase space plot once we have sketched it let us analyse the following two cases.

Example 5

The Hamilton equations for the problem:

$$\text{maximise} \quad -\frac{1}{2}\int_0^T (x^2 + u^2)\,dt$$

$$\text{subject to} \quad \dot{x} = x + u$$
$$x(0) = x(T) = a \tag{6.45}$$

have the form:

$$\left.\begin{array}{l}\dot{\lambda} = x - \lambda \\ \dot{x} = x + \lambda\end{array}\right\} \tag{6.46}$$

with $F = x - \lambda$ and $G = x + \lambda$. The trajectories for these equations can be found by the phase space method in the following steps.

(i) The λ-isocline is the radial line $x = \lambda$ (Fig. 6.15a). The sign of $\dot{\lambda}$ in the region to the left of this line can be established by evaluating $\dot{\lambda}$ at any point in this region since the sign of $\dot{\lambda}$ is common to all its points. If we choose for example $x = -1, \lambda = 0$ we find that $\dot{\lambda}$ is equal to -1 and hence we are moving downwards in this region. To the right of the isocline $\dot{\lambda}$ is positive (since, for example, $\dot{\lambda}$ is equal to $+1$ at the point with $x = +1$, $\lambda = 0$) and hence in this region we are moving upwards.

(ii) The x-isocline is the radial line $x = -\lambda$ (Fig. 6.15b). Since $\dot{x}$ is -1 at $x = -1, \lambda = 0$ we deduce that to the left of the isocline we are moving to the left on any trajectory. Similarly to the right of this isocline we move to the right.

(iii), (iv) Superimposing these results we obtain the plot of Fig. 6.15c with four regions and a pair of arrows for each region. Using this information we conclude that the trajectories are positioned as shown in Fig. 6.15d, with four sets of trajectories divided by four special trajectories emanating from or going into the equilibrium point at the origin. (Note that the points of intersection of an x-isocline and a λ-isocline are necessarily equilibrium points.)

The completed phase plot gives us the shape of all possible trajectories for the equation system (6.46). To pick out the trajectory that corresponds to the optimal solution of problem (6.45) we must impose the end-point conditions $x(0) = x(T) = a$. The trajectory we require is one that crosses the vertical line $x = a$ at time $t = 0$ and recrosses it at time T later. From Fig. 6.15d it is clear that there is a set of trajectories which cross this line twice. Which of these crosses the line after time T has elapsed is a detailed question which can only be answered by solving equations (6.46) analytically or numerically after specifying a and T. What we can say, however, is that the optimal trajectory has roughly the shape shown in Fig. 6.16a. As we move along this trajectory the state variable x first decreases

168   *Optimal control theory: the basics*

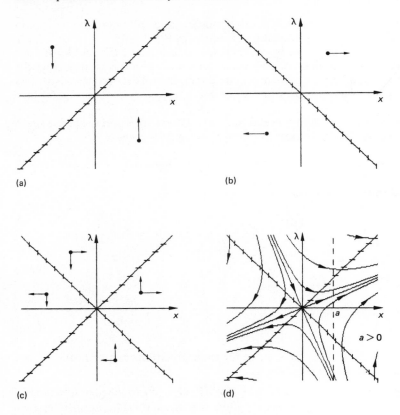

Fig. 6.15

from its value a at $t = 0$, eventually rising back to the value a at $t = T$. In the process λ, which in this example equals u, progressively rises from a negative to a positive value. The graphs of x and u against t are sketched from these observations in Figs 6.16b, c.[10]

Example 6
We found in Section 6.5 that the Hamiltonian analysis for the problem:

$$\text{maximise} \int_0^T e^{-t} u^{1/2} \, dt$$
$$\text{subject to } \dot{x} = -u, \quad x(0) = R_0, \quad x(T) = 0$$

yields the Hamilton equations:

$$\dot{\lambda} = 0, \quad \dot{x} = -e^{-2t}/(4\lambda^2).$$

We cannot apply the Phase Space Method directly to these equations since they involve t explicitly via the exponential discount factor. We can however eliminate this factor if we define a new multiplier λ_1 as $\lambda = \lambda_1 e^{-t}$. Differentiating this expression with respect to t and cancelling the exponential factor, the first of the Hamilton equations becomes:

$$\dot{\lambda}_1 = \lambda_1$$

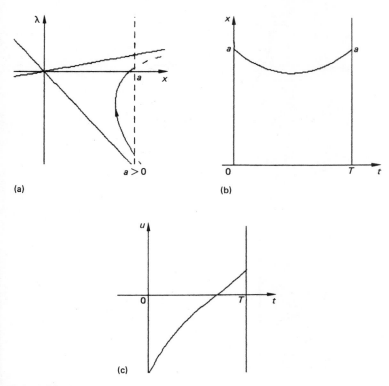

Fig. 6.16

and the second:

$$\dot{x} = -1/(4\lambda_1^2).$$

There is no explicit time dependence in these equations and hence the Phase Space Method applies.

(i) The λ_1-isocline is the horizontal line $\lambda_1 = 0$ coincident with the x-axis (Fig. 6.17a). Above this line $\dot{\lambda}_1$ is positive corresponding to upwards movement while below this line movement is downwards.

(ii) There is no x-isocline since there are no finite points satisfying the equation $G = 0$ in this case. Throughout phase space $\dot{x}$ is always negative and hence movement is always to the left (Fig. 6.17a). Following the arrow directions we obtain the phase space diagram of Fig. 6.17b.

The optimal trajectory is the one that crosses the vertical line $x = R_0$ at $t = 0$ and the λ_1-axis at $t = T$. The shape of such a trajectory is isolated in Fig. 6.17c, clearly indicating that x decreases and λ_1 increases throughout the relevant section of the trajectory. Since $u = 1/(4\lambda_1^2)$ we conclude that the phosphate production rate also decreases throughout the exploitation phase. The overall optimum with T infinite corresponds to trajectory AA (Fig. 6.17b) which has the λ_1-axis as asymptote.

In preparation for the Exercise Set let us summarise the various stages of the Phase Space Method in Procedure Box 6.3.

170   *Optimal control theory: the basics*

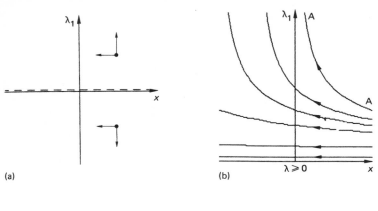

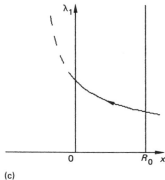

(c)

Fig. 6.17

PROCEDURE 6.3
The Phase Space Method

1. (a) Sketch the λ-isocline (look out for degeneracies with no or all points satisfying the isocline equation).
   (b) Superimpose tangents on the isocline.
   (c) Add arrows to each region defined by the isocline.
2. Repeat stage 1 for the x-isocline.
3. Superimpose the diagrams from stages 1 and 2.
4. Sketch the trajectories using the rule that no trajectories cross or touch except possibly at equilibrium points.
5. Identify the trajectories satisfying the end-point conditions at $t = 0$ and some end-point time T. Derive properties of the optimal state and control variables.

Exercise Set 3
Sketch the Phase Space Trajectories for problem (6.30) when the functions f, g are identified as follows:

	(i)*	(ii)	(iii)*	(iv)	(v)*	(vi)
f	$-(x^2+xu+u^2)$	$-\frac{1}{2}(x+u)^2$	$2e^{-t}u^{1/2}$	$2e^{-4t}u^{1/2}$	$2u^{1/2}$	$u-\frac{1}{2}u^2$
g	$x-u$	$u-x$	$3x-u$	$3x-u$	$2x(1-x)-u$	$2x(1-x)-u$

*6.7 Ramsey's Growth Model

One well known example of optimal control theory is Ramsey's growth model[11] which addresses the issue of how much of the economy's output should be in the form of consumer goods and how much in the form of capital investment such as new plant and equipment.

Let us suppose that the rate of output Q of the economy is a function of two inputs, capital K and labour L i.e.

$$Q = F(K, L).$$

A simple but not unreasonable choice for F is the Cobb–Douglas function: $F = cK^\beta L^{1-\beta}$ where c and β are constants with β satisfying the inequalities $0 < \beta < 1$. The output is used for three purposes (i) for consumption C (ii) to replace old capital equipment (assumed at a rate proportional to the current capital stock) and (iii) for new capital investment. Precisely we have:

$$Q = C + bK + \dot{K}$$

where b is the proportionality factor and $\dot{K}$ denotes the new investment. The equation to describe the economy is therefore given by:

$$C + bK + \dot{K} = cK^\beta L^{1-\beta}$$

which, on dividing by L and using the identity:

$$\dot{K}/L = (K/L)(\dot{L}/L) + \frac{d}{dt}(K/L),$$

becomes

$$C/L + b(K/L) + \frac{d}{dt}(K/L) + (K/L)(\dot{L}/L) = c(K/L)^\beta \qquad (6.47)$$

This equation can be simplified if we suppose that L is increasing exponentially, with $L = Le^{nt}$ for then we have $(\dot{L}/L) = n$. With the relabelling $x = K/L$ and $u = C/L$ equation (6.47) then becomes:

$$\dot{x} = cx^\beta - \theta x - u \qquad \text{where } \theta = b+n \qquad (6.48)$$

The optimal choice of consumption u at each point in time is determined in this model by maximising total discounted utility:

$$\int_0^T e^{-rt} u^\alpha \, dt \qquad 0 < \alpha < 1$$

over a planning period of length T subject to the dynamic equation (6.48) and appropriate

end-point conditions. The Hamiltonian in this case is given by:
$$H = e^{-rt}u^\alpha + \lambda(cx^\beta - \theta x - u)$$
and hence:
$$\frac{\partial H}{\partial u} = \alpha e^{-rt}u^{\alpha-1} - \lambda = 0, \quad \text{i.e. } u = (\alpha e^{-rt}\lambda^{-1})^{1/(1-\alpha)} \tag{6.49}$$
$$-\dot\lambda = \beta\lambda c x^{\beta-1} - \lambda\theta$$
$$\dot x = cx^\beta - \theta x - u.$$

Redefining the multiplier as $\lambda_1 = \lambda e^{rt}$ we obtain the Hamilton equations:
$$\dot\lambda_1 = \lambda_1(r + \theta - \beta c x^{\beta-1})$$
$$\dot x = cx^\beta - \theta x - (\alpha/\lambda_1)^{1/(1-\alpha)}.$$

It would clearly be difficult to solve these equations analytically and so we will use the phase space method to analyse their properties.

From these equations we find that the λ_1-isocline is the pair of lines $\lambda_1 = 0$ and $x = x^*$ where $x^* = (\beta c/(r+\theta))^{1/(1-\beta)}$ and the x-isocline is the curve with equation:
$$\lambda_1 = \frac{\alpha}{(cx^\beta - \theta x)^{1-\alpha}}.$$

This has the shape shown in Fig. 6.18a with asymptotes at $x = 0$ and $x = \bar x$ where $\bar x = (c/\theta)^{1/(1-\beta)}$. The trajectories, sketched in Fig. 6.18b, are clearly structured about the equilibrium point P at the intersection of the isoclines. (This point always exists whatever the values of the parameters.) If $x(0) = x^*$ then the optimal path is just the point P with per capita consumption and per capita capital stock remaining constant while total consumption and capital stock increase exponentially with the labour force along a 'balanced growth path'. The crucial feature of the model that generates this equilibrium point is the 'decreasing returns to scale' exhibited by our particular choice of production function F. By this we mean that the greater the input the less proportionately is the increase in production. In the Pioneer model this feature was not present because it was implicitly assumed that there was always enough good land available for cultivation. This is usually not the case. The quality of extra land brought into cultivation will gradually deteriorate leading to decreasing returns to scale.

If we start the Ramsey model with a less developed economy, $x(0) < x^*$ then we can reach level x^* along the trajectories shown in Fig. 6.18c. It takes longer to reach x^* along trajectory B compared with trajectory A since along the former trajectory one is consuming more (see equation (6.49)). On the limiting trajectory C through P it in fact takes an infinite amount of time to reach P.[12]

Finally we must check whether the inequality constraints that we have so far ignored in the analysis are satisfied by the optimal solutions of interest. There are three such constraints:
$$x \geq 0, \ u \geq 0, \ cx^\beta \geq u \quad \text{for } 0 \leq t \leq T.$$

The first two are obvious and obviously satisfied by all the trajectories in Fig. 6.18b. The third constraint states that current consumption does not exceed current production. This is true for those trajectories where the capital stock is always increasing in time since the region where the third constraint is operative lies below the curve:
$$\lambda_1 = \alpha(cx^\beta)^{(\alpha-1)}$$

(shown dotted in Fig. 6.18b) which in turn lies below the x-isocline. Entering this region on a trajectory with decreasing capital stock corresponds to eating the machinery!

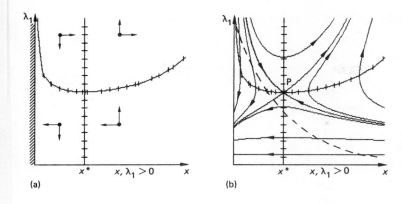

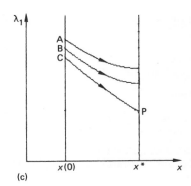

Fig. 6.18

Problem Set 6

1 Use Procedure 6.1 to solve the Discrete Control Problem (6.4) when:

	f_k	g_k	n	x_0	x_{n+1}
(i)*	$-\frac{1}{2}(x_k+u_k)^2$	u_k	3	1	1
(ii)	$-\frac{1}{2}u_k^2$	$u_k - x_k$	3	1	1
(iii)*	$-(1+u_k^2)^{1/2}$	u_k	n	0	$n+1$
(iv)	$-(x_k+\frac{1}{2}u_k^2)$	u_k	n	0	0

2 Use Procedure 6.2 to solve the Continuous Control Problem (6.30) when $T = 1$ and:

	f	g	$x(0)$	$x(T)$
(i)*	$-(x+e^u)$	u	1	$2\ln 2$
(ii)	$-(1+e^{-u})$	u	0	1
(iii)*	$-u^2$	1	0	1
(iv)	$-\frac{1}{2}(x^2+u^2)$	x	0	1

3 Sketch the Phase Space Trajectories for problem (6.30) when:

	(i)*	(ii)	(iii)*	(iv)
f	$-(x+\tfrac{1}{2}u^2)$	$-\tfrac{1}{2}(5x^2+u^2-2xu)$	$-(x+e^u)$	$e^{-t}(1-e^{-u})$
g	$u+x$	$u+x$	u	$-u$

4 (a)* (i) Show that the state variable for the problem:

$$\text{minimise} \quad \int_0^T x(1+u^2)^{1/2}\,dt$$

subject to $\dot{x}=u$, $x(0)=x(T)=a>0$

satisfies the differential equation:

$$\dot{x}=C^{-1}(x^2-C^2)^{\frac{1}{2}}, \quad C \text{ a constant.}$$

(ii) By solving this equation using the substitution $x=C\cosh\theta$ and applying the end conditions show that:

$$x = C\cosh(C^{-1}(t-\tfrac{1}{2}T))$$

where C satisfies the algebraic equation:

$$\cosh y = (2aT^{-1})y \quad \text{with } y=\tfrac{1}{2}TC^{-1}.$$

(iii) By finding the points of intersection of the graph $z=\cosh y$ and the radial line $z=(2aT^{-1})y$ find y and hence C when a/T equals 1, 0.75 and 0.5.

(This problem is the well known Soap Film Problem where the shape of a soap film suspended between two coaxial circular wires of radius a is sought. This interpretation is established in Section I of the Appendix.)

(b) (i) Show that the state variable for the problem:

$$\text{minimise} \quad \int_0^T x^{-1/2}(1+u^2)^{1/2}\,dt$$

subject to $\dot{x}=u$, $x(0)=0$, $x(1)=a>0$

satisfies the differential equation:

$$\dot{x} = ((C^2 x)^{-1}-1)^{1/2}$$

(ii) By using the substitution $x=C^{-2}\sin^2(\theta/2)$ show that:

$$t = R\theta - R\sin\theta \quad \text{with } x=R-R\cos\theta \text{ and } R=\tfrac{1}{2}C^{-2}.$$

Interpret this solution in terms of the motion of a point on a circle of radius R. (This is the Brachistochrone Problem, to find the shape of a smooth wire lying in a vertical plane along which a bead would travel between two points in minimum time. See Section I of the Appendix for details.)

(c) (i) Show that the state variable for the problem:

$$\text{minimise} \quad \int_0^T (x^2+u^2(1+h^2(x)))^{1/2}\,dt$$

subject to $\dot{x}=u$, $x(0)=a$, $x(T)=b$; $a,b>0$

satisfies the differential equation:

$$\dot{x} = \frac{x(x^2-C^2)^{1/2}}{C(1+h^2(x))^{1/2}}.$$

(ii) Solve this equation for $h = \alpha$, a constant. (Use the substitution $x = C \sec \theta$.) (This problem is an example of a Geodesic Problem, to find the shortest distance between two points on a surface, in this case a surface symmetric about the vertical axis.)

5 (a)* Show that the dual of the discrete control problem:

$$\text{maximise} \quad -\sum_{k=0}^{n} (x_k + \tfrac{1}{2} u_k^2)$$
$$\text{subject to} \quad x_{k+1} - x_k = u_k \quad \text{for } k = 0, \ldots, n$$
$$x_0 = x_{n+1} = 0$$

is given by:

$$\text{minimise} \quad \tfrac{1}{2} \sum_{k=0}^{n} \lambda_k^2$$
$$\text{subject to} \quad \lambda_k = \lambda_{k-1} + 1 \quad \text{for } k = 1, 2, \ldots, n$$

by following the procedure of Chapter 4, maximising the Lagrangean with respect to both the state and control variables. By examining its structure carefully, solve this dual and hence find the primal optimal value.

(b) Find and solve the dual of the problem:

$$\text{maximise} \quad \sum_{k=0}^{n} u_k^{1/2}$$
$$\text{subject to} \quad x_{k+1} = c(x_k - u_k), \quad c > 1$$
$$x_{n+1} = \gamma c^{n+1} x_0, \quad 0 < \gamma < 1.$$

6 (a) In the Pioneer Problem discussed in Section 6.3 determine the limit of the state and control sequences x_k, u_k when the utility exponent α tends to 1 with $\rho c < 1$.
(b)* Check your result by the following alternative method.
  (i) Show that the dual of the linear (control) program:

$$\text{maximise} \quad \sum_{k=0}^{n} \rho^k u_k$$
$$\text{subject to} \quad x_{k+1} = c(x_k - u_k); \quad x_k, u_k \geq 0$$

is given by:

$$\text{minimise} \quad cx_0 \lambda_0 - x_{n+1} \lambda_n$$
$$\text{subject to} \quad \lambda_k \geq c \lambda_{k+1} \quad \text{for } k = 0, \ldots, n-1$$
$$c \lambda_k \geq \rho \quad \text{for } k = 0, \ldots, n.$$

  (ii) By carefully considering the structure of the inequalities find the solution of the dual and hence (by the Complementary Slackness Conditions) find the optimal primal variables when $\rho c < 1$.

7 Discuss the change in the shape of the family of trajectories for the problem:

$$\text{maximise} \quad \int_0^T e^{-rt} u^{1/2} \, dt$$
$$\text{subject to} \quad \dot{x} = cx - u, \quad x(T) > x(0) > 0, \quad c > 0$$

as r increases from zero to a value greater than c. Taking this to be a continuous harvesting model describe the change in harvesting strategy as the discount factor decreases from unity.

176 *Optimal control theory: the basics*

8* (a) The optimal control problem (6.30) is said to be a problem from the Calculus of Variations if its dynamic equation has the form $\dot{x} = u$. For such a problem show by eliminating the multiplier function $\lambda(t)$ that the state variable function $x(t)$ satisfies the second order differential equation:

$$\frac{d}{dt}\left(\frac{\partial f}{\partial \dot{x}}\right) = \frac{\partial f}{\partial x}$$

where $f(x, u, t) = f(x, \dot{x}, t)$. (This equation is called the *Euler–Lagrange equation*.)

(b) Derive this equation for the Soap Film* and Brachistochrone problems of Question 4.

References

1. For further reading on the theory of difference equations see M. Casson, *Introduction to Mathematical Economics*, Nelson (1973).
2. We will not keep on writing out explicitly the range of values of k for each equation. If in doubt one can refer back to the relations (6.9), (6.10) and (6.11).
3. To actually prove this one would use the method of induction.
4. That the limit of the equation system (6.9), (6.10) and (6.11) gives the solution to the limit of the control problem (6.4) is an assertion that would have to be justified in a rigorous treatment. A more direct derivation of the equations (6.31), (6.32) and (6.33) will be given in Section J of the Appendix, although even there the treatment will not be totally rigorous.
5. The book *Differential Equations and Their Applications* by M. Braun, Springer-Verlag (1975) has an extensive discussion on first and second order differential equations.
6. For a full discussion see C. W. Clark, *Mathematical Bioeconomics*, Wiley (1976) and P. S. Dasgupta and G. M. Heal, *Economic Theory and Exhaustible Resources*, Cambridge University Press (1979).
7. The original article by H. Hotelling, 'The Economics of Exhaustible Resources', can be found in the *Journal of Political Economy*, Volume 39, pages 137–175 (1931).
8. There is also an ambiguity if both F and G are infinite at a point but we will ignore this possibility.
9. Except of course if P is an equilibrium point when the tangent is not defined.
10. Note that for u we do not have the information to determine whether its graph is bowed up (concave) or bowed down (convex) or has a more complicated shape. We only know that it has no stationary point in the interval 0 to T.
11. The original article by F. Ramsey, 'A Mathematical Theory of Saving' can be found in the *Economic Journal*, Volume 38, pages 543–559 (1928).
12. This can be shown by linearising the equations about P. The details can be found in M. D. Intriligator, *Mathematical Optimisation and Economic Theory*, Prentice-Hall (1971).

7
Optimal Control: Generalisations

7.1 Introduction

In the last chapter we discussed how to solve the basic optimal control problem with no inequality bounds on the state or control variables. Most optimal control problems of interest, however, are not of this form, and this includes some of those mentioned in the last chapter. For example in the Ramsey Growth Model the control variable, per capita consumption, must be non-negative and cannot exceed current production (unless we allow storage of goods for future consumption). Also in the Exhaustible Resource Model both the state and control variables, the current stock and production rate, are necessarily non-negative. We ignored these bounds on the assumption that they would not be binding in the optimal solution, an assumption that was checked at the end of each analysis. In this chapter we present cases where the bounds do become binding and discuss how to extend the theory to solve them. In particular in Section 7.2 we use the fact that the only change in the basic algorithm required when fixed bounds on the control variable are present is that the unconstrained problem of maximizing the Hamiltonian with respect to u is changed into a constrained maximisation problem. In Section 7.5 we discuss how to handle the less traceable case where there are bounds on the state variable.

In some of the applications discussed in the last chapter there was also some uncertainty as to what values should be chosen for the end parameters T and $x(T)$. It was not clear for example what length of time T to allow for depletion of the phosphate deposits in the Exhaustible Resource Model nor what value to take in the Ramsey Growth Model for the capital stock $x(T)$ to bequeath to future generations. One way of deciding is to choose T or $x(T)$ optimally. Indeed this is what we did in the Resource Model where we came to the conclusion that for certain demand functions T should be infinite. The optimisation with respect to T and $x(T)$ can in fact be absorbed within the main algorithm by generalising the end conditions on the state variables and multipliers. The precise conditions are stated in Section 7.4.

Two other generalisations of the basic model are also dealt with in this chapter. The first concerns the addition of integral constraints to the control problem, probably the best known example of this being the Isoperimetric Problem where one seeks the largest area within a boundary of given length. A less academic example is the minimum time problem where, typically, the objective is to minimise the time taken to achieve a given profit from a given production system. The second generalisation tackles the issue of how to deal with those systems that require more than one state variable or control variable. This extension is called Vector Control Theory.

7.2 Bounds on the Control Variable

Let us suppose that the control variable u satisfies the inequalities:

$$\underline{u} \leqslant u \leqslant \bar{u}$$

where $\underline{u}$ and $\bar{u}$, the *lower* and *upper* bounds on u, are constants. In optimising the Hamiltonian with respect to u we must take these bounds into account, i.e. the optimisation becomes a constrained optimisation problem solvable by the method described in Section 3.2.

Our first example of a control problem with bounds on u is autonomous and so we will be able to use the phase space method to explore the effect of these bounds on the behaviour of the system. (In Section 7.6 we will look at an example which is not autonomous.)

Example 1
The Hamiltonian for the problem:

$$\text{maximise} \quad -\tfrac{1}{2}\int_0^T (x^2 + u^2)\,dt$$

$$\text{subject to} \quad \dot{x} = x + u$$
$$-1 \leqslant u \leqslant 1$$
$$x(0) = a, x(T) = b$$

with $\underline{u} = -1$ and $\bar{u} = 1$ is given by:

$$H = -\tfrac{1}{2}(x^2 + u^2) + \lambda(x + u).$$

If we consider H to be a function of u with x and λ as parameters then its graph is a parabola with peak located at $u = \lambda$. The maximum of H will depend on whether this peak

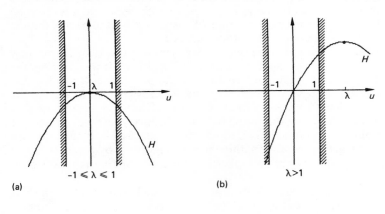

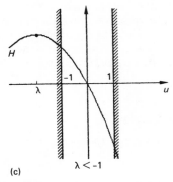

Fig. 7.1

7.2 Bounds on the control variable

lies in or outside the feasible region, and hence on the value of the parameter λ. There are three possibilities to consider.

(a) If $-1 \leq \lambda \leq 1$ then the maximum occurs at the peak $u = \lambda$ since it lies in the feasible region (Fig. 7.1a).
(b) If $\lambda > 1$ then the maximum occurs on the upper boundary $u = 1$ since the peak lies to the right of the feasible region (Fig. 7.1b).
(c) If $\lambda < -1$ then the maximum occurs at the lower boundary $u = -1$ (Fig. 7.1c).

The form of the Hamilton equations:

$$\left. \begin{array}{l} -\dot{\lambda} = \lambda - x \\ \dot{x} = x + u \end{array} \right\} \quad (7.1)$$

will as a consequence depend on which of the three ranges is being considered.

(a) With $-1 \leq \lambda \leq 1$ and $u = \lambda$, equations (7.1) reduce to

$$\dot{\lambda} = x - \lambda; \quad \dot{x} = x + \lambda.$$

The trajectories for these equations, sketched in Fig. 7.2a, are valid only in the region of phase space defined by $-1 \leq \lambda \leq 1$. The trajectories in this region are in fact identical to those sketched in Fig. 6.15d where we considered the same problem, except for the bounds.

(b) With $\lambda > 1$ and $u = 1$ equations (7.1) reduce to

$$\dot{\lambda} = x - \lambda; \quad \dot{x} = x + 1.$$

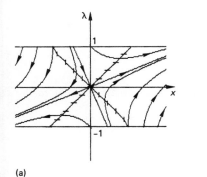

(a)

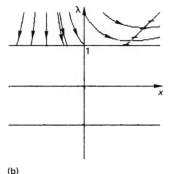

(b)

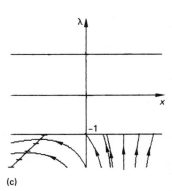

(c)

Fig. 7.2

The corresponding trajectories, constructed from the isoclines $x = -1$, $x = \lambda$ in the region $\lambda > 1$, are shown in Fig. 7.2b.
(c) With $\lambda < -1$ and $u = -1$ the equations:
$$\dot{\lambda} = x - \lambda; \quad \dot{x} = x - 1$$
generate the trajectories shown in Fig. 7.2c for the region $\lambda < -1$.

To obtain the complete phase space diagram (Fig. 7.3) we smoothly join together these trajectory sections, using the fact that at each point on the boundary the trajectory possesses a tangent. This is because there is no jump in the optimal value of u in the transition from a boundary to internal maximum of H and hence no jump in the derivatives of the Hamilton equations.

Comparison with Fig. 6.15d shows how the bounds on u distort the phase space diagram.

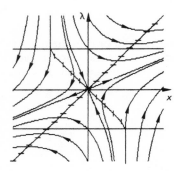

Fig. 7.3

As a second example we consider a problem where the Hamiltonian is linear in the control variable.

Example 2
The Hamiltonian for the problem:

$$\text{maximise} \quad \int_0^T u \, dt$$

$$\text{subject to} \quad \dot{x} = 2x - u; \quad 0 \leqslant u \leqslant 1$$

$$x(0) = x(T) = a$$

has the linear form:

$$H = u + \lambda(2x - u) = u(1 - \lambda) + 2\lambda x.$$

The position of its maximum with respect to u depends on the orientation of its straight line graph.

(a) When $\lambda < 1$ the maximum of H is taken at the upper bound $u = 1$ since the gradient is positive (Fig. 7.4a).
(b) When $\lambda > 1$ the maximum of H is taken at the lower bound $u = 0$ since the gradient is now negative (Fig. 7.4b).
(c) When $\lambda = 1$ the graph is horizontal and hence all points in the interval $0 \leqslant u \leqslant 1$ are optimal (Fig. 7.4c).

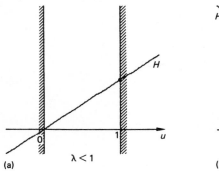
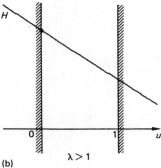

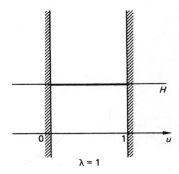

Fig. 7.4

To sketch the phase space trajectories we have to consider the Hamilton equations:

$$\dot{\lambda} = -2\lambda; \quad \dot{x} = 2x - u \tag{7.2}$$

for each of these three cases.

(a) With $\lambda < 1$ and $u = 1$ the isoclines are the lines $\lambda = 0$ and $x = \frac{1}{2}$, which also act as solutions to the equations (7.2). The trajectories for this region are sketched in Fig. 7.5a.
(b) With $\lambda > 1$ and $u = 0$ the trajectories are structured about the single isocline $x = 0$ (Fig. 7.5b).
(c) On the boundary $\lambda = 1$ between the two phase space regions the two sets of trajectories connect up to form continuous trajectories (Fig. 7.5c). The trajectories will, however, not possess tangents on the boundary because in crossing the boundary there is a sudden switch in the control variable from value 0 to value 1, generating discontinuities in the derivatives of the Hamilton equations. (The switch is instantaneous because from equations (7.2) it is clear that on the boundary λ is non-zero.) This sudden switch between control variable bounds explains the name *Bang-Bang* given to this type of behaviour.

In general the bounds $\underline{u}, \bar{u}$ will depend on t and x. If they depend on t we can no longer use the phase space method and an analytic solution should be sought. If they depend on x

182 Optimal control: generalisations

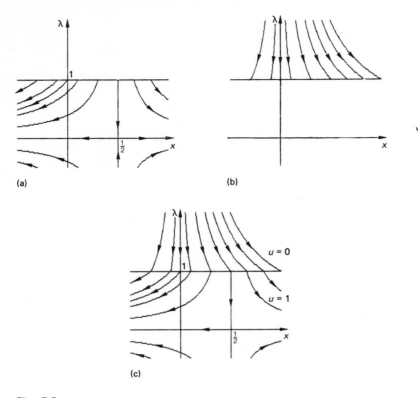

Fig. 7.5

then the multiplier function equation changes form through the addition of terms involving the derivative of the bounding functions with respect to x. The best way to tackle this case is to add the bounds to the Hamiltonian with new multiplier functions. Examples will be given in the Problem Set.

Exercise Set 1
Sketch the phase space trajectories for the continuous optimal control problem when the functions f, g and bounds $\underline{u}$, $\bar{u}$ are chosen to be:

	(i)*	(ii)	(iii)*	(iv)	(v)*	(vi)
f	$-(x+\frac{1}{2}u^2)$	$-(x+\frac{1}{2}u^2)$	$-x^2$	$-x^2$	$2u^{1/2}$	u
g	$x+u$	$x-u$	u	$x+u$	$2x(1-x)-u$	$2x(1-x)-u$
$\underline{u}, \bar{u}$	0, 2	0, 2	$-1, 1$	0, 1	$0, \frac{1}{4}$	0, 1

*7.3 An Application in Renewable Resource Theory[1]

The fishing cooperative of a small island state wants to determine its options for exploiting the fishing grounds around the island. As a first step it has built a simple control model with the state variable, $x(t)$, denoting the level of the fish stock present in the fishing grounds at time t and the control variable, $u(t)$, the size of the fleet actually fishing at that

7.3 An application in renewable resource theory

time. The dynamic equation for the problem models the growth of the fish stock. If there is no fishing and the stock is low then that stock is assumed to grow according to the differential equation:

$$\dot{x} = c_0 x$$

where c_0 is the growth factor. The fishing grounds however can only support a certain stock, say $\bar{x}$, and hence as x grows towards $\bar{x}$ the growth factor c_0 must decrease until it is zero in the limit. This saturation effect can be modelled by the form $c_0 = c(\bar{x} - x)$ where c is a constant. When fishing is in progress the rate of growth of x is decreased by the rate F at which the fish are caught, which in this case is found to be proportional to both the number of fish and the number of boats fishing, i.e.

$$F = Axu.$$

Hence the dynamic equation for the fish stock is given by:

$$\dot{x} = c(\bar{x} - x)x - Axu \tag{7.3}$$

The cooperative's objective is to maximise the total discounted profit over a planning period of length T. If p_0 is the (supposed constant) price of landed fish and costs are incurred at a rate $w_0 u$, proportional to the number of boats used, then the objective is to:

$$\text{maximise} \int_0^T e^{-rt}(p_0 Axu - w_0 u)\,dt$$

subject to equation (7.3), the control bounds $0 \leqslant u \leqslant \bar{u}$ (where $\bar{u}$ denotes the maximum number of boats available), the condition $x \leqslant \bar{x}$ (which, as we will see, will automatically be satisfied) and a government imposed condition that the stock at the end of the planning period must not be less than at the beginning.

The Hamiltonian for this problem is linear in u and hence the system is subject to the Bang-Bang phenomenon. Precisely, with the multiplier redefined as $\lambda_1 = \lambda e^{rt}$, we have:

$$He^{rt} = u((p_0 - \lambda_1)Ax - w_0) + \lambda_1 c(\bar{x} - x)x \tag{7.4}$$

together with the Hamilton equations:

$$\dot{x} = c(\bar{x} - x)x - Axu \tag{7.3}$$

$$\dot{\lambda}_1 = 2\lambda_1 c(x - \tfrac{1}{2}\bar{x}) + uA(\lambda_1 - p_0) + r\lambda_1 \tag{7.5}$$

With w_0 and r put equal to zero for simplicity, the coefficient of u in H is just $(p_0 - \lambda_1)Ax$. Hence if $\lambda_1 > p_0$ then u is at its lower bound, zero, and if $\lambda_1 < p_0$ then u is at its upper bound, $\bar{u}$.

The structure of the phase space trajectories depends critically on the magnitude of $\bar{u}$, the size of the available fleet. We first study the case where the fleet is sufficiently large that it can catch fish faster than they can breed i.e. when $\bar{u}$ is such that $\dot{x} < 0$ for all $0 < x < \bar{x}$ when $u = \bar{u}$.

(i) If $\bar{u} > c\bar{x}A^{-1}$ then the isoclines for the two regions $\lambda_1 > p_0$, $\lambda_1 < p_0$ are as shown in Fig. 7.6a, the curved isocline in the lower region being part of a branch of the hyperbola

$$\lambda_1(x - x^*) = (\tfrac{1}{2}\bar{u}A\, p_0 c^{-1}) \tag{7.6}$$

where $x^* = \tfrac{1}{2}(\bar{x} - \bar{u}A/c)$. The trajectories are sketched in Fig. 7.6b and are clearly structured about the point P on the boundary $\lambda_1 = p_0$. This point is in fact an equilibrium point since with $\lambda_1 = p_0$, $x = \tfrac{1}{2}\bar{x}$, $r = 0$ and $u = \tfrac{1}{2}c\bar{x}A^{-1}$ equations (7.3), (7.5) give $\dot{x}$ and $\dot{\lambda}_1$ as zero. P represents the ideal fishing situation where a fraction $\tfrac{1}{2}c\bar{x}A^{-1}\bar{u}^{-1}$ of the fleet is employed to cream off the natural increase in the fish stock which is at its maximum.

184 *Optimal control: generalisations*

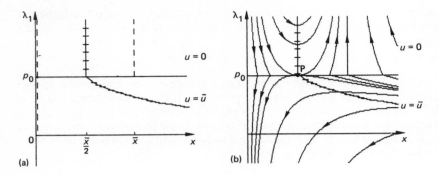

Fig. 7.6

If the initial stock of fish is less than $\frac{1}{2}\bar{x}$ and the condition $x(0) = x(T)$ is imposed then the optimal strategy consists of three stages (Fig. 7.7) if T is sufficiently large. In the first the fish stock is allowed to reach the ideal level $\frac{1}{2}\bar{x}$ in the shortest possible time (i.e. with no fishing taking place at all). One then stays at the ideal level as long as possible, leaving it only to return to the initial level at the end of the planning period. This last step is also

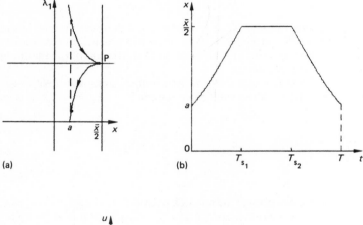

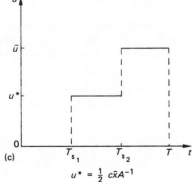

$$u^* = \tfrac{1}{2} c\bar{x}A^{-1}$$

Fig. 7.7

7.3 An application in renewable resource theory   185

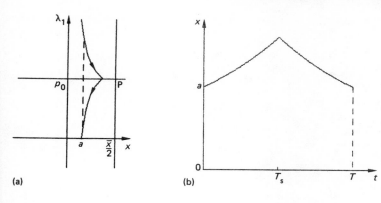

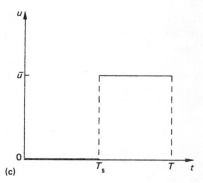

Fig. 7.8

achieved in the shortest possible time with all the fleet employed. If T is reduced sufficiently then there is no time to reach the ideal state before returning to the given end state (Fig. 7.8).

In summary we can say that for low T we have a Bang-Bang situation (Fig. 7.8) with control switching instantaneously from $u = 0$ to $\bar{u}$, while for large enough T (Fig. 7.7) we have what we might call a Bank-Sit-Bang situation where we stay for a time on the 'knife-edge' between the extreme positions where the control is fully on or off (with the linear graph of H with respect to u horizontal).

These observations might lead the government to revise its end-point condition to read as:

$$x(T) = \tfrac{1}{2}\bar{x}$$

for $T > T_{s1}$, (Fig. 7.7b) so that the fish stock is left at a level which is sustainable.

(ii) We lose the Bang-Sit-Bang structure when the fleet is much smaller, with $\bar{u} < \tfrac{1}{2}c\bar{x}A^{-1}$. The phase space structure in this case is shown in Fig. 7.9. There is no longer an equilibrium point P on the boundary since there is no longer a solution to the equations (7.3), (7.5) with $\dot{x} = \dot{\lambda} = 0$ and $u \leqslant \bar{u}$. The equilibrium point has shifted to a point Q where the hyperbola (7.6) intersects a new section of the x-isocline. At Q the fleet is fully employed, with fishing and breeding in balance. If one is not initially at Q then the trajectories of Fig. 7.9 suggest that one should move as close to Q for as long as the final end-point condition will allow.

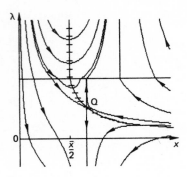

Fig. 7.9

The fishing example provides us with a clear interpretation for the multiplier λ_1 and the Hamiltonian H. The multiplier λ_1 is the so-called *imputed price* that values the fish in the sea, of interest to someone who might wish to take over the cooperative as a going concern. We see from Figs 7.7 and 7.8 that we do not fish if the value of the fish in the sea is greater than it is on shore ($\lambda_1 > p_0$) and we do fish if the converse is true ($\lambda_1 < p_0$). At the equilibrium point P (Fig. 7.6b) with $\lambda_1 = p_0$ the fish at sea and on the dock are valued equally. The first term of the Hamiltonian for this problem:

$$He^{rt} = p_0 Axu + \lambda_1(c(\bar{x}-x)x - Axu)$$
$$= p_0 F + \lambda_1 \dot{x}$$

gives the rate of profit from the fish caught and its second the rate of increase in the value of the fish stock. H therefore measures the rate of increase in the total assets of the Cooperative.

7.4 Terminal Conditions

In some of the examples we have discussed $x(T)$ or T or both are not specified beforehand but are to be chosen optimally. This can be done either as a second step after the main algorithm or as part of it. In the latter case, discussed here, we have to amend Procedure 6.2 by replacing the end-point condition on $x(T)$ by a generalised set of conditions which we derive in three stages.

(a) T fixed, $x(T)$ to be chosen optimally

The problem to be solved can be specified as:

$$\text{maximise} \quad \int_0^T f(x, u, t)\,dt + h(x(T)) \tag{7.7}$$

$$\text{subject to} \quad \dot{x} = g(x, u, t)$$
$$x(0) = a, \quad x(T) \text{ to be chosen optimally.}$$

The function h added to the objective functional measures the value of the final state $x(T)$. For a company that has been expanding its productive capacity over a given period the integral in (7.7) would represent its total profits in that period and the h function the value of the total assets of the company at the end of that period. Similarly for a community making a living from a continuously harvestable crop the integral would represent, as before, the total utility while the h function would value the possible resource bequests that could be left for future generations.

The optimal choice for the final state value: $x_T \equiv x(T)$ in problem (7.7) can be obtained by imposing an end-point condition on $\lambda_T \equiv \lambda(T)$. Precisely, we replace condition $x_T = b$ (given) in Procedure 6.2 by the condition:

$$\lambda_T = \frac{dh}{dx_T} \tag{7.8}$$

leaving the rest of the algorithm as it is. The derivation of this condition is quite subtle and is left to Section J of the Appendix where we discuss the variational derivation of the Hamilton equations. As an example of the use of equation (7.8) let us consider a problem in geometry chosen for its algebraic simplicity.

Example 3
The problem

$$\text{maximise} \quad -\int_0^T (1+u^2)^{1/2} \, dt - \tfrac{1}{2}(x_T-1)^2$$

$$\text{subject to} \quad \dot{x} = u$$

$$x(0) = 0, \quad x_T \text{ to be chosen optimally}; \; T = 1$$

with asset function $h = -\tfrac{1}{2}(x_T-1)^2$ has as its Hamiltonian the function:

$$H = -(1+u^2)^{1/2} + \lambda u$$

which generates the equations:

$$\frac{\partial H}{\partial u} = -u(1+u^2)^{-1/2} + \lambda = 0 \qquad \text{i.e. } u = \lambda(1-\lambda^2)^{-1/2}$$

$$-\dot{\lambda} = \frac{\partial H}{\partial x} = 0 \qquad \text{i.e. } \lambda = A, \text{ (a constant)}$$

$$\dot{x} = u = A(1-A^2)^{-1/2} \qquad \text{i.e. } x = A(1-A^2)^{-1/2}t + B.$$

The constants of integration A and B are obtained from the end-point conditions:

$$x(0) = 0, \; \lambda_T = \frac{dh}{dx_T} = -(x_T-1), \; T = 1$$

using equation (7.8). The first condition gives $B = 0$ and the second yields the equation:

$$\lambda_T = A = -A(1-A^2)^{-1/2} + 1 = -x_T + 1,$$

an algebraic equation for A with unique solution at the intersection of the curve $z = A(1-A^2)^{-1/2}$ and the line $z = 1 - A$ (Fig. 7.10a). In fact $A \simeq 0.5$ and hence the solution to our problem is (to a good approximation) the radial line

$$x(t) = 0.5t$$

terminating at the point $(x, t) = (0.5, 1)$ (Fig. 7.10b).

As an interpretation of this problem imagine that a schoolmaster, who firmly believes in the marriage of brawn and brain, is organising a cross-country race which finishes at the

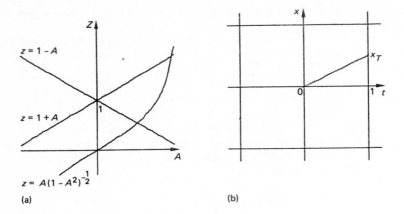

Fig. 7.10

origin of the 'map' shown in Fig. 7.10b. Instead of specifying the route the schoolmaster tells the participants that they can start at any point on a road running North–South along the grid at $t = 1$ and then choose any path along the flat terrain to the finishing post. (Note that t denotes a distance rather than time in this example.) However, he imposes a penalty of $-h(x_T) = \frac{1}{2}(x_T - 1)^2$ to add to the time taken to complete the run, the magnitude of this penalty depending on the position x_T along the road at which a participant may wish to start. To find the best route we first note that the length of a path $x(t)$ between points $(1, x_T)$ and $(0, 0)$ is given by $l = \int_0^1 (1 + \dot{x}^2)^{1/2} dt$ with $x(0) = 0$ (see Section I of the Appendix) and hence the total time taken along this path, with penalty included, is given by:

$$l/v + \tfrac{1}{2}(x_T - 1)^2$$

where v is the running speed along the path. This is just the objective function of Example 3 with $u = \dot{x}$, $v = 1$ and a sign change to convert the minimisation into a maximisation problem.

(b) x_T fixed, T to be chosen optimally

To solve the problem:

$$\text{maximise} \quad \int_0^T f(x, u, t)\,dt + h(T)$$

subject to $\quad \dot{x} = g(x, u, t)$
$\qquad\qquad\quad x(0), x(T)$ given; $\quad T$ to be chosen optimally

we add to the condition on $x(T)$ in Procedure 6.2 the end-point condition:

$$H_T + \frac{dh}{dT} = 0 \tag{7.9}$$

where by H_T we mean the Hamiltonian evaluated at $t = T$ for the optimal state, control and multiplier functions. This result is discussed in Section J of the Appendix. Such a problem would arise, for example, if the road in the last example ran East–West rather than North–South and if the penalty were $\tfrac{1}{2}(T-1)^2$. We know the optimal solution by symmetry (we simply reverse the x and t axes in Fig. 7.10b) but we will derive it using equation (7.9).

Example 4
To solve the problem:

$$\text{maximise} \quad -\int_0^T (1+u^2)^{1/2}\,dt - \tfrac{1}{2}(T-1)^2$$

$$\text{subject to} \quad \dot{x} = u$$

$$x(0) = 0,\; x_T = 1;\quad T \text{ to be chosen optimally}$$

we have to find the constants A, B in the optimal functions:

$$\lambda = A,\; x = A(1-A^2)^{-1/2}t + B$$

by using the conditions:

$$x(0) = 0,\; x_T = 1,\; H_T + \frac{dh}{dT} = 0.$$

From the first we find that $B = 0$ and from the second:

$$A(1-A^2)^{-1/2}T = 1 \tag{7.10}$$

Since

$$H = -(1+u^2)^{1/2} + \lambda u = -(1-A^2)^{1/2}$$

on substituting for u in terms of λ we have from equation (7.9):

$$-(1-A^2)^{1/2} - (T-1) = 0 \tag{7.11}$$

Eliminating T from equations (7.10), (7.11) we find that A is the solution to the equation $1 + A = A(1-A^2)^{-1/2}$. By solving graphically (Fig. 7.10a), we find that $A \simeq 0.9$ and hence $T \simeq 0.5$ with $x = 2t$ as a good approximation to the optimal solution.

(c) x_T **and** T **both to be chosen optimally subject to a Terminal Condition**

In the problem:

$$\text{maximise} \quad \int_0^T f(x, u, t)\,dt + h(x_T, T)$$

$$\text{subject to} \quad \dot{x} = g(x, u, t),\; x(0) \text{ given} \tag{7.12}$$

$$k(x_T, T) = 0$$

$$x_T, T \text{ to be chosen optimally}$$

the only information given on the values of x_T and T is that they must satisfy the equation $k(x_T, T) = 0$, the *terminal condition* for the problem. In the cross-country example this equation would indicate the location of the road which need no longer lie East–West or North–South nor for that matter even be straight.

To solve problem (7.12) we first 'free' the parameters x_T, T by absorbing the terminal condition in the asset function h, with a multiplier μ to form a new asset function:

$$\varphi = h + \mu k.$$

With x_T and T now free we impose both optimality conditions (7.8) and (7.9).

190 *Optimal control: generalisations*

The precise algorithm is displayed in Procedure Box 7.1. It includes as special cases T fixed with x_T chosen optimally and x_T fixed with T chosen optimally.

PROCEDURE 7.1
To solve problem (7.12)

1 Form the Hamiltonian $H = f + \lambda g$ and solve the Hamiltonian system:

   maximise $H(u)$

   $$-\dot{\lambda} = \frac{\partial H}{\partial x}$$

   $$\dot{x} = \frac{\partial H}{\partial \lambda}.$$

2 Form the new asset function:

   $$\varphi(x_T, T) = h(x_T, T) + \mu k(x_T, T).$$

   Determine the constants of integration, the Lagrange multiplier μ and time T by solving the algebraic equations:

   $x(0)$ given

   $k(x_T, T) = 0$

   $$\lambda_T = \frac{\partial \varphi}{\partial x_T}$$

   $$H_T + \frac{\partial \varphi}{\partial T} = 0.$$

   (The end values x_T, λ_T are obtained by evaluating at $t = T$ the solutions $x(t)$, $\lambda(t)$ obtained in stage 1.)

The next example illustrates the use of this Procedure.

Example 5
1 The problem:

$$\text{maximise} \quad -\int_0^T (1+u^2)^{1/2} \, dt$$

$$\text{subject to} \quad \dot{x} = u \tag{7.13}$$

$$x(0) = 0, \; x_T + T = 1$$

$$T, x_T \text{ to be chosen optimally}$$

has, as before, the general solution:

$$x(t) = A(1-A^2)^{-1/2} t + B, \quad \lambda(t) = A, \quad H(t) = -(1-A^2)^{1/2} \tag{7.14}$$

2 Since $h = 0$ and $k = x_T + T - 1$ the function φ is given by:

$$\varphi = \mu(x_T + T - 1)$$

where μ is a Lagrange multiplier.
To find A, B, μ, T we have to solve the equations:

$$x(0) = 0 \qquad (7.15)$$
$$x_T + T = 1 \qquad (7.16)$$
$$\lambda_T = \mu \qquad (7.17)$$
$$-(1 - A^2)^{1/2} + \mu = 0 \qquad (7.18)$$

where

$$x_T = A(1 - A^2)^{-1/2} T + B \qquad (7.19)$$
$$\lambda_T = A \qquad (7.20)$$

from (7.14). The solution to these equations can be obtained using the following argument.
(i)   (7.15) in (7.14) gives $B = 0$.
(ii) (7.20) in (7.17) and (7.18) gives $A = (1 - A^2)^{1/2}$, i.e. $A = 2^{-1/2} = \mu$.
(iii) Hence from equation (7.19) we obtain $x_T = T$ and from equation (7.16) $x_T = T = \frac{1}{2}$.

The solution to the problem (7.13) is therefore the radial line $x(t) = t$ terminating when $t = \frac{1}{2}$ and $x = \frac{1}{2}$. The problem is in fact the rather simple one of finding the shortest path from the line $x + t = 1$ to the origin with no penalty incurred.

It should be noted that equations (7.8) and (7.9) are only necessary conditions, i.e. not all solutions correspond to a maximum. However in most of the problems we shall be discussing it will be clear from the context which (if any) of the solutions correspond to the maximum.

Exercise Set 2
1 Using equation (7.8) solve Problem (7.7) with $T = 1$, $h = 0$ and the following choices for functions f, g and parameter a.

	(i)*	(ii)	(iii)*	(iv)
f	$-(x + \frac{1}{2}u^2)$	$-\frac{1}{2}(x + u)^2$	$-\frac{1}{2}(x^2 + u^2)$	$-x(1 + u^2)^{1/2}$
g	u	$u - x$	$u + x$	u
a	2	1	0	2

Use the results of Exercise Set 2 of Chapter 6 and Question 4(a) of Problem Set 6 where appropriate.

2 (a)* Using equation (7.9) and the maximum principle show that in the optimal solution to the Resource Problem (Section 6.5):

$$\text{maximise} \quad \int_0^T e^{-rt} u D(u) \, dt$$

subject to $\quad \dot{x} = -u, \quad x(0) = R_0, \quad x(T) = 0, \quad T$ to be chosen optimally

it must in general be true that $u_T = 0$ and $\lambda_T e^{-rt} = D(0)$ if $D(0)$, $D'(0)$ are finite and positive.
(b) Using this result solve the problem of part (a) where $R_0 = r = 1$ and
(i)* $D = (1 - e^{-u})$ (ii) $1 - u$ for $0 \leq u < 1$, $D = 0$ for $u > 1$.

192 *Optimal control: generalisations*

3 In this problem we find the shortest distance from the point $(0, 1)$ to given terminal curves. Using Procedure 7.1 solve the problem:

$$\text{minimise} \quad \int_0^T (1+u^2)^{1/2} \, dt$$

subject to $\dot{x} = u, \quad x(0) = 1, \quad x_T, T$ to be chosen optimally
$\qquad\qquad k(x_T, T) = 0$

where (i)* $k = T^2 - x_T$ (ii) $k = x_T - T$
     (iii)* $k = (T-2)^2 + x_T^2 - 1$ (iv) $k = 2T^2 + (x_T - 1)^2 - 1$.

7.5 Further Generalisations

In this section we discuss three further generalisations of the basic optimal control model. In the first we consider integral constraints, in the second inequality bounds on the state variable and in the third multiple state and control variables.

(a) Integral Constraints

One of the problems faced by a group of farmers setting up a horticultural business in a temperate country is protecting their seedlings against the elements. After some discussion they have decided to construct out of a new type of reinforced plastic a set of long buildings with a shape that will enclose a given volume at minimum cost (Fig. 7.11a). In cross-section (Fig. 7.11b) their problem is to find the curve $x = x(t)$ of minimum length that encloses a given area A, i.e. they wish to solve the problem:

$$\text{minimise} \quad \int_0^T (1+u^2)^{1/2} \, dt$$

subject to $\dot{x} = u$

$$\int_0^T x \, dt = A \qquad (7.21)$$

$x(0) = x(T) = 0, \quad T$ and A given.

In solving this problem for the farmers we can free ourselves from the integral constraint (7.21) by absorbing its integrand in the Hamiltonian using a new multiplier v, a constant. Precisely we have[2]:

$$H = -(1+u^2)^{1/2} + \lambda u + vx$$

We now apply Procedure 6.2 to the Hamiltonian H carrying through the multiplier v as a parameter. At the end of the analysis we will determine v by imposing the integral constraint (7.21):

$\dfrac{\partial H}{\partial u} = -u(1+u^2)^{-1/2} + \lambda = 0$ i.e. $u = \lambda(1-\lambda^2)^{-1/2}$

$-\dot{\lambda} = v$ i.e. $\lambda = -vt + A$

$\dot{x} = u = (A - vt)/(1 - (A-vt)^2)^{1/2}$ i.e. $x = v^{-1}(1 - (A-vt)^2)^{1/2} + B$ (7.22)

Applying the end-point conditions we find that $A/v = \tfrac{1}{2}T$, $B = -(v^{-2} - \tfrac{1}{4}T^2)^{1/2}$ and

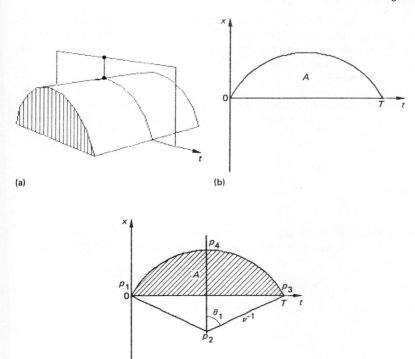

Fig. 7.11

hence, on rearranging equation (7.22), that:

$$(x - B)^2 + (t - \tfrac{1}{2}T)^2 = v^{-2}$$

The optimal curve is therefore a segment of a circle centred at $(\tfrac{1}{2}T, B)$ with radius v^{-1}. To find v we eliminate x from (7.21) using equation (7.22):

$$\int_0^T x \, dt = \int_0^T \{(v^{-2} - (t - \tfrac{1}{2}T)^2)^{1/2} - (v^{-2} - \tfrac{1}{4}T^2)^{1/2}\} \, dt = A.$$

Using the substitution $t = \tfrac{1}{2}T + v^{-1} \sin \theta$ this integral can be evaluated as:

$$A = v^{-2}(\theta_1 - \sin \theta_1 \cos \theta_1)$$

where $\theta_1 = \sin^{-1} \tfrac{1}{2}Tv$ is the angle defined in Fig. 7.11c. This equation can in fact be established more directly by evaluating the area A in Fig. 7.11c as the difference of the circle sector $P_1 P_2 P_3 P_4$ and the triangle $P_1 P_2 P_3$. For a given A and T this equation can be solved graphically or numerically to find v.

(b) Inequality Bounds on the State Variable

Let us now consider some farmers who are cultivating the coastal region of a hot mediterranean country. It seldom rains on the coast so the regional government has built a dam in the nearby mountains to provide a reliable supply of water for their crops.[3] The problem faced by the dam operators is how to ration the water given the likely rainfall pattern over, say, the following T days. To model the situation let $x(t)$ denote the amount of water in the dam, $u(t)$ the rate at which the water is released to the farmers and $r(t)$ the rate at which the rainfall is collected in the dam at time t. The rate of increase of water in the dam is equal to the rate at which water flows in, r, minus the rate at which it flows out, u, i.e.

$$\dot{x} = r - u.$$

If the objective is to satisfy the needs of the farmers as measured by the integral $\int_0^T u^{1/2} dt$ then the optimisation problem to be solved by the dam operators can be written as:

$$\text{maximise} \quad \int_0^T u^{1/2} dt$$

$$\text{subject to} \quad \dot{x} = r - u$$
$$x(0) = x(T) = a$$

if we suppose that they are to have as much water in the dam at the end of the planning period as at the beginning. (For the moment we are ignoring possible inequality bounds on the variables.)

The standard Hamiltonian analysis gives in this case:

$$H = u^{1/2} + \lambda(r - u)$$

$$\frac{\partial H}{\partial u} = \tfrac{1}{2} u^{-1/2} - \lambda = 0 \qquad \text{i.e.} \quad u = (4\lambda^2)^{-1} \tag{7.23}$$

$$-\dot{\lambda} = \frac{\partial H}{\partial x} = 0 \qquad \text{i.e.} \quad \lambda = A_0 \text{ (a constant)} \tag{7.24}$$

$$\dot{x} = r - (4A_0^2)^{-1} \tag{7.25}$$

From equations (7.23) and (7.24) it is clear that the water is to be released at a constant rate $u_0 \equiv (4A_0^2)^{-1}$, its value being determined by integrating the dynamic equation (7.25) over the planning period:

$$x(T) - x(0) = \int_0^T r \, dt - u_0 T = 0 \tag{7.26}$$

For example if the rainfall has the pattern shown in Fig. 7.12a, with no rain falling in the first quarter of the period, then we deduce from equation (7.26) that $u_0 = \tfrac{1}{2} r_0$ and from equation (7.25) that the amount of water in the dam at any time t is given by:

$$x = a + \int_0^t r \, dt - u_0 t$$

with the graph shown in Fig. 7.12b. It is clear from this graph that if a, the initial amount of water in the dam, is low enough then the dam at some stage will become dry. In this eventuality we have to consider the more general problem:

$$\text{maximise} \quad \int_0^T u^{1/2} dt$$

$$\text{subject to} \quad \dot{x} = r - u; \quad x(t) \geq 0 \qquad \text{for } 0 \leq t \leq T$$
$$x(0) = x(T) = a > 0.$$

We can free ourselves of the state variable inequality by absorbing it into the Hamiltonian

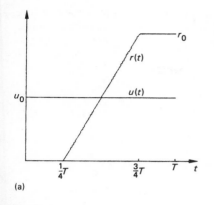 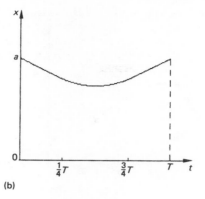

Fig. 7.12.

with a multiplier function $\mu(t)$. The Hamiltonian is now:

$$H = u^{1/2} + \lambda(r - u) + \mu x$$

with equations:

$$\frac{\partial H}{\partial u} = \tfrac{1}{2}u^{-1/2} - \lambda = 0 \tag{7.27a}$$

$$-\dot{\lambda} = \mu \tag{7.27b}$$

$$\dot{x} = r - u \tag{7.27c}$$

There are two possibilities. Either the state variable inequality is binding or it is not.

(i) In the latter case we have $\mu = 0$ and the solution of these equations is as before with λ equal to a constant.

(ii) In the former case we have $x = 0$ and hence $r = u$ from (7.27c), $\lambda = \tfrac{1}{2}r^{-1/2}$ from (7.27a) and $\mu = \tfrac{1}{4}r^{-3/2}\dot{r}$ from (7.27b). But μ must be non-negative, being the multiplier of an inequality constraint and hence $\dot{r}$ must also be non-negative. This means that the dam should only be allowed to run dry on a rising rainfall.

The functions u, x, λ will be continuous functions of time formed by appropriately piecing together solutions of the types (i) and (ii). Since initially and finally the constraint $x \geq 0$ is not binding (since $a > 0$) u must start and end as a solution of type (i). The solution for u can only switch to a solution of type (ii) when $r = u$ and hence for the given rainfall profile the graph of u must be as shown in Fig. 7.13a, with the dam dry between switching points T_1 and T_2. The switching time T_1 and constant u_0 can be found by imposing the conditions that, at T_1, r equals u and the dam is dry, i.e. the area A_1 in Fig. 7.13a is equal to the original amount of water in the dam. The second switching time T_2 and constant u_1 can be found from the conditions that, at T_2, r equals u and the net inflow over period T is zero, i.e. areas A_1 and A_2 in Fig. 7.13a are equal. The graph of x determined from equation (7.27b) is shown in Fig. 7.13b and that for λ in Fig. 7.13c.

With a control inequality constraint x and λ remain continuous (provided u is bounded). This is not always the case for λ in the presence of state inequality constraints. With such a constraint, $S(x, t) \leq 0$, it can be shown that the changes in the Hamiltonian ΔH and the multiplier $\Delta\lambda$ as the constraint becomes binding satisfy the 'jump conditions':

$$\Delta H = v\frac{\partial S}{\partial t}; \qquad \Delta\lambda = -v\frac{\partial S}{\partial x}$$

196 *Optimal control: generalisations*

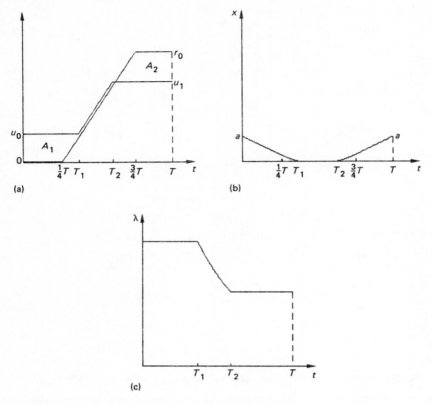

Fig. 7.13

when there is just one state variable (v is an additional multiplier). For the dam problem the first of these conditions together with the maximum condition establish the continuity of u and λ as the dam becomes dry (with $v = 0$ from the second condition).

The jump conditions can be established by a simple extension of the arguments that led to the terminal conditions (7.8), (7.9). The details together with the generalisation to more than one state variable can be found in Section J of the Appendix and in the texts mentioned in reference 4.

(c) Vector Control Theory

Let us suppose that the regional government has built a second dam further up the mountain to generate hydroelectricity by allowing water to cascade down into the first dam (Fig. 7.14). In this situation we have two variables x_1 and x_2 denoting the amounts of water in the dams and two variables u_1 and u_2 denoting the rates at which water is released from the dams. The dynamic equations describing their operation are given by:

$$\left. \begin{array}{l} \dot{x}_1 = r_1 - u_1 + u_2 \\ \dot{x}_2 = r_2 - u_2 \end{array} \right\} \tag{7.28}$$

where r_1, r_2 are the rates at which the rainfall is collected by the two dams. A simple

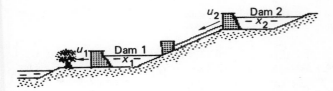

Fig. 7.14

generalisation for the objective functional is:

$$\int_0^T (a_1 u_1^{1/2} + a_2 u_2^{1/2}) \, dt \tag{7.29}$$

where a_1 and a_2 weight the importance to the farmers of water and electricity. The optimisation problem is therefore to maximise (7.29) subject to the dynamic equations (7.28) and end-point and inequality conditions. (For simplicity we suppose that the rainfall is such that none of the inequality conditions becomes binding, i.e. neither dam runs dry or for that matter overflows.)

The new feature of this problem is that there are now two state variables and two control variables but fortunately the basic Hamiltonian algorithm can easily be generalised to handle this situation. We introduce a multiplier for each dynamic equation, maximise the Hamiltonian with respect to each control variable and generate a set of Hamilton equations, one for each multiplier function and state variable. Precisely, we have:

$$H = a_1 u_1^{1/2} + a_2 u_2^{1/2} + \lambda_1(r_1 - u_1 + u_2) + \lambda_2(r_2 - u_2)$$

and the equation system:

$$\begin{cases} \dfrac{\partial H}{\partial u_1} = \tfrac{1}{2} a_1 u_1^{-1/2} - \lambda_1 = 0 & \text{i.e.} \quad u_1 = a_1^2/4\lambda_1^2 \\[2pt] \dfrac{\partial H}{\partial u_2} = \tfrac{1}{2} a_2 u_2^{-1/2} + \lambda_1 + \lambda_2 = 0 & \text{i.e.} \quad u_2 = a_2^2/4(\lambda_2 - \lambda_1)^2 \end{cases}$$

$$\begin{cases} -\dot{\lambda}_1 = \dfrac{\partial H}{\partial x_1} = 0 & \text{i.e.} \quad \lambda_1 = A_1 \text{ (a constant)} \\[2pt] -\dot{\lambda}_2 = \dfrac{\partial H}{\partial x_2} = 0 & \text{i.e.} \quad \lambda_2 = A_1 + A_2 \text{ (a constant)} \end{cases}$$

$$\dot{x}_1 = r_1 - u_{10} + u_{20} \qquad \text{where} \quad u_{i0} = a_i^2/4A_i^2 \quad \text{for } i = 1, 2.$$
$$\dot{x}_2 = r_2 - u_{20}$$

Again we see that the flow rates are constant. To find their values we integrate the dynamic equations:

$$x_1(T) - x_1(0) = R_1 - u_{10}T + u_{20}T$$
$$x_2(T) - x_2(0) = R_2 - u_{20}T$$

where $R_i = \int_0^T r_i \, dt$ for $i = 1, 2$. If the initial and final state values are equal for each dam then:

$$u_{20} = R_2/T \quad \text{and} \quad u_{10} = (R_1 + R_2)/T.$$

The optimal flow rate from the upper dam is therefore equal to the average rainfall into that dam and the optimal flow rate for the lower dam is equal to the sum of the average rainfalls into the two dams.

Vector Control Theory is the fifth and last generalisation of the basic optimal control problem (6.33) that we will consider. Before considering more complicated applications let us summarise in Procedure Box 7.2 the various techniques we have used in this chapter.

PROCEDURE 7.2
Changes to Procedure 6.2 resulting from amendments to the basic optimal control problem (6.30)

1 If there are fixed bounds on the control variable then replace '$\frac{\partial H}{\partial u} = 0$' with
   $$\underset{\underline{u} \leq u \leq \bar{u}}{\text{maximise}} \quad H(u).$$
2 (a) If T is fixed but x_T is to be chosen optimally then replace x_T given by
   $$\lambda_T = \frac{dh}{dx_T}$$ where h denotes the asset function.
   (b) If x_T is fixed but T is to be chosen optimally then add condition:
   $$H_T + \frac{dh}{dT} = 0.$$
   (c) If x_T and T are to be chosen optimally subject to a terminal condition then refer to Procedure 7.1.
3 If there is an integral constraint: $\int_0^T h(x, u, t) dt \geq A$ replace H with $\bar{H} \equiv H + vh$, where v is a new multiplier, a constant. Determine v by imposing the integral constraint at the end of the analysis.
4 If there is a state variable inequality constraint: $S(x) \leq 0$ then: replace H with $\bar{H} \equiv H - \mu S$ where μ is a new multiplier function. Use the Complementary Slackness Condition that $\mu = 0$ when $S(x) < 0$ and $S(x) = 0$ when $\mu > 0$, together with the appropriate jump conditions.
5 (a) If there are n state variables then introduce a multiplier function $\lambda_i(t)$ for the ith dynamic equation $\dot{x}_i = g_i$ $(i = 1, \ldots, n)$ and construct the Hamiltonian:
   $$H = f + \sum_{i=1}^{n} \lambda_i g_i.$$
   Derive and solve the Hamilton equations:
   $$-\dot{\lambda}_i = \frac{\partial H}{\partial x_i}; \quad \dot{x}_i = \frac{\partial H}{\partial \lambda_i} \text{ for each } i.$$
   (b) If there is more than one control variable then maximise the Hamiltonian with respect to each of them.
For more general cases, see Section J of the Appendix.

Exercise Set 3
1 Solve the following problems:
   (i)* maximise x_T
   subject to $\int_0^T (1 + u^2)^{1/2} dt = 2^{1/2}, \quad \dot{x} = u, \quad x(0) = 0, \quad T = 1$

(ii)   maximise T

subject to $\quad \int_0^T (1+u^2)^{1/2} dt = 2^{1/2}, \quad \dot{x} = u, \quad x(0) = 0, \quad x(T) = 1$

(iii)*   maximise T

subject to $\quad \int_0^T (1+u^2)^{1/2} dt = 1, \quad \dot{x} = u, \quad x(0) = 1, \quad x_T = T$

(iv)   maximise T

subject to $\quad \int_0^T (1+u^2)^{1/2} dt = 2, \quad \dot{x} = u, \quad x(0) = 1, \quad x_T^2 + (T-2)^2 = 1$

2   (i)*   minimise $\quad \int_0^1 u^2 dt$

subject to $\quad \dot{x}_1 = x_2, \quad \dot{x}_2 = u$

$\quad\quad\quad\quad\quad x_1(0) = x_2(0) = 0; \quad x_1(1) = 1, \quad x_2(1) = 1$

(ii)   minimise $\quad \int_0^1 (x_1 + \tfrac{1}{2} u^2) dt$

subject to $\quad \dot{x}_1 = x_2, \quad \dot{x}_2 = u$

$\quad\quad\quad\quad\quad x_1(0) = x_2(0) = 0; \quad x_1(1) = 1, \quad x_2(1) = 1$

3   Show that the optimal solution for the problem:

maximise $\quad \int_0^T x \, dt$

subject to $\quad \dot{x} = u, \quad \int_0^T (1+u^2)^{1/2} = l, \quad x(0) = x(T) = 1, \quad l \leq \tfrac{1}{2} \pi T$

is given by:

$$x = (v^2 - (t - \tfrac{1}{2} T)^2)^{1/2} - (v^2 - \tfrac{1}{4} T^2)^{1/2}$$

where v is the multiplier for the integral constraint. Show by choosing T optimally that the graph of x against t is a semicircle. (This problem is essentially the Isoperimetric Problem mentioned in the Introduction and discussed further in Section 7.7.)

7.6   Further Applications

Having developed the techniques let us now apply them to solve two important problems, one an investment problem in Energy Conservation and the other a minimum time problem in the Fishing Industry.

A   Energy Conservation

Consider a household or company which wishes to reduce its energy consumption by installing extra energy saving equipment, for example insulation or control devices Let $x(t)$ denote the stock of energy saving equipment at time t and suppose the amount of energy used with this equipment is given by:

$$E(x) = E_0 e^{-x/c}$$

where E_0 and c are constants. Clearly energy costs will be reduced by buying more

Optimal control: generalisations

equipment but that will involve higher capital costs. The right balance between the two can be determined by minimising total (discounted) costs over a given planning period, i.e. minimising:

$$\int_0^T e^{-rt}(q_0 \dot{x}(t) + p(t)E(x(t)))\,dt$$

where the first term corresponds to the capital costs (with q_0 being the assumed constant price of capital equipment) and the second to the energy costs (with $p(t)$ the energy price). Introducing the variable $u = \dot{x}$ and supposing the price of energy increases exponentially as $p = p_0 e^{\alpha t}$, we derive the control problem:

$$\text{minimise} \quad \int_0^T e^{-rt}(q_0 u + p_0 E_0 e^{(\alpha t - x/c)})\,dt$$

subject to $\dot{x} = u$

$0 \leqslant u \leqslant \bar{u}$

$x(0) = a$, $x(T)$ to be chosen optimally

where $\bar{u}$ models any financial limit on capital spending that may be operative. This is a non-autonomous problem, with linear Hamiltonian and an optimal end-point condition. Its Hamiltonian is in fact:

$$H = u(\lambda - q_0 e^{-rt}) - p_0 E_0 e^{((\alpha - r)t - x/c)}$$

yielding the Hamilton equations:

$$-\dot{\lambda} = p_0 E_0 c^{-1} e^{((\alpha - r)t - x/c)} \tag{7.30}$$

$$\dot{x} = u \tag{7.31}$$

The optimal value for u depends on the sign of the coefficient of u in H.

(i) If $\lambda > q_0 e^{-rt}$ then $u = \bar{u}$ and $x = \bar{u}t + A$ where A is a constant.
(ii) If $\lambda < q_0 e^{-rt}$ then $u = 0$ and $x = B$, where B is a constant.
(iii) If $\lambda = q_0 e^{-rt}$ then substitution for λ in equation (7.30) yields, on rearrangement,

$$x = c \ln\left(\frac{p_0 E_0}{rcq_0}\right) + c\alpha\, t \tag{7.32}$$

and from (7.31) $u = c\alpha$. There is therefore a singular (knife-edge) solution provided $c\alpha < \bar{u}$.

The final step is to piece together these solutions to form continuous functions x, λ satisfying the end-point condition $\lambda_T = 0$ for optimality of x_T. To show how to do this let us suppose that initially $\lambda > q_0$ then situation (i) holds with λ proportional to $e^{(\alpha - r - \bar{u}/c)t}$ according to equation (7.30). If $\bar{u} > c\alpha$ then eventually λ falls to value $q_0 e^{-rt}$ (Fig. 7.15) and there is a switch to a solution of type (iii) with $\lambda = q_0 e^{-rt}$. To satisfy the end-point condition $\lambda_T = 0$ we must again switch (point Q in Fig. 7.15a) this time to a type (ii) solution with $\lambda < q_0 e^{-rt}$. The switching times T_1, T_2 are determined by the continuity of λ at P and Q. The corresponding graph for x is shown in Fig. 7.15b. The singular solution (7.32) defines what we might call a balanced investment path with annualised capital cost equalling fuel savings. Initially we have too little energy saving equipment and hence we should spend as much as is available to get onto path (7.32). We eventually leave this path because of the finite planning period. Investment made after T_2 will not pay for itself. We note finally that if there is no limit on capital spending, i.e. $\bar{u} = \infty$, then we move onto the path (7.32) immediately (Fig. 7.15c) generating a discontinuity in the x function[5].

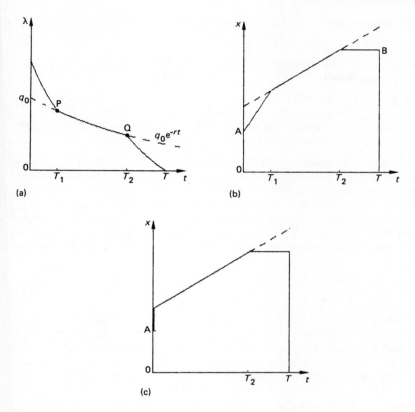

Fig. 7.15

B The Minimum Time Problem

Suppose that the fishing cooperative discussed in Section 7.3 wishes to achieve a given profit B in minimum time, i.e. it wishes to:

minimise T

subject to $\dot{x} = c_0 x - Axu$, c_0 a constant

$$\int_0^T e^{-rt} p_0 Axu \, dt = B$$

$$0 \leqslant u \leqslant \bar{u}, \quad x(0) = x(T) = a$$

ignoring population saturation. This problem has an integral constraint, involves an optimal end-point decision and is linear in u. The consequent Bang–Bang structure will considerably simplify the solution procedure since we will be able to solve the dynamic equation directly with u equal to either its lower or upper bound. The role of the λ equation will be to give the correct order in which to piece together the solutions for x into a continuous function.

For definiteness let us take $A = c_0 = \bar{u} = B = 2$, $p_0 = a = 1$ and $r = 0$, then the Hamiltonian becomes after rearrangement:

$$H = 2x(v - \lambda)u + 2\lambda x$$

and the Hamilton equations:
$$-\dot{\lambda} = 2\lambda(1-u) + 2vu, \quad \dot{x} = 2x(1-u)$$
where v is the constant multiplier for the integral constraint. Maximising H with respect to u we identify three situations.

(i) If $v < \lambda$ then $u = 0$, $\dot{x} = 2x$ and $\dot{\lambda} = -2\lambda$.
(ii) If $v > \lambda$ then $u = 2$, $\dot{x} = -2x$ and $\dot{\lambda} = -4v + 2\lambda$.
(iii) There is no solution when $v = \lambda$, since with $\dot{\lambda} = 0$ (v being a constant) the first of the Hamilton equations cannot be satisfied with a non-zero multiplier v.

In (i) the dynamic equation yields $x = A e^{2t}$ and in (ii) $x = B e^{-2t}$. Neither of these solutions satisfy the end-point conditions and hence we will have to construct x by piecing together solutions of type (i) and (ii) to form a continuous function x. Our phase space analysis of Section 7.3 suggested that a solution of type (i) should precede that of type (ii) as in Fig. 7.16. This can be verified independently by noting that λ is always negative and hence the only way to achieve a switch is to start with $\lambda > v$ in which case there is only one switch (Fig. 7.16). Applying the end-point conditions we find that $A = 1$ and $B = e^{2T}$ and hence:
$$\begin{aligned} x &= e^{2t} & \text{for } 0 \leq t \leq T_1 \\ x &= e^{2(T-t)} & \text{for } T_1 \leq t \leq T \end{aligned} \quad (7.33)$$
where T_1 is the switching time. At $t = T_1$ these two expressions are equal by continuity. This holds when $T_1 = T/2$ (Fig. 7.16), as is otherwise clear from symmetry. The minimum time T is determined by substituting solution (7.33) in the integral constraint:
$$\int_0^T 2xu \, dt = \int_{\frac{1}{2}T}^T 4e^{2(T-t)} \, dt = 2e^T - 2 = 2.$$
The minimum time for achieving the given profit is therefore $T = \ln 2$.

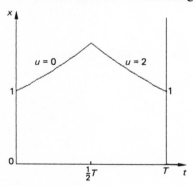

Fig. 7.16

It is perhaps surprising that we have been able to solve the problem without having to use the end-point condition (7.9). This is because of the singular nature of the equation system in Bang–Bang situations. We would, however, need this condition to find the multiplier v.

7.7 The Calculus of Variations

In this and the previous chapter we have discussed several of the classical optimisation problems solved by Mathematicians in the 17th and 18th centuries. Probably the most famous of these is the Isoperimetric problem where we seek the maximum area that can be bounded by a curve of given length. Legend has it that this is the problem that a Phoenician Princess, Dido, had to solve on her arrival on the African coast after fleeing from her homeland in fear that her brother would kill her.[6] In response to her entreaties for help in starting a new life for herself and her entourage she was offered as much land as could be surrounded by a bullock's hide (suitably cut up into strips). It is not reported whether Dido identified the maximum area to be that surrounded by a circle but out of that early settlement grew the powerful city state of Carthage.

The question of who obtained the first mathematically acceptable solution to this problem is not easy to resolve. Both Jacob and Johannes Bernoulli claimed to have been the first and quarrelled ferociously over the issue with Jacob accusing his brother of stealing his solution.[7] Certainly the most elegant solution came later from the 19-year old Italian, Lagrange. His method contained the elements of what became known as the Calculus of Variations. This theory is concerned with problems of the form:

$$\text{maximise} \int_0^T f(x, \dot{x}, t) \, dt$$

i.e. control problems with dynamic equation $\dot{x} = u$. Lagrange, in essence, argued that if $x = x^*(t)$ is an optimal function and $x(t) = x^*(t) + \theta y(t)$ is any other function (with parameter θ and $y(t)$ an arbitrary differentiable function), then:

$$\frac{dJ}{d\theta} = 0 \quad \text{at} \quad \theta = 0 \tag{7.34}$$

where

$$J(\theta) = \int_0^T f(x^* + \theta y, \dot{x}^* + \theta \dot{y}, t) \, dt.$$

Otherwise we could find an improved solution with θ non-zero. Put in words, equation (7.34) states that variations in the objective functional due to variations in the state function are, to first order, zero at the optimum. From (7.34) Lagrange derived the differential equation:

$$\frac{d}{dt}\left(\frac{\partial f}{\partial \dot{x}}\right) = \frac{\partial f}{\partial x} \tag{7.35}$$

as a necessary condition for optimality. The details of the derivation are given in Section J of the Appendix. (You may recall that we obtained this equation from the Hamilton equations in Problem Set 6.) Equation (7.35) is known as the Euler–Lagrange equation and is a second order differential equation for $x(t)$. For autonomous systems (where there is no explicit dependence on t) it can be reduced to a first order differential equation.

Further studies of problems in the Calculus of Variations brought to light certain pathologies in the behaviour of optimal solutions; in particular solutions with corners and discontinuities were found. These pathologies were deeply disturbing to those who believed that, in the best of all possible worlds, everything should be perfectly smooth. They remained somewhat of a mystery until the development of Hamiltonian theory at the end of the 19th century. In this new formalism these particular pathologies are seen to arise in general from multiple or infinite solutions in the maximisation of the Hamiltonian with respect to u. We examine these possibilities in turn.

(i) Corners[4]

Corners correspond to discontinuities in $\dot{x}$ and can be generated when there is switching between different optima in the maximisation of H with respect to u. An example is provided by the Bang–Bang phenomenon discussed in Section 7.2. In this case the multiple solutions occur when the linear graph of the Hamiltonian has zero gradient. Corners, however, can occur with nonlinear Hamiltonians as the following example shows.

The problem:

$$\left.\begin{aligned}\text{minimise} \quad & \int_0^1 (1-u^2)^{1/2}\, dt \\ \text{subject to} \quad & \dot{x}=u, \quad -1 \leqslant u \leqslant 1 \\ & x(0)=0, \quad x(1)=\tfrac{1}{2}\end{aligned}\right\} \qquad (7.36)$$

has as its Hamiltonian the function:

$$H = -(1-u^2)^{1/2} + \lambda u$$

with the graph shown in Fig. 7.17a. The maximum of H is achieved at either the upper or lower bound when $\lambda \neq 0$ but at both bounds when $\lambda = 0$ (Fig. 7.17b). In the former case the Hamilton equations:

$$\dot{x} = u, \quad \dot{\lambda} = 0$$

yield the solution:

$$\lambda = A, \quad x = ut + B \qquad (A, B \text{ constants}) \qquad (7.37)$$

where u is $+1$ when $A > 0$ and -1 when $A < 0$. Neither of these solutions satisfy the end-point conditions (7.36) and hence the required solution can only come from the degenerate case with $\lambda = 0$, which yields solutions of the form (7.37) with u *either* $+1$ or -1. We can piece together such solutions in an infinite number of ways to satisfy the end conditions (Fig. 7.17c), necessarily creating corners at each 'join'. This set of solutions is in fact otherwise obvious from the form of the objective functional which takes its minimum value of zero if u is either $+1$ or -1 throughout the integration range. This is clearly also the case with the problem:

$$\text{minimise} \quad \int_0^1 (1-u^2)^2\, dt$$

$$\text{subject to} \quad \dot{x} = u, \quad x(0) = 0, \quad x(1) = \tfrac{1}{2}$$

which illustrates the fact that inequality bounds are not necessary in generating corners.

(ii) Discontinuities

Discontinuities in the x function can be viewed as infinities in its derivative u in the case of problems from the Calculus of Variations. An example occurred in the Energy Conservation model of Section 7.6 when there was no upper bound on u and hence the optimal value of u was infinite when the Hamiltonian, linear in u, had positive gradient. A second example is provided by the version of the Isoperimetric problem discussed in Exercise Set 3:

$$\text{maximise} \quad \int_0^T x\, dt$$

$$\text{subject to} \quad \dot{x} = u, \quad x(0) = x(T) = 0$$

$$\int_0^T (1+u^2)^{1/2}\, dt = l, \quad T \text{ fixed.}$$

7.7 The calculus of variations

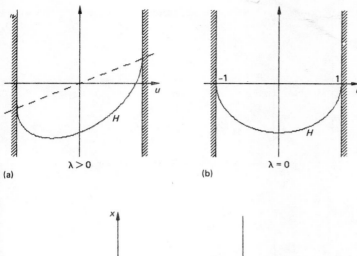

Fig. 7.17

For low values of l ($<\frac{1}{2}\pi T$) the solution for x was found to be a segment of a circle (Fig. 7.18a). With $l = \frac{1}{2}\pi T$ the segment becomes a semicircle. (With T chosen optimally the solution will in fact always be a semicircle.) With T fixed and $l > \frac{1}{2}\pi T$ the analysis might suggest that the optimal solution corresponds to the 'bulging' circle segment shown in Fig. 7.18c. This is in fact not optimal because the areas overlapping the vertical boundary lines do not contribute to the objective functional. The optimal solution is in fact a semicircle 'resting on two vertical walls' (Fig. 7.18d) corresponding to infinite values of the derivative $\dot{x} = u$. The optimal value of u in the maximisation of the Hamiltonian for this problem:

$$H = x + \lambda u - v(1 + u^2)^{1/2}$$

is in fact only finite (Fig. 7.19a) when $v > \lambda > 0$. If $\lambda \geqslant v > 0$ then the optimal value is taken at infinity (Fig. 7.19b).

A more dramatic discontinuity arises in the Soap Film Problem discussed in Question 4a of Problem Set 6:

$$\text{minimise} \quad \int_0^T x(1+u^2)^{1/2} \, dt$$

$$\text{subject to} \quad \dot{x} = u, \quad x(0) = x(T) = a > 0$$

which describes the shape taken by a soap film suspended between two coaxial circular

206 Optimal control: generalisations

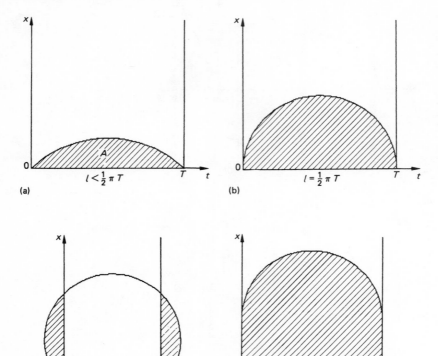

Fig. 7.18

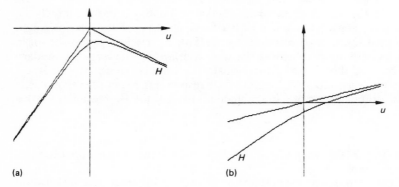

Fig. 7.19

wires of radius a, distance T apart (Fig. 7.20a). Variable $x(t)$ measures the radius of the film at distance t along the axis. The shape was found to be hyperbolic:

$$x = C \cosh \frac{1}{C}(t - \tfrac{1}{2}T)$$

7.7 The calculus of variations

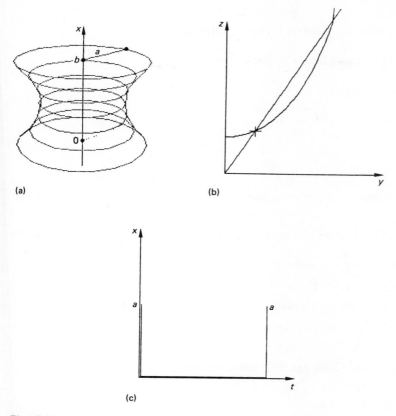

Fig. 7.20

where C is a constant satisfying the algebraic equation:

$$\cosh y = \left(\frac{2a}{T}\right) y \quad \text{with} \quad y = \tfrac{1}{2}T/C.$$

Its solution is obtained as the lower intersection of the graph of $z = \cosh y$ and the radial line $z = (2a/T)y$ (Fig. 7.20b). As the separation T between the wires increases the gradient of the radial line in Fig. 7.20b decreases until at some point the algebraic equation ceases to have a solution. The physical explanation is that the soap film has collapsed onto the end circles and the axis, corresponding to the discontinuous solution for x shown in Fig. 7.20c. The discontinuities at $t = 0, T$ are generated by an infinite solution in the maximisation of H when $x \leqslant \lambda$ and the solution $x = 0$ for $0 < t < T$ by the binding of the state inequality constraint $x \geqslant 0$ essential in a complete formulation of the problem because of the definition of x as a radius. In the Problem Set we will examine these 'singularities' more closely using the phase space approach[8].

Problem Set 7

1   Sketch the phase space trajectories for the control problem

$$\text{maximise} \int_0^T f \, dt \text{ subject to } \dot{x} = g \text{ and } \underline{u} \leq u \leq \bar{u}$$

for the following choices of f, g, $\underline{u}$, $\bar{u}$:

	(i)*	(ii)	(iii)*	(iv)
f	$-(x^2 + xu + u^2)$	$-\frac{1}{2}(x+u)^2$	$2u$	$2u$
g	$x - u$	$u - x$	u	$x^2 - u$
$\underline{u}, \bar{u}$	0, 1	$-1, 1$	0, 1	$-1, 1$

2   (a) Use the Hamilton method to show that there are optimal solutions for the problem:

$$\text{minimise} \int_0^1 f(x, u) \, dt \text{ subject to } \dot{x} = g(x, u), \; x(0) = 0, \; x(1) = \tfrac{1}{2} \quad (7.38)$$

that possess corners when $g = u$ and

(i)* $f = x^2(u-1)^2$ (ii) $f = (1-u^2)^2$.

(b) Repeat part (a) when the constraint $-1 \leq u \leq 1$ is applied and

(i)* $f = -(1+u^2)^{1/2}$ (ii) $f = (1-u^2)^{1/2}$

(iii)* $f = -|u|$ (iv) $f = |u|$.

(c) Show that the problem (7.38) has discontinuities when

(i)* $f = \tfrac{1}{2}x^2$, $g = \tfrac{1}{2}u^2$

(ii) $f = x^2 u^2$, $g = u$.

(d) Sketch the phase space trajectories for the Soap Film and Isoperimetric Problems:

(i)* minimise $\int_0^T x(1+u^2)^{1/2} \, dt$ subject to $\dot{x} = u, \; x(0) = x(T) = a$

(ii) maximise $\int_0^T x \, dt$ subject to $\begin{cases} \dot{x} = u, \; \int_0^T (1+u^2)^{1/2} \, dt = l, \\ x(0) = x(T) = a \end{cases}$

and identify the trajectories corresponding to the discontinuous solutions discussed in Section 7.7.

3   (a) Solve the following vector control problems:

(i)* minimise T, subject to $\begin{cases} \dot{x}_1 = x_2, \; \dot{x}_2 = u; \; -1 \leq u \leq 1 \\ x_1(0) = -1, \; x_2(0) = 1, \; x_1(T) = x_2(T) = 0. \end{cases}$

(ii) maximise $\int_0^1 (1 + u_1^2 + u_2^2)^{1/2} \, dt$

subject to $\begin{cases} \dot{x}_1 = u_1, \; \dot{x}_2 = u_2, \; x_1(0) = x_2(0) = 0, \\ x_1(1) = x_2(1) = 1. \end{cases}$

(This second problem is to find the shortest distance between the origin and the point (1 1 1) in the 3-dimensional space of the variables t, x_1, x_2.)

(b) A control problem with integral constraint $\int_0^T h(x, u, t)\,dt = A$ can be transformed into a vector control problem with no integral constraint. Precisely if $x_1 = x$ and $x_2 = \int_0^t h\,dt$ then the dynamic equation for x_2 is obtained on differentiation as:

$$\dot{x}_2 = h$$

with $x_2(0) = 0$, $x_2(T) = A$. Use this technique to solve the problems:

(i)* maximise x_T, subject to $\begin{cases} \dot{x} = u, \int_0^T (1+u^2)^{1/2}\,dt = 2^{1/2}, \\ x(0) = 0, T = 1. \end{cases}$

(ii) the Isoperimetric Problem.

4 (a) (i) Solve the single dam problem of Section 7.5 when $T = 4$, $r_0 = 1$, $a = \tfrac{1}{2}$ and the rainfall pattern is as shown in Fig. 7.21a.
(ii) Repeat part (i) for a period of length $2T$ with the rainfall pattern of Fig. 7.21a repeating itself over the second half of the extended period.
(iii) Repeat part (i) when the rainfall follows the monsoon-like pattern shown in Fig. 7.21b.
(b) (i) Derive the Hamilton equations for the single dam problem when the additional constraint $x \leqslant M$ is imposed, where M is the maximum capacity of the dam.
(ii) Solve this problem for the rainfall pattern of Fig. 7.21a with $T = 4$, $r_0 = 1$, $a = c = \tfrac{1}{2}$.

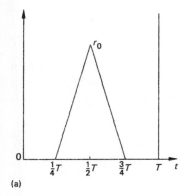
(a)
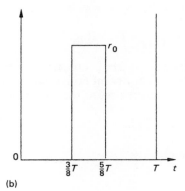
(b)

Fig. 7.21

5 (a)* In the fishing model of Section 7.3 we did not include one feature which is of particular interest, namely the possibility of the fish stock becoming extinct if it falls below some critical level, e say. To see how this affects the behaviour of the system sketch the phase space trajectories for the problem:

$$\text{maximise} \int_0^T xu\,dt$$

subject to $\quad \dot{x} = 2(x-e)(1-x) - xu, \qquad 0 \leqslant u \leqslant 3$

where $e = \tfrac{1}{4}$.
(b) Discuss the effect on the phase space trajectories of Fig. 7.6b and Fig. 7.9 of a non-zero rate of interest r (but $e = 0$).

210 *Optimal control: generalisations*

6 If the control bounds depend on the state variable x then we should include these bounds in the Hamiltonian, each with a new multiplier equal to zero when the corresponding constraint is not binding.

(a) (i) Use this procedure to show that the Hamilton equations for the problem:

$$\text{maximise} \quad -\frac{1}{2}\int_0^T u^2\,dt$$

subject to $\quad \dot{x} = u$

$$u \leqslant x; \quad x(0) = 1, \quad x(T) = 2$$

are given by $\dot{x} = x$, $\dot{\lambda} = x - \lambda$ (after elimination of the multiplier) when on the boundary $u = x$.

(ii) Sketch the phase space trajectories for the problem and obtain the analytic solution when $T = 3/4$. For what values of T are there solutions?

(b) Repeat part (a) for the problem:

$$\text{maximise} \quad \int_0^T u\,dt$$

subject to $\quad \dot{x} = 2x - u$

$$-1 \leqslant u \leqslant x; \quad x(0) = 1, \quad x(T) = 2$$

where the boundary equations are given by $\dot{\lambda} = -(\lambda + 1)$, $\dot{x} = x$ and $T = 1/2$.

7 Two towns A and B are distance 100 km apart with town B at an altitude 2 km above A. The land between the two towns is well approximated by the curve:

$$e(t) = 1 - \cos\left(\frac{\pi}{100}t\right) \quad \text{(Fig. 7.22)}$$

relating altitude to distance. The problem is how to build a road from A to B at minimum cost with the gradient at any point no greater in magnitude than $1/40$. If the equation for the road is $x = x(t)$ and the cost of the road is proportional to:

$$\int_0^{100} (\dot{x}(t) - \dot{e}(t))^2\,dt$$

find $x(t)$ for minimum cost.

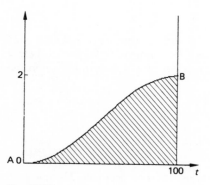

Fig. 7.22

8 (a)* Discuss the relationship between the solutions of the two problems:

maximise $\int_0^T f(x, u, t) \, dt$ minimise $\int_0^T h(x, u, t) \, dt$

subject to $\dot{x} = g(x, u, t)$ subject to $\dot{x} = g(x, u, t)$

$\int_0^T h(x, u, t) \, dt \leq A$ $\int_0^T f(x, u, t) \, dt \geq B$

(b) Repeat part (a) for the problems:

maximise $\int_0^T f(x, u, t) \, dt$ minimise T

subject to $\dot{x} = g(x, u, t)$ subject to $\dot{x} = g(x, u, t)$
$T = A$ $\int_0^T f(x, u, t) = B$

References

1 See reference 6 of Chapter 6 and the article by A. M. Spence 'Blue Whales and Applied Control Theory' in *Systems Approaches and Environmental Problems*, edited by H. W. Gottinger and published by Vandenhoeck and Ruprecht (Rotterdam) (1975).
2 The full justification for this is to go back to the Lagrangean for the problem, obtained as the continuous limit of the discrete Lagrangean of Section 6.2. For the basic optimal control problem we have:

$$L = \int_0^T (f - \lambda(\dot{x} - g)) \, dt$$

$$= \int_0^T (H - \lambda \dot{x}) \, dt$$

and for a problem with an integral constraint equivalent to $\int_0^T h \, dt \geq A$ we have:

$$L = \int_0^T (H - \lambda \dot{x}) \, dt + v \left(\int_0^T h \, dt - A \right)$$

$$= \int_0^T ((H + vh) - \lambda \dot{x}) \, dt - vA.$$

The effective Hamiltonian is therefore $H + vh$ with $v \geq 0$.
3 For an alternative approach to this 'dam' problem see P. Whittle, *Optimization under Constraints*, Wiley (1971) and M. E. El-Hawary and G. S. Christonsen, *Optimal Economic Operation of Electric Power Systems*, Academic Press (1979), and for further references: M. S. Bazaraa and C. M. Shetty, *Nonlinear Programming*, Wiley (1979). Following Whittle we will use the term 'dam' to denote both the civil engineering construction and the reservoir contained by it.
4 See A. E. Bryson and Y-C. Ho. *Applied Optimal Control*, John Wiley (1975) or G. Knowles, *An Introduction to Applied Optimal Control*, Academic Press, (1981).
5 To see the advantage of the Hamiltonian method read the article by P. P. Craig and J. Reeds, 'On Optimal Investment Strategies for Energy Conservation', in *Energy*, Volume 5, pages 1–11 (1980).

6 For the story of Queen Dido see J. R. Newman, *The World of Mathematics*, George Allen and Unwin (1961).
7 For background reading in the development of Optimisation Theory in the 17th and 18th centuries see E. T. Bell, *Men of Mathematics*, Simon and Schuster (1937).
8 It should be noted that the state equation is not defined at corners and discontinuities of a solution. Technically we say that we are looking for solutions, $x(t)$, in the space of functions that are piecewise differentiable, i.e. differentiable at all but a finite number of points.

8
Postscript

8.1 Introduction

We have discussed many of the essential ideas in Optimisation Theory in the last seven chapters but of necessity much has been omitted. In particular we have restricted attention almost entirely to those situations where the decision variables take a continuous range of values and so we have been able to use the full power of the differential and integral calculus. However, many problems, especially in Operational Research, are not of this form. Indeed in the microcomputer example of Chapter 5 the decision variables denoting the numbers of 'Super' and 'Standard' models to produce are restricted to integer values. Often the integer constraints can be ignored if all we want is an approximate answer and the optimal values of the integer variables are large in magnitude. When this is not the case—for example when the variable relates to the number of nuclear power stations to build or the number of supertankers to order—the integer constraint cannot safely be neglected.

In many cases of interest the integer constraint implies that the system being modelled can only exist in a finite number of 'states'. Such a problem can be represented by a graph defined as a set of nodes connected by a set of arcs. Examples are shown in Figs 8.1 and 8.5. The nodes represent the states of the system and the arcs the transitions that can take place between those states. Associated with each arc is a 'weight' equal to the penalty incurred in making that transition and each path through the graph is associated with a total penalty equal to the sum of the penalties for each constituent arc. The optimisation problem is to find the path through the graph with least penalty. If we interpret penalty as a measure of distance then the search is for the shortest path through the graph. Two examples are analysed in the next section. The first is a trucking problem where we seek the shortest route to a particular destination and the second is a 'capital replacement' problem where the concern is with how frequently a particular item of capital equipment should be replaced. In our example the item is a domestic washing machine. In other cases of interest, instead of a penalty we have a 'benefit' (e.g. a profit) and hence the problem is to find the longest path on the graph. Examples will be given in Section 8.2.

Another limitation to our discussion has been the fact that we have, with one or two exceptions, only discussed deterministic models, namely those without an element of uncertainty (or, in more technical language, stochastic variation). The major exception was the commodity dealing problem at the end of Chapter 2 where a food processing company wished to reduce the risk involved in forward contracting. Also, the sensitivity analysis in linear programming (Chapter 5) and the linear regression discussed in Chapter 1 can be viewed as attempts to handle variability in the data. There were other models that we discussed where we should have included uncertainty to achieve an acceptable level of realism. For example in the dam problem of Chapter 7 it is seldom the case that we have total confidence in our rainfall predictions. Releasing water now on the expectation of future rainfall could lead to unfortunate results if that expectation is not fulfilled. Also in the Pioneer Problem of Chapter 6 we ignored the possibility of variable weather conditions in assuming that the growth factor c did not change from year to year. Bad harvests could, however, jeopardise the consumption calculations. Incorporating uncertainty into an optimisation problem is usually a non-trivial modelling exercise and the formulation obtained is usually significantly more difficult to solve than its deterministic counterpart[1]. We give some indication of why this is so in Section 8.4 where we discuss the Pioneer Problem in an uncertain climate.

The finite state problems of Section 8.2, the optimal control problems of Chapters 6 and 7 and the stochastic optimisation problem in Section 8.4 can all be viewed as sequential decision problems in which the optimisation problem can be decomposed into a sequence of simpler optimisation problems. This approach is implemented by an algorithm called *Dynamic Programming*[2] which solves for the optimal variables and the optimal value of the problem recursively. Its main advantage is in solving finite state and stochastic programs although it can sometimes be of use as an alternative to the Hamiltonian method. The details of the Dynamic Programming algorithm and its derivation from Bellman's *Principle of Optimality*[2] can be found in Section 8.3.

Another question we have not had time to address is the practicability of many of our algorithms. The main purpose of these algorithms has been to solve a carefully selected set of examples on which to build an understanding of the essentials of the theory. The question of general applicability was therefore not of primary concern. In 'real-life' problems, however, we would have to augment our working with numerical algorithms using a computer to find the solutions. Numerical optimisation is a large and growing area of study, too large for us to cover in this text. Many numerical algorithms, however, have evolved from the theory and algorithms that we have discussed and hence this text can be viewed as a prerequisite to the study of the numerical aspects of optimisation[3].

8.2 Finite State Systems[4]

To introduce the topic of finite state systems let us consider the following routing problem.

Example 1
A road transport company in Bordeaux has taken on a contract to transport wine in bulk containers from a local producer to a bottling plant in Vienna. The operations manager of the company has the task of deciding which route the trucks should take and, after much discussion, opts for the route with shortest distance. The identification of this route does not appear to be a particularly challenging problem. The possible routes are simply listed, the distance for each determined and the shortest picked out. The number of possible

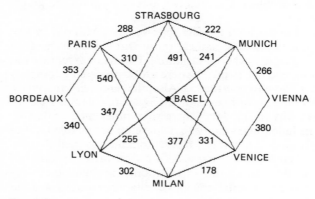

Fig. 8.1

routes, however, is quite large and the challenge is in finding an efficient way of sorting through them. Our strategy will be to reduce the problem to a sequence of simpler subproblems where we find the shortest distances from intermediate cities to the final destination. Precisely, for the routes passing through the seven cities shown in Fig. 8.1 we (a) first find the shortest distance from Strasbourg, Basel and Milan to Vienna (b) use this information to find the shortest distances from Paris and Lyon to Vienna (c) use this information to find the shortest distance from Bordeaux to Vienna.

For subproblem (a) we have the following distances in miles:

Strasbourg–Munich–Vienna	$222 + 266 = 488$	(shorter)
Strasbourg–Venice–Vienna	$491 + 380 = 871$	
Basel–Munich–Vienna	$241 + 266 = 507$	(shorter)
Basel–Venice–Vienna	$331 + 380 = 711$	
Milan–Munich–Vienna	$377 + 266 = 643$	
Milan–Venice–Vienna	$178 + 380 = 558$	(shorter)

We represent the shorter routes by the paths shown in Fig. 8.2.

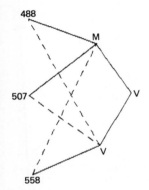

Fig. 8.2

For subproblem (b) we have the distances:

Paris to Vienna

via Strasbourg	$288 + 488 = 776$	(shortest)
via Basel	$310 + 507 = 817$	
via Milan	$540 + 558 = 1098$	

Lyon to Vienna

via Strasbourg	$347 + 488 = 835$	
via Basel	$255 + 507 = 762$	(shortest)
via Milan	$302 + 558 = 860$	

choosing the shorter routes from the intermediate cities to Vienna. The shortest routes from Paris and Lyon to Vienna are shown in Fig. 8.3.

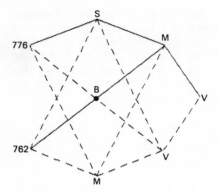

Fig. 8.3

For subproblem (c) we have the distances:

Bordeaux to Vienna
via Paris $353 + 776 = 1129$
via Lyon $340 + 762 = 1102$

The shortest overall distance is therefore 1102 miles taking the route:

Bordeaux–Lyon–Basel–Munich–Vienna (Fig. 8.4).

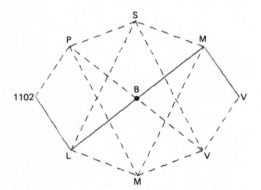

Fig. 8.4

The advantage of this method over the obvious method of listing all possible routes is now clear. We are building the optimal route from sections that are themselves optimal from intermediate points. Hence all routes with sections that are not optimal are automatically excluded from consideration. For example the route:

Bordeaux–Lyon–Basel–Venice–Vienna

is eliminated because the section:

Basel–Venice–Vienna

is not the shortest route from Basel to Vienna.

There is a great variety of problems in Operational Research that can be reduced to the problem of finding the shortest path on a graph. As an example consider the following capital replacement problem.

Example 2

A family, planning to move abroad in three years' time because of a new posting within the Diplomatic Service, is wondering whether it should keep its 1-year old washing machine for the duration of that period or trade it in for another one in the meantime. The family has decided that if ever it does trade it in then it will be for a new one and the decision will be made at the beginning of a year.

This problem can be represented by the graph of Fig. 8.5. Each node represents a possible 'state of the system'. For example node D corresponds to the family having a 3-year old machine at the end of the second year and node J to having a 1-year old machine at the end of the third year. The arcs joining the nodes correspond to the possible decisions that can be taken. For example at node D the 3-year old machine can either be kept for another year (arc DG) to become a 4-year old machine or traded-in for a new machine (arc DJ) which at the end of the year becomes a 1-year old machine. The expected cost incurred over the year for each course of action is noted against the relevant arc and is termed the 'weight' of that arc. For example the expected cost over the year resulting from the decision represented by arc DJ is £392, obtained as:

(purchase price) − (trade-in price) − (expected repair costs in the first year).

For the decision represented by arc DG on the other hand the £171 denotes just the expected annual repair costs for a 3-year old machine.

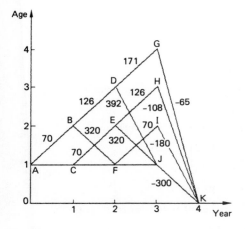

Fig. 8.5

At the end of three years the machine that they then have is sold, a decision that is represented by the arcs leading to node K with zero age representing the absence of a machine and the negative cost the income gained from the sale. A path on the graph from the starting node A to the final node K corresponds to a possible sequence of decisions implying a total cost obtained by adding all the weights on the arcs of the path. For example for path ABDJK the total cost is £288. We can find the sequence of decisions that yields the path with least cost by using the sequential method introduced in Example 1. Precisely:

(a) we find the least cost strategy from year 2 to 4 for each of the possible ages of the machine at year 2.

$$\begin{array}{ll} D \to G \to K & 171 - 65 = 106 \\ D \to J \to K & 392 - 300 = 92 \end{array} \quad \text{(shorter)}$$

$$\begin{array}{ll} E \to H \to K & 126 - 108 = 18 \quad \text{(shorter)} \\ E \to J \to K & 320 - 300 = 20 \end{array}$$

$$\begin{array}{ll} F \to I \to K & 70 - 180 = -110 \quad \text{(shorter)} \\ F \to J \to K & 200 - 300 = -100 \end{array}$$

The least cost paths from D, E and F are shown in Fig. 8.6.

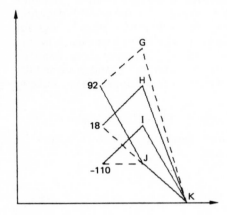

Fig. 8.6

(b) we find the least cost paths from year 1 using the information from (a).

$$\begin{array}{ll} B \to D \to K & 126 + 92 = 218 \\ B \to F \to K & 320 - 110 = 210 \end{array} \quad \text{(shorter)}$$

$$\begin{array}{ll} C \to E \to K & 70 + 18 = 88 \quad \text{(shorter)} \\ C \to F \to K & 200 - 110 = 90 \end{array}$$

The least cost paths from B and C are shown in Fig. 8.7.

(c) we find the least cost path from year 0 using the information from (b).

$$\begin{array}{ll} A \to B \to K & 70 + 210 = 280 \\ A \to C \to K & 200 + 88 = 288 \end{array}$$

The least cost path is that shown in Fig. 8.8.

$$A \to B \to F \to I \to K$$

The machine is therefore to be traded in after 1 year and then kept until departure.

As we mentioned, the 'capital replacement problem' is just one of many types of Operational Research problem that can be represented by a graph, with the nodes

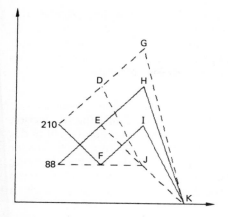

Fig. 8.7

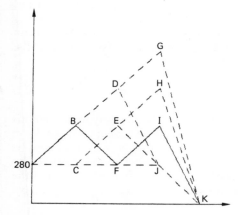

Fig. 8.8

describing the possible states of the 'system' and the arcs the possible transitions between those states. To show the variety of such Finite State Systems we give three further examples.

(i) *The Allocation Problem*[4]

A company wishes to determine how best to spend its advertising budget between different TV companies covering different regions of the country. Spending in regions of high density of population might seem preferable but in less densely populated areas the advertising rates are much lower. A node of the graph for this problem corresponds to the amount of money remaining at a given stage of the negotiations after contracts with some of the TV companies have already been signed. The arcs from this node correspond to the amounts (in multiples of £10 000 say) that could be spent in the current negotiation and the arc weights the expected increased sales resulting from the contract. The problem is to maximise total increased sales, i.e. to find the longest path on the graph.

(ii) *A Capital Investment Problem*[5]

A company is experiencing great demand for its products and is wondering whether it should open up new factories to meet that demand. The problem is that the capital costs are high and it may be that the new facilities will not pay for themselves, certainly if sales decline later. The nodes of the graph for this problem correspond to the productive capacity that is available in a given year and the arcs to the decision to invest or not to invest in a new factory. The arc weights correspond to the net profit in a year allowing for capital costs. The problem again is to find the longest path on the graph.

(iii) The Production–Inventory Problem[6]

It is often the case that goods are produced in batches and then stored until eventual sale rather than being produced in response to individual orders. Certainly the production costs will be less in this strategy but this will be offset by increased storage costs. The nodes of the graph for this problem correspond to the number of goods in store in a particular week (say), the arcs to the decisions on how many to produce in that week and the arc weights to the cost incurred, from production and storage. The problem is to minimise total cost over a given planning period, i.e. to find the shortest path on the graph, under a given demand pattern.

8.3 Dynamic Programming[2]

The problems discussed in the previous section are called sequential decision problems, each decision in the sequence changing the state of the system with some penalty incurred. We denote the state just before the kth decision by x_k, the state after the decision by x_{k+1} and the penalty incurred $f_k(x_k, x_{k+1})$ (Fig. 8.9a). The problem is to minimise the total penalty in changing from state x_0 to x_{n+1}, i.e.

$$\text{minimise} \sum_{k=0}^{n} f_k(x_k, x_{k+1}) \tag{8.1}$$

by choosing the intermediate states $x_1, x_2, \ldots, x_n$ appropriately. The way we solved this problem was to work with the function $F_k(x_k)$ which denotes the minimum total penalty in changing from state x_k to state x_{n+1}. An optimal path from state x_k to x_{n+1} must also be optimal from state x_{k+1} (Fig. 8.9b); otherwise we could find a path with lower penalty by taking the optimal path from x_{k+1} to x_{n+1}. This observation is often referred to as the *Principle of Optimality*. The total penalty on an optimal path from x_k to x_{n+1} is given by:

$$F_k(x_k) = f_k(x_k, x_{k+1}^*) + F_{k+1}(x_{k+1}^*)$$

where x_{k+1}^* denotes the optimal choice for x_{k+1} found by sorting through the finite set of

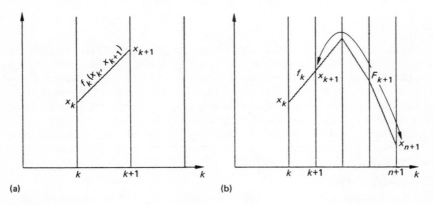

Fig. 8.9

states x_{k+1} to find the one where the penalty:

$$f_k(x_k, x_{k+1}) + F_{k+1}(x_{k+1})$$

is least, i.e.

$$F_k(x_k) = \min_{x_{k+1}} \{f_k(x_k, x_{k+1}) + F_{k+1}(x_{k+1})\} \quad \text{for } k = 0, \ldots, n-1 \quad (8.2)$$

This relation is usually referred to as a *Dynamic Program* and defines a recursive scheme whereby one can successively calculate the optimal penalties $F_k(x_k)$ starting with the identification:

$$F_n(x_n) = f_n(x_n, x_{n+1})$$

and ending with $F_0(x_0)$ as the required minimum penalty for the original problem (8.1).

If instead of minimising cost the objective is to, say, maximise profit then the minimisation operation in equation (8.2) is simply replaced by a maximisation. A Dynamic Program can also be solved in the Forward direction starting with $F_1(x_1) = f_0(x_0, x_1)$ and ending with $F_{n+1}(x_{n+1})$ as the required optimal value of the objective function.

The Dynamic Program (8.2) was derived on the assumption that the set of possible states for a given k was finite in number. This is an unnecessary restriction and the program (8.2) can be used as an alternative method of solving separable optimisation problems of the form (8.1) where x_k are interpreted as variables taking a continuous range of values. As an example let us reconsider the basic Pioneer Problem:

$$\text{maximise} \quad \sum_{k=0}^{n} u_k^{1/2}$$

$$\text{subject to} \quad x_{k+1} = c(x_k - u_k)$$

which, after elimination of the control variables, has the form:

$$\text{maximise} \quad \sum_{k=0}^{n} (x_k - c^{-1} x_{k+1})^{1/2}$$

with $f_k(x_k, x_{k+1}) = (x_k - c^{-1} x_{k+1})^{1/2}$. In the equivalent Dynamic Programming formulation we have:

$$F_k(x_k) = \max_{x_{k+1}} \{(x_k - c^{-1} x_{k+1})^{1/2} + F_{k+1}(x_{k+1})\}.$$

With $n = 2$, $x_0 = 1$, $x_3 = 4$ and $c = 2$, the first stage of the calculation with $k = 1$ gives:

$$F_1(x_1) = \max_{x_2} \{(x_1 - \tfrac{1}{2} x_2)^{1/2} + F_2(x_2)\} \quad (8.3)$$

where $F_2(x_2) = (x_2 - \tfrac{1}{2} x_3)^{1/2} = (x_2 - 2)^{1/2}$. Since x_2 is now a continuous variable the maximisation can be performed using the differential calculus. In fact under the constraints $\tfrac{1}{2} x_3 \leq x_2 \leq 2 x_1$, necessary for the definition of the square root terms, the maximum occurs at the interior point $x_2 = \tfrac{1}{3}(2 + 4x_1)$ with:

$$F_1(x_1) = 3^{1/2}(x_1 - 1)^{1/2}$$

on substitution in equation (8.3) (Fig. 8.10a). The second stage of the calculation with $k = 0$ gives:

$$F_0(1) = \max_{x_1} \{(1 - \tfrac{1}{2} x_1)^{1/2} + 3^{1/2}(x_1 - 1)^{1/2}\}$$

with solution $x_1 = 13/7$ and $F_0(1) = (7/2)^{1/2}$. The process is represented in the two diagrams of Fig. 8.10.

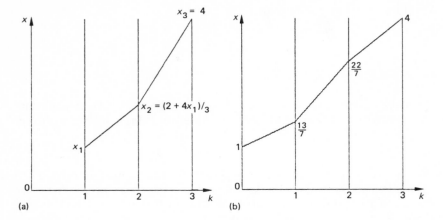

Fig. 8.10

We can write a basic discrete optimal control problem directly in Dynamic Programming form without having to substitute out the control variable. This is achieved by replacing x_{k+1} in equation (8.2) using the dynamic equation:

$$x_{k+1} = x_k + g_k(x_k, u_k)$$

to obtain:

$$F_k(x_k) = \max_{u_k} \{\tilde{f}_k(x_k, u_k) + F_{k+1}(x_k + g_k(x_k, u_k))\} \tag{8.4}$$

where:

$$\tilde{f}_k(x_k, u_k) = f_k(x_k, x_k + g_k(x_k, u_k))$$

is the summand in the objective function for the optimal control problem.

The Dynamic Programming approach can also be used in the limit of a continuous optimal control problem. The continuous analogue of equation (8.4) is determined by taking the limit of (8.4) as the time interval h of the discrete problem decreases to zero. To prepare for this limiting operation we first multiply the component functions $\tilde{f}, g$ by h and subtract $F_{k+1}(x_k)$ from both sides of the equation:

$$-(F_{k+1}(x_k) - F_k(x_k)) = \max_{u_k} \{\tilde{f}_k(x_k, u_k)h + (F_{k+1}(x_k + hg_k) - F_{k+1}(x_k))\}$$

The final step before taking the limit is to linearise the function F_{k+1} about x_k. A linear approximation to a general differentiable function $p(x)$ is given by:

$$p(x + l) \simeq p(x) + lp'(x),$$

the approximation being better the smaller the magnitude of l. Geometrically we are approximating the graph of $p(x)$ by its tangent at x (Fig. 8.11). If we apply this linearisation with $p = F_{k+1}$ and $l = hg_k$ we obtain:

$$-(F_{k+1}(x_k) - F_k(x_k)) \simeq \max_{u_k} \left\{ \tilde{f}_k(x_k, u_k)h + hg_k \frac{\partial F_{k+1}}{\partial x_k} \right\}.$$

In the limit we obtain:

$$-\frac{\partial F}{\partial t} = \max_u \left\{ f(x, u, t) + \frac{\partial F}{\partial x} g(x, u, t) \right\} \tag{8.5}$$

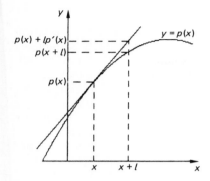

Fig. 8.11

with $F(x, t)$ as the limit of $F_k(x_k)$, dropping the tilde on the f function for convenience. If u^* denotes the optimal value of u in the maximisation in equation (8.5) then:

$$-\frac{\partial F}{\partial t} = f(x, u^*, t) + \frac{\partial F}{\partial x} g(x, u^*, t).$$

This equation is called the *Hamilton–Jacobi equation*[7] and is a partial differential equation for the function $F(x, t)$ since it involves the partial derivatives of F with respect to x and t. Its right hand side bears a close resemblance to the Hamiltonian with $\partial F/\partial x$ playing the role of the Lagrange multiplier.

As an example let us consider the continuous harvesting model:

$$\text{maximise} \quad \int_0^T u^{1/2} \, dt$$
$$\text{subject to} \quad \dot{x} = cx - u$$

with $f = u^{1/2}$ and $g = cx - u$ for which we have the equation:

$$-\frac{\partial F}{\partial t} = \max_u \left\{ u^{1/2} + \frac{\partial F}{\partial x}(cx - u) \right\}.$$

Maximisation yields $u^* = \frac{1}{4}(\partial F/\partial x)^{-2}$. Hence the Hamilton–Jacobi equation in this case reads as:

$$-\frac{\partial F}{\partial t} = \frac{1}{4}\left(\frac{\partial F}{\partial x}\right)^{-1} + cx\left(\frac{\partial F}{\partial x}\right)$$

on substituting for u^*. There are special techniques for solving partial differential equations of Hamilton–Jacobi type just as there are for solving the ordinary differential equations of Chapter 6. We do not have the space here to discuss these techniques but there are several texts where a full discussion can be found.[8]

The Dynamic Programming and Hamiltonian approaches look at the optimal control problem in complementary ways. In the former the primary object of interest is the optimal value of the objective function $F(x, t)$ from a starting point (x, t). The optimal paths are determined at a second stage by going back and determining the optimal control variable u^* and hence $\dot{x}$ from the dynamic equation. In the Hamiltonian approach, on the other hand, the primary object of interest is the optimal path $x(t)$ with the optimal objective function value being determined at a second stage of the analysis.

8.4 Stochastic Programming[9]

To examine some of the difficulties in incorporating uncertainty into an optimisation problem let us re-examine the problem faced by the Pioneer Community in planning its consumption when the climate is so unreliable that the assumption of a constant growth rate factor c is no longer tenable. The simplest response the Community can make would be to replace c by its expected value and to calculate the consumption path as before. This however could lead to difficulties since in some years the harvest may be so bad that the planned consumption might be greater than the actual harvest for that year. There would also be no guarantee that the final agreed stock level x_{n+1} could be achieved.

There are various ways they could tackle this end-point problem. If the harvests are variable but always good enough to enable the end-point value to be achieved then they could impose inequality constraints on the consumption path to guarantee it. If this is not possible they could introduce a probabilistic end-point condition stating that a given final level is to be achieved with a certain probability. A further possibility is to remove the end-point condition altogether and add an 'asset' function to the objective function to value the end state.

Such amendments to the deterministic optimal control problem do not address one of the deeper issues in modelling uncertainty in planning problems—that the amount of information that is available changes with time. If the Community, for example, wishes in year zero to make its decisions about the consumption path then it only has available (at best) the expected level of the harvest in any year but by the time the consumption decision has to be implemented that level is known with certainty. What the Community really needs is a set of decision rules which say how much they should set aside for consumption given the size of the harvest in that year. The following example shows how this can be done.

For the deterministic Pioneer Problem:

$$\text{maximise} \quad \sum_{k=0}^{2} u_k^{1/2} + \varphi(x_3)$$

$$\text{subject to} \quad x_{k+1} = c(x_k - u_k),$$

with asset function $\varphi = x_3^{1/2}$ we have the 'Stochastic' dynamic program:

$$F_k(x_k) = \max_{u_k} \{u_k^{1/2} + E(F_{k+1}(c(x_k - u_k)))\} \quad \text{for } k = 0, 1, 2$$

where E denotes the expected value (i.e. average) of the expression in the following brackets. We solve this equation recursively with the following steps.

(i) After the harvest in year 3 there is a known amount of corn x_3 and this yields $x_3^{1/2}$ in Utility from the asset function φ.

(ii) After the harvest in year 2 there is a known amount of corn x_2 to divide between consumption and seedcorn but it is not certain what the yield on the seedcorn will be at the next harvest. With this uncertainty the consumption u_2 is determined by:

$$F_2(x_2) = \max_{u_2} \{u_2^{1/2} + E(c(x_2 - u_2))^{1/2}\}$$

$$= \max_{u_2} \{u_2^{1/2} + \alpha(x_2 - u_2)^{1/2}\}$$

where $\alpha = E(c^{1/2})$. The solution is obtained as:

$$u_2 = x_2/(1 + \alpha^2) \tag{8.6}$$

and $\quad F_2(x_2) = x_2^{1/2}(1 + \alpha^2)^{1/2}.$

(iii) If the harvest in the first year is x_1 then the corresponding consumption decision satisfies:
$$F_1(x_1) = \max_{u_1} \{u_1^{1/2} + (1+\alpha^2)^{1/2} E(c(x_1 - u_1)^{1/2})\}$$
with solution:
$$u_1 = x_1/(1+\alpha^2+\alpha^4) \tag{8.7}$$
and $F_1(x_1) = x_1^{1/2}(1+\alpha^2+\alpha^4)^{1/2}$.

(iv) Finally the initial decision in year zero with stock level unity is determined by:
$$F_0(1) = \max_{u_0} \{u_0^{1/2} + (1+\alpha^2+\alpha^4)^{1/2}\alpha(1-u_0)^{1/2}\}$$
with solution:
$$u_0 = x_0/(1+\alpha^2+\alpha^4+\alpha^6) \tag{8.8}$$
and $F_0(1) = x_0^{1/2}(1+\alpha^2+\alpha^4+\alpha^6)^{1/2}$

In summary equations (8.6), (8.7) and (8.8) give the decision rules required. Once the harvests are known the optimal consumption values can be determined.

One practical way that is used to cope with the uncertainty is to build up a store of corn to be used in years of bad harvest. Try to model this feature and determine how each harvest should be divided between seedcorn, present consumption and precautionary storage.

8.5 The Theory in Perspective

Finally, in this last section, let us go back over the material we have covered in the eight chapters of this book and remind ourselves of how the various topics are related to each other.

For a finite number n of continuously valued decision variables the general optimisation problem, often called the general *mathematical program*, has the form:
$$\left. \begin{array}{l} \text{maximise} \quad f(\mathbf{x}) \\ \text{subject to} \quad g_j(\mathbf{x}) \leqslant 0 \quad \text{for} \quad j = 1, \ldots, m \end{array} \right\} \tag{8.9}$$

This includes the case of a mixture of equality and inequality constraints since any equality constraint can be written as a pair of inequality constraints:

$g = 0$ is equivalent to $g \leqslant 0$ and $g \geqslant 0$.

We have considered three important special cases of (8.9), the linear, the quadratic and the geometric programs. For both the linear and quadratic cases the constraints are linear in the (usually) non-negative variables, while the objective function is linear for the former and quadratic for the latter. The geometric program is somewhat different in form, since its objective function and constraints are sums of products of powers of the variables, the so-called posynomials.

In Chapter 3 we used the Lagrange method to derive two sets of necessary conditions for a solution to (8.9). The Lagrange equations gave the positions of the unconstrained stationary points of the Lagrangean and the complementary slackness conditions reimposed the constraints to find the constrained optimum (Section 3.6). When the variables are all non-negative these conditions become the Kuhn–Tucker conditions. If the objective function f is concave and the constraint functions convex then (with some further qualification) the Kuhn–Tucker conditions are sufficient to determine the optimum, i.e. all solutions of the Kuhn–Tucker conditions are optimal. From these

226 Postscript

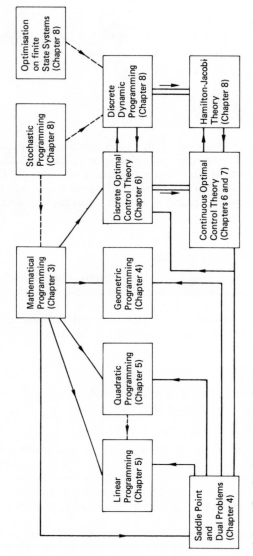

Fig. 8.12

conditions we developed in Chapter 5 an efficient algorithm, the Simplex Algorithm, for finding the optimum of a linear program. We first established that the optimum is taken in general at a vertex on the boundary of the feasible region. The algorithm searches for this vertex by following a path on the boundary along which the Lagrangean and the objective function progressively increase in value. This algorithm was adapted to solve a quadratic program by searching for a feasible vertex in the extended space of the decision variables and multipliers.

When the appropriate concavity and convexity conditions are satisfied we can construct from the Lagrangean a second optimisation problem called the dual whose variables are the multipliers of the original (primal) problem. These problems are, in a sense described in Chapter 4, different cross-sections of a larger problem called the Saddle Point Problem where we search for a saddle point of the Lagrangean in the extended space of the primal and dual variables. One implication of this structure is that from the solution of, say, the dual one can deduce the solution of the primal. If therefore the dual is easier to solve it might be preferable to solve the primal by first solving the dual. This is the case with geometric programming where the dual, in many cases of interest, is of zero dimension with the optimum being determined solely by the constraints.

The discrete optimal control problem is also a particular case of (8.9) with the variables divided into two sets—the state and control variables, the control variables determining the state variables through the dynamic equation. The solution is obtained by solving a pair of difference equations defined in terms of the Hamiltonian function which is derived directly from the Lagrangean. Being in effect a sequential decision problem the discrete optimal control problem can also be solved by Dynamic Programming where the optimisation is carried out as a sequence of 1-variable optimisation problems rather than simultaneously in all variables.

The Dynamic Programming Algorithm can easily be adapted to solve sequential problems where the variables take a discrete set of values. The optimisation is achieved by sorting through a finite set of states at each stage to find the optimum. Such problems can be translated into the shortest path problem on a graph, with the nodes representing the finite number of states that the system can take. Dynamic Programming can also be used to handle Stochastic programs where the amount of information available is time dependent. This is because the optimisation is sequential and hence all the available information can be used at each stage.

Besides discussing these various types of optimisation problems involving a finite number of variables we also took the opportunity (in Chapters 6 and 7) to discuss one class of problem where that number was infinite. This is the continuous optimal control problem obtained as the limit of the discrete case as the interval between system changes decreases to zero. Like the discrete case there are two ways of solving this problem—either by solving a set of 'ordinary' differential equations for the optimal state and multiplier variables, $x(t)$ and $\lambda(t)$, (the Hamilton equations) or by solving a partial differential equation (the Hamilton–Jacobi equation) for the optimal value of the objective functional $F(x, t)$ with initial state (x, t). The former is in general the more tractable approach although the latter has proved useful in solving particular types of problem especially in the Physical Sciences[8].

We have summarised these relationships in the flowchart of Fig. 8.12.

References

For further reading refer to:
1 G. Hadley, *Nonlinear and Dynamic Programming*, Addison-Wesley (1972).
2 R. E. Bellman and S. E. Dreyfus, *Applied Dynamic Programming*, Princeton University Press (1962).

3   M. S. Bazaraa and C. M. Shetty, *Nonlinear Programming: Theory and Algorithms*, Wiley (1979).
4   T. B. Boffey, *Graph Theory in Operations Research*, Macmillan (1982).
5   G. L. Nemhauser and Z. Ullman, 'Discrete Dynamic Programming and Capital Allocation', *Management Science*, Volume 15 (1969), pages 494–505.
6   R. L. Schultz (editor), *Applications of Management Science*, Volume 1, JAI Press (1981).
7   M. D. Intriligator, *Mathematical Optimization and Economic Theory*, Prentice-Hall (1971) and H. Goldstein, *Classical Mechanics*, Addison-Wesley (1964).
8   C. Lanczos, *The Variational Principles of Mechanics*, University of Toronto Press (1960) and R. Courant and D. Hilbert, *Methods of Mathematical Physics*, Volume 2, Interscience (1966).
9   O. L. R. Jacobs, *An Introduction to Dynamic Programming*, Chapman and Hall (1972).

Appendix

A Linear Simultaneous Equations

The equations:

$$x_1 + 2x_2 - 3x_3 = 1$$
$$3x_1 - 2x_2 = -5$$
$$2x_1 - x_2 - x_3 + 2x_4 = 0.5$$

are all examples of a *linear equation*, whose general form for n variables is:

$$a_1 x_1 + a_2 x_2 + \ldots + a_n x_n = b \tag{A.1}$$

where $a_1, \ldots, a_n$ are its coefficients and b is the constant term. Geometrically, a linear equation defines a line in a plane when $n = 2$ (Fig. A.1a), and a plane in three-dimensional space when $n = 3$ (Fig. A.1b). A set of m linear equations in n variables is said to have a solution if we can find finite values for $x_1, x_2, \ldots, x_n$ which simultaneously satisfy all m equations. For example, the pair of equations:

$$x_1 + 2x_2 = 5$$
$$x_1 - x_2 = -1 \tag{A.2}$$

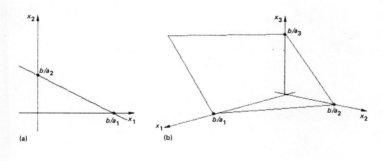

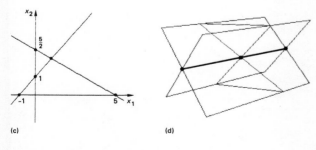

Fig. A.1

230 Appendix

has $x_1 = 1, x_2 = 2$ as a solution. There is usually only one solution when the number of equations equals the number of variables. For two and three variables this corresponds to the fact that two lines in a plane and three planes in a three-dimensional space have only one point in common (Figs A.1c, d). There are, however, exceptions to this rule—we can have no solutions or an infinite number of solutions. For example, the pair of equations:

$$2x_1 - x_2 = 5$$
$$-4x_1 + 2x_2 = a$$

has no (finite) solution when, for example, $a = 3$, since the corresponding lines are parallel and distinct (Fig. A.2a). When $a = -10$, however, the two lines are parallel and coincident (Fig. A.2b), so that any point on the coincident lines satisfies both equations. As an example in three variables consider:

$$\begin{aligned} x_1 + x_2 + x_3 &= 6 \\ 2x_1 - x_2 + x_3 &= 3 \\ -4x_1 + 5x_2 + x_3 &= b. \end{aligned} \qquad \text{(A.3)}$$

When $b = 3$ the corresponding planes intersect in a straight line rather than a point, giving an infinite number of solutions. If b takes any other value then the third plane will not intersect the line of intersection of the first two, and hence there will be no (finite) solution. The equations are then said to be *inconsistent*.

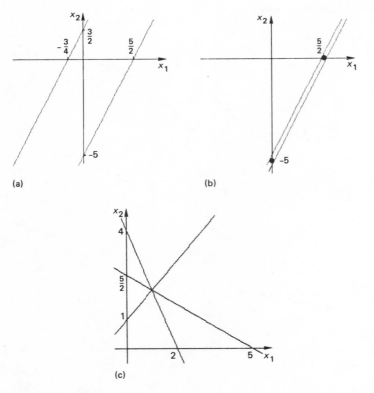

Fig. A.2

The reason for the infinite number of solutions with $b = 3$ is that one of the equations is redundant, containing no additional information. With this value of b we can obtain the third equation from the other two by subtracting 3 times the second equation from 2 times the first. The third equation is said to be *linearly dependent* (on the others). If we remove the dependent equation we obtain two equations in three variables—not enough to determine the three variables uniquely.

Linear dependence is an essential condition for there to be a solution when there are more equations than variables, otherwise the equations will necessarily be inconsistent. For example, the equations:

$$x_1 + 2x_2 = 5$$
$$x_1 + x_2 = -1$$
$$2x_1 + x_2 = b$$

have a solution only if $b = 4$, in which case the third equation can be written as the sum of the first two. Geometrically, $b = 4$ is the condition that the line defined by the third equation contains the point of intersection of the first two (Fig. A.2c).

B Summation Notation

General expressions such as the left hand side of (A.1) can be written more simply as:

$$a_1 x_1 + a_2 x_2 + \ldots + a_n x_n = \sum_{i=1}^{n} a_i x_i$$

by introducing the summation operator $\sum_{i=1}^{n}$, which instructs us to add together the n terms $a_i x_i$ where the index i can use any label for the indexing of the elements in the sum. For example:

$$\sum_{i=1}^{n} a_i x_i = \sum_{j=1}^{n} a_j x_j = \sum_{k=1}^{n} a_k x_k.$$

If we are considering a set of m such equations, then we have to introduce a second subscript to distinguish the coefficients of the various equations. If a_{ji} denotes the coefficient of x_i in the jth equation, then that equation

$$a_{j1} x_1 + a_{j2} x_2 + \ldots + a_{jn} x_n = b_j$$

can be written in summation notation as:

$$\sum_{i=1}^{n} a_{ji} x_i = b_j \tag{A.4}$$

If we now wish to consider the expression:

$$M = p_1 b_1 + p_2 b_2 + \ldots + p_m b_m = \sum_{j=1}^{m} p_j b_j$$

then, using (A.4), we can write it as a double summation:

$$M = \sum_{j=1}^{m} p_j b_j = \sum_{j=1}^{m} p_j \left(\sum_{i=1}^{n} a_{ji} x_i \right) \tag{A.5}$$

which is shorthand for

$$
\begin{aligned}
M = & p_1(a_{11}x_1 + a_{12}x_2 + \ldots + a_{1n}x_n) \\
& + p_2(a_{21}x_1 + a_{22}x_2 + \ldots + a_{2n}x_n) \\
& + \ldots \\
& + p_m(a_{m1}x_1 + a_{m2}x_2 + \ldots + a_{mn}x_n).
\end{aligned}
$$

If we evaluate M by adding the terms in each column and then adding these column sums, we can rewrite M as:

$$M = \sum_{i=1}^{n} \left(\sum_{j=1}^{m} p_j a_{ji} \right) x_i \tag{A.6}$$

Comparison of (A.5) and (A.6) shows that the order of the summations does not matter, and hence we can write the double sum more symmetrically as

$$M = \sum_{i=1}^{n} \sum_{j=1}^{m} p_j a_{ji} x_i.$$

For typographical convenience, we shall usually omit the range of the index when its values are clear from the context. For example the last expression might be written:

$$M = \sum_{i} \sum_{j} p_j a_{ji} x_i.$$

C Matrices

An $n*m$ matrix A is a rectangular array of elements written in n rows and m columns. For example

$$\begin{pmatrix} 3 & 0 & 2 \\ 2 & -1 & 5 \end{pmatrix}, \begin{pmatrix} 5 \\ 4 \\ 1 \end{pmatrix}, \quad (0 \ 2 \ 0)$$

are $2*3$, $3*1$ and $1*3$ matrices respectively. Two matrices are *equal* when they have the same number of rows and the same number of columns and all corresponding elements are equal. The *transpose* of a matrix A, written as A^T, is the $m*n$ matrix formed from A by interchanging its rows and columns. For example, the following matrices are equal to those defined above.

$$\begin{pmatrix} 3 & 2 \\ 0 & -1 \\ 2 & 5 \end{pmatrix}^T, \quad (5 \ 4 \ 1)^T, \quad \begin{pmatrix} 0 \\ 2 \\ 0 \end{pmatrix}^T.$$

An $n*1$ matrix is known as a *column vector*, and a $1*m$ matrix as a *row vector*. By a vector we shall usually mean a column vector; a row vector being considered as the transpose of a column vector.

The *scalar product* of two vectors **a** and **x** is defined only if they contain the same number of elements. If

$$\mathbf{a}^T = (a_1 a_2 \ldots a_n) \quad \text{and} \quad \mathbf{x}^T = (x_1 x_2 \ldots x_n)$$

then their scalar product $\mathbf{a}^T\mathbf{x}$ is defined as:

$$(a_1 \ldots a_n) \begin{pmatrix} x_1 \\ \vdots \\ x_n \end{pmatrix} = \mathbf{a}^T\mathbf{x} = \sum_{i=1}^{n} a_i x_i.$$

By symmetry this is also equal to $\mathbf{x}^T\mathbf{a}$. For example, the scalar product of the vectors $(1 \quad 2)^T$ and $(x_1 \quad x_2)^T$ is given by:

$$(1 \quad 2) \begin{pmatrix} x_1 \\ x_2 \end{pmatrix} = x_1 + 2x_2.$$

We can define the product of an $n*m$ matrix A and an m element vector $\mathbf{x}$ in terms of scalar products as:

$$A\mathbf{x} = \begin{pmatrix} \mathbf{a}_1^T \\ \mathbf{a}_2^T \\ \cdot \\ \mathbf{a}_n^T \end{pmatrix} \mathbf{x} = \begin{pmatrix} \mathbf{a}_1^T\mathbf{x} \\ \mathbf{a}_2^T\mathbf{x} \\ \ldots \\ \mathbf{a}_n^T\mathbf{x} \end{pmatrix} \quad (A.7)$$

where $\mathbf{a}_j^T$ denotes the row vector defining the jth row of A. For example if

$$A = \begin{pmatrix} 1 & 2 \\ 1 & -1 \\ 2 & 1 \end{pmatrix}$$

then

$$A\mathbf{x} = \begin{pmatrix} 1 & 2 \\ 1 & -1 \\ 2 & 1 \end{pmatrix} \begin{pmatrix} x_1 \\ x_2 \end{pmatrix} = \begin{pmatrix} x_1 + 2x_2 \\ x_1 - x_2 \\ 2x_1 + x_2 \end{pmatrix}$$

with $\mathbf{a}_1^T = (1 \quad 2)$, $\mathbf{a}_2^T = (1 \quad -1)$ and $\mathbf{a}_3^T = (2 \quad 1)$.

The linear equation

$$\sum_{i=1}^{n} a_i x_i = b$$

can be written as the scalar product:

$$\mathbf{a}^T\mathbf{x} = b$$

and the jth of a set of m linear equations

$$\sum_{i=1}^{n} a_{ji} x_i = b_j$$

as

$$\mathbf{a}_j^T\mathbf{x} = \mathbf{h}_j$$

where the ith element of the jth vector $\mathbf{a}_j^T$ is denoted by a_{ji}. From (A.7) we see that these equations can be written as a single matrix equation

$$A\mathbf{x} = b$$

where $\mathbf{b}$ is the vector $(b_1, b_2, \ldots, b_m)^T$. For example, the equations
$$x_1 + 2x_2 = 5, \quad x_1 - x_2 = -1, \quad 2x_1 + x_2 = 4$$
can be written as:
$$\begin{pmatrix} 1 & 2 \\ 1 & -1 \\ 2 & 1 \end{pmatrix} \begin{pmatrix} x_1 \\ x_2 \end{pmatrix} = \begin{pmatrix} 5 \\ -1 \\ 4 \end{pmatrix}.$$

The double sum (A.5) can also be written as a matrix product if we take the scalar product of (A.4) with the vector $(p_1\ p_2\ \ldots\ p_n)^T$. This gives
$$\sum_j p_j b_j = \mathbf{p}^T \mathbf{b} = \mathbf{p}^T (A\mathbf{x}) = \mathbf{p}^T A \mathbf{x}.$$

D Limits of Functions

In Chapter 1 we have to determine the behaviour of functions at infinity in order to decide where the overall maximum or minimum occurs. To do this we need to understand the behaviour at infinity of the most commonly occurring functions of one and two variables.

(a) Functions of one variable

We restrict our attention to polynomial, exponential and logarithmic functions, since most of the single variable functions discussed in the first chapter are constructed from them.

(i) *Polynomial functions* An nth order polynomial has the form:
$$f(x) = a_n x^n + a_{n-1} x^{n-1} + \ldots + a_1 x + a_0 \qquad (a_n \neq 0)$$
where $a_0, \ldots, a_n$ are its coefficients and n is a positive integer. As x tends to $+\infty$ the behaviour of f is determined by that of its highest power, since all other terms tend to zero relative to it. The behaviour of f at infinity therefore depends on the sign of a_n and whether n is even or odd. For example, for the polynomial
$$-x^6 + 2x^2 + 1$$
the dominating term is $-x^6$ and hence the polynomial tends to $-\infty$ as x tends to $+\infty$. We write this as:
$$-x^6 + 2x^2 + 1 \to -\infty \text{ as } x \to +\infty.$$
Similarly: $x^3 - x + 1 \to \infty \quad \text{as } x \to \infty$
$$\to -\infty \text{ as } x \to -\infty$$
since the x^3 term dominates. We can use the same technique to determine the behaviour at infinity of rational functions (i.e. ratios of two polynomials). For example, for the function
$$f = (x-3)/(x^2+1)$$
x dominates the numerator and x^2 the denominator. The function therefore behaves like $x/x^2 = 1/x$, and hence $f \to 0$ as $x \to \infty$.

(ii) *The exponential function* The exponential function, written as $\exp x$ or e^x, increases in value faster than any positive power of x as x tends to plus infinity and decreases to zero

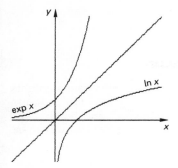

Fig. A.3

faster than any negative power of x as x tends to minus infinity (Fig. A.3), i.e.

$$\frac{\exp x}{x^n} \to \infty \quad \text{as } x \to \infty \quad (n > 0)$$

and

$$\frac{\exp x}{|x|^{-n}} \to 0 \quad \text{as } x \to -\infty \quad (n > 0).$$

These dominance properties extend to the exponential of polynomial and rational functions. For example:

$$x^5 \exp(-2x) = \frac{\exp(-2x)}{x^{-5}} \to 0 \quad \text{as } x \to \infty$$

since the numerator and denominator both tend to zero but the numerator at a much faster rate. Also

$$x^{-4} \exp x^2 = \frac{\exp x^2}{x^4} \to \infty \quad \text{as } x \to -\infty$$

since the numerator and denominator both tend to ∞ but the numerator at a much faster rate.

For more complicated combinations of exponential and rational functions we can also find the limit using dominance arguments. For example

(α) $f = \exp(x^3)/(1 + \exp x)$ behaves like $\exp(x^3)/\exp x = \exp(x^3 - x)$, and this, in turn, like $\exp(x^3)$ since x^3 dominates x. Consequently $f \to \infty$ as $x \to \infty$.
(β) For $f = \exp(x^2 + 1)/(x^{20} + 1)$, the dominating terms are $\exp(x^2)$ and x^{20}, and so f behaves like $\exp(x^2)/x^{20}$, which $\to \infty$ as $x \to \infty$.
(γ) $(x^3 - x + 1)\exp(-2x) \to 0$ as $x \to \infty$, since it behaves like $x^3 \exp(-2x)$.

(iii) *The logarithmic function* The natural logarithmic function, written $\ln x$ or $\log_e x$, is the inverse of the exponential function (Fig. A.3), and hence is not defined for negative x. We have:

$$\ln x \to \infty \quad \text{as } x \to \infty$$

and $\quad \ln x \to -\infty \quad \text{as } x \to 0.$

Being the inverse of the exponential function, the logarithm function is dominated *by* powers of x. Previously:

$$(\ln x)/x^n \to 0 \qquad \text{as } x \to \infty \qquad (n > 0)$$
$$(\ln x)/x^{-n} = x^n \ln x \to 0 \qquad \text{as } x \to 0 \qquad (n > 0).$$

When the logarithm is a function of a polynomial expression, then its behaviour at infinity is determined by the dominating terms and the identity $\ln x^m = m \ln x$. For example, $f = \ln(x^6 + x^4 + 1)/(x^2 + 2x + 2)$ behaves like $(\ln x^6)/x^2 = (6\ln x)/x^2$ and hence $f \to 0$ as $x \to \infty$. In this case the composite logarithmic function is defined for all x and we can see that $f \to 0$ as $x \to -\infty$ also.

(b) Functions of Two Variables

Assessing the behaviour at infinity of functions of two variables is more difficult than in the one variable case because there are an infinite number of ways of reaching infinity and, in general, the value of the function will depend on the actual path taken. Due to the subtlety of the problem we will restrict our attention to multinomial expressions involving sums of products of powers of the two variables. Even with this restriction there are no easy rules that will work on every function, so we must content ourselves with indicating strategies that will often prove useful.

Let us restrict our problem, for the moment, to finding the maximum of the function, and consider three cases.

(i) If we can show that the function takes an infinite value *somewhere* at infinity, then our search is ended, since this must correspond to the overall maximum of the function. This most certainly will be the case if the function takes either of the following two forms (though these do not cover all possibilities).

(α) The highest power of one of the variables has a positive coefficient for some value of the other variable. Consider, for example, $f = x_1^4 x_2 - x_1^3 x_2$. The highest power of x_1 is x_1^4 and its coefficient is x_2. So for any $x_2 > 0$ ($x_2 = 1$, say) we have $f(x_1, 1) = x_1^4 - x_1^3 \to \infty$ as $x_1 \to +\infty$ since the first term dominates. Consequently $f \to \infty$ along $x = 1$.

(β) The highest power of one of the variables is odd and has a negative coefficient for some value of the other variable. Suppose $f = 2x_1 x_2^3 - 3x_1^2 x_2^5$. Here, the highest power of x_2 is x_2^5 with coefficient $-3x_1^2$, which is negative for all $x_1 \neq 0$. Consequently, if we take $x_1 = 1$ (say) $f(1, x_2) = 2x_2^3 - 3x_2^5 \to \infty$ as $x \to -\infty$ along $x_1 = 1$.

(ii) If the function takes the value $-\infty$ at all points at infinity, then the maximum must be attained at an internal point. This situation is more difficult to check. However, if we can rewrite the function as minus the sum of even powers, at least one of which is infinite at any point at infinity, then the function must tend to minus infinity everywhere at infinity. This is true, for example, for the function $f = -x_1^2 - x_2^4$, since at infinity either x_1 or x_2 or both are infinite. Similarly $f = -x_1^2 + x_1 x_2 - x_2^2 = -(x_1 - x_2/2)^2 - 3x_2^2/4 \to -\infty$, since if x_2 tends to infinity then $f \to -\infty$ through the second term, and if x_1 tends to infinity then $f \to -\infty$ through the first term unless $x_1 - x_2/2$ stays finite, in which case x_2 must tend to infinity and $f \to -\infty$ through the second term.

(iii) If neither of the above cases holds, then the function will tend to a finite limit somewhere at infinity. For example, $f = -(x_1^2 + (x_1 x_2 - 1)^2)$ stays finite as $x_2 \to \infty$ along $x_1 x_2 = c$ since in this case $x_1 \to 0$. Consequently the maximum value of f is zero along $x_1 x_2 = 1$ at infinity. As you can imagine, handling functions of this subtlety requires some care and for this reason none of our examples falls into this category.

If we need to find the minimum of a function, all we need do is to find the maximum of

minus the function. For example, to find the minimum of
$$f = x_1^3 x_2 - x_1^4 x_2^2$$
we consider $f' = -f = -x_1^3 x_2 + x_1^4 x_2^2$. The highest power of x_1, x_1^4, has coefficient x_2^2, which is always positive for $x = 0$, so by $(i\alpha) f' \to \infty$ as $x_1 \to \infty$ along any line $x_2 = c$ ($c \neq 0$). Hence $f \to -\infty$ in this direction. The minimum of f is therefore taken at infinity.

E Geometric Vectors

A two-component row vector $\mathbf{v} = (a, b)$ can be represented geometrically by a point Q in the x, y plane with coordinates $x = a$, $y = b$ or alternatively by the directed line segment *from* the origin O *to* the point Q (Fig. A.4a), called the *geometric vector* of $\mathbf{v}$. The scalar multiple $\lambda \mathbf{v}$ of $\mathbf{v}$ is the row vector with components:

$$\lambda(a, b) = (\lambda a, \lambda b).$$

If $\lambda > 0$ then the geometric vector of $\lambda \mathbf{v}$ is obtained from that of $\mathbf{v}$ by scaling its length by a factor λ, keeping the direction unchanged (Fig. A.4b). If $\lambda < 0$ then the length is scaled by a factor $|\lambda|$ and the direction reversed (Fig. A.4c). Conversely, it can be shown (by similar triangles) that two geometric vectors that are collinear (in that they lie along the same radial line) are scalar multiples of each other, i.e.

$$\mathbf{v}_1 = \lambda \mathbf{v}_2$$

where $\mathbf{v}_1$, $\mathbf{v}_2$ are the corresponding row vectors and $\lambda > 0$ if they point in the same direction and $\lambda < 0$ if in opposite directions.

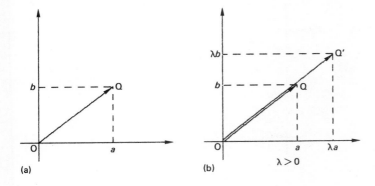

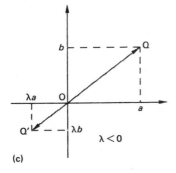

Fig. A.4

238   Appendix

The geometric vector corresponding to the sum $v_1 + v_2 = (a+c, b+d)$ of the two row vectors $v_1 = (a, b)$ and $v_2 = (c, d)$ is the directed line segment lying along the diagonal of the parallelogram defined by v_1, v_2 (Fig. A.5a). More generally, the geometric vector corresponding to the linear combination

$$\lambda_1 v + \lambda_2 v = \lambda_1(a, b) + \lambda_2(c, d) = (\lambda_1 a + \lambda_2 c, \lambda_1 b + \lambda_2 d)$$

of row vectors is obtained by scaling the geometric vectors of v_1, v_2 by factors λ_1 and λ_2 and then adding the resulting vectors by 'parallelogram addition' (Figs A.5b, c). The radial lines defined by the geometric vectors of v_1, v_2 define four regions in the plane if they are not collinear. Which of these regions the vector $\lambda_1 v_1 + \lambda_2 v_2$ lies in depends on the signs of λ_1, λ_2 as indicated in Fig. A.5d. Conversely, a vector lying in one of these regions can be written in the form $\lambda_1 v_1 + \lambda_2 v_2$ with λ_1, λ_2 having the requisite signs.

The scalar product of two geometric vectors is defined in terms of their row vectors (a, b), (c, d) as

$$v_1 v_2^T = (a\ b)(c\ d)^T = ac + bd.$$

This can be expressed in terms of the lengths $|v_1|, |v_2|$ and angles θ_1, θ_2 of their geometric vectors (Fig. A.6a). In fact:

$$v_1 v_2^T = |v_1||v_2|\cos\theta_1 \cos\theta_2 + |v_1||v_2|\sin\theta_1 \sin\theta_2$$
$$= |v_1||v_2|(\cos\theta_1 \cos\theta_2 + \sin\theta_1 \sin\theta_2)$$
$$= |v_1||v_2|\cos\theta$$

where θ is the angle between the vectors.

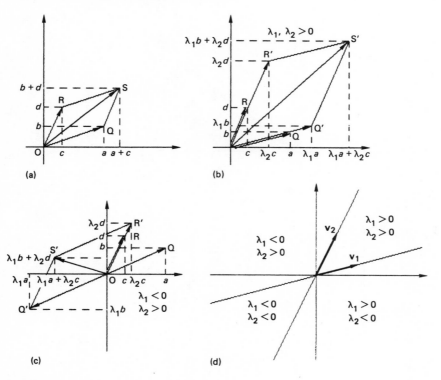

Fig. A.5

Gradient Vectors

Until now, our geometric vectors have been defined relative to the origin of coordinates, though this is not essential. They can be defined relative to any point. This observation is relevant to our discussion in Chapter 3, where we introduce the notion of a gradient vector

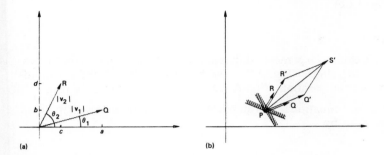

Fig. A.6

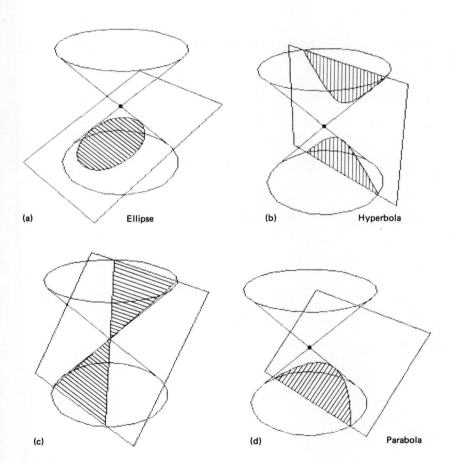

Fig. A.7

240 Appendix

relative to an arbitrary point P, the direction of this vector giving the direction of steepest ascent on a surface at P and its magnitude the value of that steepest gradient. The discussion concerns the relationship between several such gradient vectors when P is a constrained stationary point or a vertex on the boundary of the feasible region. The relationship between the vectors is studied in terms of scalar multiples and sums with all vectors relative to P (Fig. A.6b).

F Conic Sections

If we slice through a double cone with planes at different orientations we obtain four types of curves of intersection—the ellipse (Fig. A.7a), the hyperbola (Fig. A.7b), a pair of straight lines (Fig. A.7c) and the parabola (Fig. A.7d). Let us consider each in turn.

(i) *The ellipse* The standard equation for an ellipse when its axes coincide with those of the coordinate system is given by

$$ax_1^2 + bx_2^2 = c \qquad (A.8)$$

where a, b and c are coefficients with common sign. The equations:

$$2x_1^2 + x_2^2 = 2$$
$$2x_1^2 + 2x_2^2 = 2,$$

for example, both define ellipses, the second being the particular case of a circle of radius 1 (Figs A.8a, b).

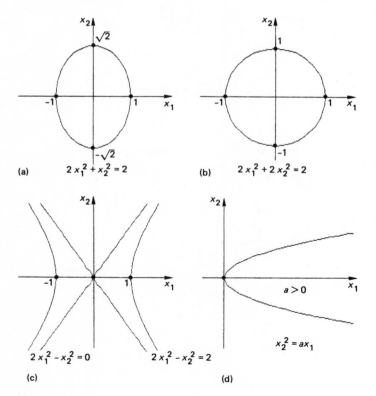

Fig. A.8

F Conic sections

(ii) *The hyperbola and line pair* Equation (A.8) defines a hyperbola with axes coincident with the coordinate axes when a and b have opposite signs and $c \neq 0$. In the case $c = 0$, the equation represents a pair of straight lines, the asymptotes to the hyperbola. For example the equation

$$2x_1^2 - x_2^2 = 2$$

defines a hyperbola intersecting the x_1 axis when $x_1 = \pm 1$ and with asymptotes

$$2x_1^2 - x_2^2 = 0, \quad \text{i.e. the two radial lines } x_2 = \pm x_1\sqrt{2} \text{ (Fig. A.8c)}.$$

(iii) *The parabola* The standard equation for a parabola with its axis along the x_1 axis and passing through the origin (Fig. A.8d) is given by

$$x_1 = ax_2^2.$$

If the axis of the parabola coincides instead with the x_2 axis then the equation is

$$x_2 = ax_1^2.$$

In many cases of interest the conic is not orientated relative to the coordinate axes as described above. The origin could be in a different position, and the axes arbitrarily rotated (Fig. A.9a). A shift of origin without rotation of the axes corresponds to the addition of linear terms to the equation. The shift can be determined by the algebraic device of 'completing the square'. For the equation

$$2x_1^2 + x_2^2 - 4x_1 + 2x_2 = 1$$

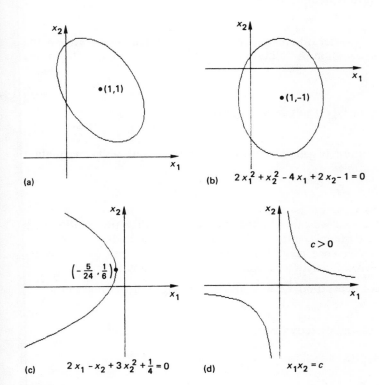

Fig. A.9

the result of completing the square is
$$2(x_1 - 1)^2 + (x_2 + 1)^2 = 4.$$
This is an ellipse centred at the point $(1, -1)$ (Fig. A.9b).

Similarly the equation
$$2x_1 - x_2 + 3x_2^2 + 1/2 = 0$$
can be rewritten as
$$(x_1 + 5/24) = (-3/2)(x_2 - 1/6)^2$$
revealing a parabola centred at the point $(-5/24, 1/6)$ with its axis parallel to the x_1 axis (Fig. A.9c).

If the axes are rotated as well then 'cross product' terms are introduced into the equation. To determine the amount of rotation requires special techniques that we have not time to discuss. We will, however, make reference to the special hyperbola with equation
$$x_1 x_2 = c$$
which has the coordinate axes as asymptotes (Fig. A.9d).

G Differential Equations

We shall restrict attention to equations of the form
$$dx/dt = F(x, t) \tag{A.9}$$
This is an example of a first order differential equation, since the first derivative is the highest that occurs in the equation. A solution is any function $x(t)$ which satisfies the equation. For example,
$$dx/dt = x - \exp(-2t) \tag{A.10}$$
with $F = x - \exp(-2t)$, is of type (A.9), and has a solution
$$x = \exp t + \tfrac{1}{3}\exp(-2t)$$
as can be verified by substitution. In general, there is an infinity of solutions to an equation of the form (A.9). For example
$$x = A \exp t + \tfrac{1}{3}\exp(-2t)$$
satisfies (A.9) whatever the value of A.

It is not always the case that we can find analytic expressions for the solution to a differential equation, so we often have to resort to numerical methods. In this book, however, we have for convenience restricted attention to three types of first order differential equation for which analytic solutions can usually be obtained.

(i) When $F(x, t)$ does not depend on x, i.e. $F(x, t) = f(t)$, equation (A.9) becomes
$$dx/dt = f(t)$$
and the solutions can be found directly by integration:
$$x = \int \frac{dx}{dt} dt = \int f(t) dt + A$$
where A is an arbitrary constant called the constant of integration. For example the

solutions of the equation
$$dx/dt = t - 1$$
are given by
$$x = \int (t-1)dt + A = t^2/2 - t + A.$$

(ii) When $F(x, t) = f(t)/g(x)$ then the equation is said to be *separable*. Its solutions can be obtained by multiplying through by $g(x)$ and then integrating.
$$G(x) \equiv \int g(x)dx = \int g(x)\frac{dx}{dt}dt = \int f(t)dt + A.$$
For example, the solutions to the equation
$$dx/dt = (t-1)/x$$
where $f(t) = t - 1$ and $g(x) = x$, are given by
$$\int x \, dx = \int (t-1)dt + A,$$
i.e. $\quad x^2/2 = t^2/2 - t + A,$
or, more simply,
$$x = \pm (t^2 - 2t + 2A)^{1/2}.$$

(iii) Equation (A.9) is said to be *linear* if it has the form
$$dx/dt = g(t)x + f(t).$$
Equation (A.10), for example, is linear with $g(t) = 1$ and $f(t) = \exp(-2t)$. Linear equations can be solved by multiplying through by a function called an *integrating factor*:
$$\mu = \exp\left(-\int g(t)dt\right)$$
to obtain, on rearrangement,
$$\mu dx/dt - \mu g(t)x = \mu f(t) \tag{A.11}$$
This particular choice for μ allows the left hand side of this equation to be expressed as the derivative of μx, since $d\mu/dt = -\mu g(t)$. From this observation we can solve (A.11) by integration
$$\mu x = \int \frac{d(\mu x)}{dt} dt = \int \mu f(t) \, dt + A.$$
As an example, consider again equation (A.10) for which
$$\mu = \exp\left(-\int 1 dt\right) = \exp(-t).$$
On multiplying the equation by μ we obtain
$$\exp(-t)dx/dt - \exp(-t)x = -\exp(-t)\exp(-2t)$$
i.e. $\quad \dfrac{d}{dt}(x\exp(-t)) = -\exp(-3t)$

which on integration yields

$$x \exp(-t) = -\int \exp(-3t)\,dt + A = \exp(-3t)/3 + A$$

or, equivalently

$$x = A\exp t + \exp(-2t)/3$$

as we previously found.

Some differential equations are both separable and linear and hence there is a choice of solution technique. For example, the equation

$$dx/dt = \alpha x \qquad (\alpha \text{ a constant})$$

can be solved by dividing through by x and integrating:

$$\frac{1}{x}\frac{dx}{dt} = \alpha$$

yielding

$$\ln|x| = \alpha t + A, \qquad x \neq 0 \tag{A.12}$$

The modulus must be included in the logarithm if the formula is to hold for x negative. Taking exponentials in (A.12), we obtain

$$|x| = A_1 \exp(\alpha t) \qquad \text{with } A_1 = \exp A$$

or, more simply

$$x = A_2 \exp(\alpha t) \qquad \text{with } A_2 = \pm A_1 \tag{A.13}$$

Considered as a linear equation, we multiply by the integrating factor $\exp(-\int \alpha\,dt) = \exp(-\alpha t)$, to obtain

$$\exp(-\alpha t)dx/dt - \alpha x \exp(-\alpha t) = 0.$$

Hence

$$d(x\exp(-\alpha t))/dt = 0,$$

and, on integrating,

$$x\exp(-\alpha t) = A.$$

Multiplying by $\exp(\alpha t)$ we have

$$x = A\exp(\alpha t) \tag{A.14}$$

The two solutions (A.13), (A.14) agree except for the special solution $x = 0$ which is excluded in (A.12) because division by zero is not permitted. If the modulus were omitted from (A.12) then half the solutions would be omitted since A_1 is necessarily positive.

A more extensive discussion of these and other solution techniques and a justification that all solutions are obtained by the techniques can be found in many textbooks on Differential Equation Theory—for example the book by M. Braun, *Differential Equations and Their Applications*, Springer-Verlag (1975).

One final note. The labelling of the variables in the equations will depend on the context, and hence in using these techniques a simple relabelling will often be necessary.

H Hyperbolic Functions

From the exponential function we can construct two new useful functions

$$\cosh x = (\exp x + \exp(-x))/2; \quad \sinh x = (\exp x - \exp(-x))/2 \tag{A.15}$$

with the graphs shown in Fig. A.10a, b. (N.B. sinh is usually pronounced 'shine'). From these definitions it is clear that the cosh function is even

$$\cosh(-x) = \cosh x \quad \text{with } \cosh 0 = 1 \qquad (A.16)$$

and the sinh function odd

$$\sinh(-x) = -\sinh x \quad \text{with } \sinh 0 = 0 \qquad (A.17)$$

The properties of these functions are all derived from those of the exponential function. For example, since $d(\exp(ax))/dx = a\exp(ax)$ we deduce that

$$d(\cosh x)/dx = \sinh x \quad \text{and} \quad d(\sinh x)/dx = \cosh x \qquad (A.18)$$

From the exponent law

$$\exp(a)\exp(b) = \exp(a+b)$$

it can be shown with a little algebra that the sinh and cosh functions satisfy the following addition rules

$$\cosh(a+b) = \cosh a \cosh b + \sinh a \sinh b \qquad (A.19)$$

$$\sinh(a+b) = \sinh a \cosh b + \cosh a \sinh b \qquad (A.20)$$

From these rules we can derive three important special cases:

(i) with $x = a = -b$ in (A.19) and using (A.16) and (A.17) we obtain the relation

$$\cosh^2 x - \sinh^2 x = 1 \qquad (A.21)$$

(ii) with $x = a = b$ in (A.19) we have

$$\cosh 2x = \cosh^2 x + \sinh^2 x$$
$$= 2\cosh^2 x - 1 \quad \text{using (A.21)}$$
$$= 1 + 2\sinh^2 x \quad \text{using (A.21)}$$

(iii) with $x = a = b$ in (A.20) we have

$$\sinh 2x = 2\sinh x \cosh x.$$

The properties of the sinh and cosh functions appear to be very similar to those of the sine and cosine functions. In fact the former can be obtained from the latter by changing the sign whenever the square of a sine is encountered. We can relate these two pairs of functions in terms of the geometry of the plane. A point on the circle $x_1^2 + x_2^2 = R^2$ can be parameterised in terms of an angle θ, where $x_1 = R\cos\theta$, $x_2 = R\sin\theta$, since $\cos^2\theta + \sin^2\theta = 1$. A point on the hyperbola $x_1^2 - x_2^2 = R^2$ on the other hand can be parameterised in terms of the 'angle' u, where $x_1 = R\cosh u$, $x_2 = R\sinh u$, since $\cosh^2 u - \sinh^2 u = 1$. It is for this reason that the cosh and sinh functions are referred to as *hyperbolic functions*. As with the trigonometric functions we can extend the family of hyperbolic functions by defining

$$\tanh x = \sinh x/\cosh x; \quad \coth x = 1/\tanh x$$

$$\text{sech } x = 1/\cosh x; \quad \text{cosech } x = 1/\sinh x.$$

The reason that hyperbolic functions are of importance to us is that in general phase space trajectories are hyperbolic in shape when the objective functional is quadratic and the dynamic equation linear in the state and control variables. The hyperbolic nature of the trajectories shows up in the solution of the Hamilton equations where we have to evaluate integrals using hyperbolic substitution. To evaluate the integral

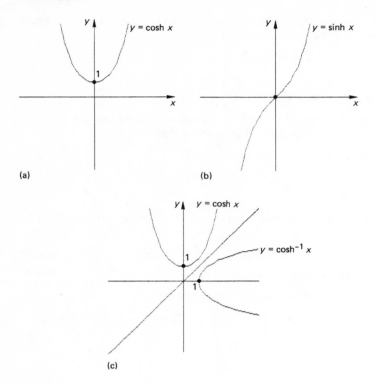

Fig. A.10

$I = \int (x^2 - a^2)^{-1/2} \, dx$, for example, we substitute $x = a \cosh u$ to obtain

$$I = \int (x^2 - a^2)^{-1/2} \, dx = \int (a \sinh u)^{-1} a \sinh u \, du = \int du = u$$

using (A.18) and (A.21). Since $u = \cosh^{-1} x/a$, where $\cosh^{-1}$ denotes an inverse function for the cosh function (Fig. A.10c) we conclude that $I = \cosh^{-1} x/a$ or $x = a \cosh I$.

The substitution $x = a \cosh u$ is also useful for the integral

$$\int (x^2 - a^2)^{+1/2} \, dx,$$

and $x = a \sinh u$ for the integrals

$$\int (x^2 + a^2)^{\pm 1/2} \, dx.$$

I Some Classical Problems

In Chapters 6 and 7 we discussed several of the 'classical' optimisation problems solved by mathematicians of the 17th and 18th centuries. These included the Geodesic, Soap-film, Brachistochrone and Isoperimetric Problems. In this section of the Appendix we show how these problems are derived from their geometric or mechanical origins.

I Some classical problems

(a) The Geodesic (or shortest distance) Problem

Consider the curve $y(x)$ joining two points A, B in the plane of the variables x, y (Fig. A.11a). A short section of this curve approximated by a line segment has length

$$ds = \sqrt{dx^2 + dy^2} = \sqrt{1 + (dy/dx)^2}\, dx.$$

Its total length is obtained by summing these 'elementary' lengths and taking the limit as $dx \to 0$. The total length is therefore the integral

$$\int_a^b (1 + (dy/dx)^2)^{1/2}\, dx \qquad (A.22)$$

The problem of finding the shortest curve between A and B is therefore:

minimise $\quad \int_a^b (1 + u^2)^{1/2}\, dx$

subject to $\quad dy/dx = u; \quad y(a), y(b)$ given.

With the relabelling $x \to t$, $y \to x$ and with $a = 0$, $b = 1$, this is the problem solved in Exercise Set 2 of Chapter 6, where we found the required curve to be a straight line. Expression (A.22) also occurs in a simplified version of the Isoperimetric Problem where we find the maximum area between the curve $y(x)$, with $y(a) = y(b) = 0$, and the x axis

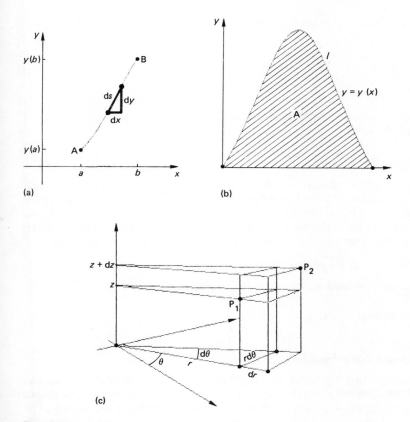

Fig. A.11

subject to the curve being of fixed length l (Fig. A.11b). The area is $\int_a^b y \, dx$ (supposing y always to be non-negative) and the length is just (A.22). Hence the Isoperimetric Problem has the form

$$\text{maximise} \quad \int_a^b y \, dx$$

$$\text{subject to} \quad \int_a^b (1+u^2)^{1/2} \, dx = l, \quad dy/dx = u.$$

A less trivial 'geodesic' problem is that of finding the shortest path between two points on a surface in three dimensions. For simplicity let us suppose that the surface is symmetrical about the vertical axis, for example a sphere or a cone. We use the so-called cylindrical polar coordinates (r, θ, z) (Fig. A.11c) to describe a point in this three-dimensional space. Suppose neighbouring points P_1, P_2 with coordinates (r, θ, z) and $(r + dr, \theta + d\theta, z + dz)$ lie on the path between two end-points A and B. The distance between P_1 and P_2 approximating the length of curve between P_1 and P_2 is given by

$$ds = \sqrt{r^2 \, d\theta^2 + dr^2 + dz^2} = \sqrt{r^2 + (dr/d\theta)^2 + (dz/d\theta)^2} \, d\theta.$$

The total length of the curve is therefore

$$\int \sqrt{r^2 + (dr/d\theta)^2 + (dz/d\theta)^2} \, d\theta.$$

If the surface has equation $z = f(r)$, then $dz/d\theta = (df/dr)(dr/d\theta)$ and so the shortest path solves the problem

$$\text{minimise} \quad \int \sqrt{r^2 + u^2(1 + h^2(r))} \, d\theta$$

$$\text{subject to} \quad dr/d\theta = u$$

where $h = df/dr$. With an obvious change of notation this is the problem solved in Question 4c of Problem Set 6.

(b) The Soap-film Problem

Imagine a soap film stretched between two identical coaxial circular rings distance b apart (Fig. A.12a). The shape formed by the film will (ignoring gravity) be symmetric about the axis and, according to the laws of physics, will be the surface of minimum area with this symmetry. Let r denote the radius of the film a distance x along the axis then a slice of the film (Fig. A.12b) between cross-sections x and $x + dx$ will have thickness ds, where

$$ds = \sqrt{dx^2 + dr^2} = \sqrt{1 + (dr/dx)^2} \, dx$$

and circumference $2\pi r$. The total surface area is therefore given by:

$$2\pi \int_0^b r(1 + (dr/dx)^2)^{1/2} \, dx.$$

(c) The Brachistochrone Problem

Imagine a bead moving along a smooth wire in a vertical plane. This wire joins the two points A and B and has equation $y = y(x)$ (Fig. A.13). The time dt taken to move a distance ds on the wire is given by $dt = ds/v$, where v is the velocity, which can be shown to be proportional to $\sqrt{y}$ when the bead starts from rest at A, the origin of the coordinate

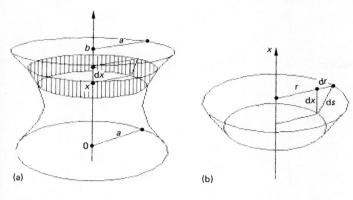

Fig. A.12

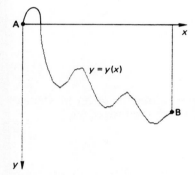

Fig. A.13

system. Hence the total time taken to traverse the wire from A to B is given by:

$$J = \int ds/v = \int_0^b \frac{\sqrt{1+(dy/dx)^2}}{c\sqrt{y}} \, dx$$

where c is the constant of proportionality. The Brachistochrone Problem is the problem of finding the curve $y(x)$ which minimises J and was solved by Hamilton's method in Question 4b of Problem Set 6.

J The Chain Rule and The Variational Principle

Two functions $x_1(t)$ and $x_2(t)$ of a variable t define a curve in the x_1, x_2 plane, each point on the curve being specified by the coordinates $(x_1(t), x_2(t))$ for some value of t. Such a curve is said to be defined *parametrically*, with t as the parameter. As we saw in Section H, the unit circle $x_1^2 + x_2^2 = 1$ can be defined in this way with $x_1 = \cos t, x_2 = \sin t$, where t is the angle made with the x_1 axis (Fig. A.14a).

Suppose we are moving on a surface $z = F(x_1, x_2)$ along a parametrically defined curve $(x_1(t), x_2(t))$, then for a given value of t we are at altitude $z = F(x_1(t), x_2(t))$ (Fig. A.14b). What is of interest to us is how z changes with t, and in particular the value of the derivative dz/dt. To determine this we return to the result obtained in Question 6 of Problem Set 1,

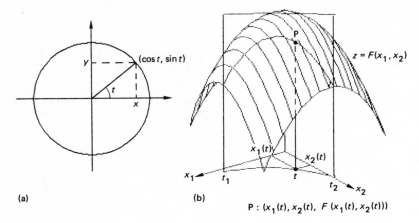

Fig. A.14

where we showed that

$$\frac{\partial F}{\partial x_1} dx_1 + \frac{\partial F}{\partial x_2} dx_2 \tag{A.23}$$

is an approximation to the change in altitude dz if we move along the path with changes dx_1, dx_2 in x_1 and x_2. If the corresponding change in t is dt then we obtain

$$\frac{dz}{dt} = \frac{\partial F}{\partial x_1} \frac{dx_1}{dt} + \frac{\partial F}{\partial x_2} \frac{dx_2}{dt}$$

by dividing this approximation by dt and taking the limit as $dt \to 0$. This result generalises straightforwardly to an n-dimensional space with

$$\frac{dz}{dt} = \sum_{i=1}^{n} \frac{\partial F}{\partial x_i} \frac{dx_i}{dt} \tag{A.24}$$

In some cases of interest the functions x_i depend on more than one variable, i.e. $x_i = x_i(t_1, t_2, \ldots, t_m)$, and in this situation (A.24) generalises further to:

$$\frac{\partial z}{\partial t_j} = \sum_{i=1}^{n} \frac{\partial F}{\partial x_i} \frac{\partial x_i}{\partial t_j} \qquad j = 1, \ldots, m \tag{A.25}$$

This relation is called the *Chain Rule* and can be used to establish several important results referred to in the text.

(a) Constant Hamiltonian

Let us take F to be the Hamiltonian H and $x_1 = x$, $x_2 = u$, $x_3 = \lambda$, $x_4 = t$, then (A.24) gives:

$$\frac{dz}{dt} = \frac{\partial H}{\partial x} \frac{dx}{dt} + \frac{\partial H}{\partial \lambda} \frac{d\lambda}{dt} + \frac{\partial H}{\partial u} \frac{du}{dt} + \frac{\partial H}{\partial t} \frac{dt}{dt}.$$

With x, u and λ optimal we must have

$$\frac{\partial H}{\partial u} = 0, \quad \frac{\partial H}{\partial x} = -\frac{d\lambda}{dt}, \quad \frac{\partial H}{\partial \lambda} = \frac{dx}{dt}$$

and hence, on substitution, it is clear that

$$\frac{dz}{dt} = \frac{\partial H}{\partial t}.$$

If H is autonomous then $\partial H/\partial t = 0$ and hence z, the value of the Hamiltonian, is constant.

(b) Multiplier Sensitivity

As a second application we establish the sensitivity result

$$\frac{\partial f^*}{\partial b_k} = \lambda_k^*$$

for the equality constrained optimisation problem:

maximise $f(x)$

subject to $g_j(x) = b_j$ for $j = 1, \ldots, m$

where λ_k^* denotes the optimal value of the kth multiplier and f^* the optimal value of the objective function. We take F in (A.25) to be the optimal value of the Lagrangean:

$$L^* = f(x^*) - \sum_j \lambda_j^* (g_j(x^*) - b_j)$$

which, like the optimal values x_i^*, λ_j^*, depends on the constraint parameters b_k. To apply (A.25) we identify t_k as b_k and the x variables as the set x_i^*, λ_j^* and b_k. Precisely, we have

$$\frac{\partial z}{\partial b_k} = \sum_i \frac{\partial L^*}{\partial x_i^*} \frac{\partial x_i^*}{\partial b_k} + \sum_j \frac{\partial L^*}{\partial \lambda_j^*} \frac{\partial \lambda_j^*}{\partial b_k} + \sum_l \frac{\partial L^*}{\partial b_l} \frac{\partial b_l}{\partial b_k}$$

$$= 0 - \sum_j (g_j - b_j) \frac{\partial \lambda_j^*}{\partial b_k} + \lambda_k^* \quad \text{since } \frac{\partial L^*}{\partial x_i^*} = 0$$

$$= \lambda_k^* \quad \text{since } g_j = b_j.$$

But always at the optimum $z = L^* = f^*$ and hence

$$\frac{\partial f^*}{\partial b_k} = \lambda_k^* \tag{A.26}$$

as required. This result also holds in the presence of inequality constraints. The argument leading to (A.26) still applies if the kth constraint is binding and holds trivially with $\lambda_k^* = 0$ if not.

(c) The Variational Method

The Chain Rule (A.25) can also be used to establish the Hamiltonian system of equations for the basic continuous optimal control problem. In analogy with the discrete case the Lagrangean for this problem is given by

$$L = \int_0^T (f(x,u,t) - \lambda(\dot{x} - g(x,u,t))) \, dt$$

$$= \int_0^T (H(x,u,\lambda,t) + \dot{\lambda}x) \, dt - (\lambda_T x_T - \lambda_0 x_0) \tag{A.27}$$

on integrating the second term by parts. Let us denote optimal state and control functions for a given multiplier function $\lambda(t)$ by $x^*(t)$ and $u^*(t)$ and generate from them a family of state and control functions by:

$$x(t) = x^*(t) + \theta_1 y(t) \tag{A.28}$$
$$u(t) = u^*(t) + \theta_2 v(t)$$

where θ_1 and θ_2 are parameters and y and v continuously differentiable functions with $y(0) = y(T) = 0$. These functions define variations about the optimal solutions ($\theta_1 = \theta_2 = 0$) (Fig. A.15a,b) and allow us to test the optimality of these solutions. In fact for θ_1, θ_2 not both zero the Lagrangean must not be greater than at $\theta_1 = \theta_2 = 0$ for a maximum problem. Necessary conditions for this are that

$$\frac{\partial L}{\partial \theta_1} = \frac{\partial L}{\partial \theta_2} = 0 \quad \text{at } \theta_1 = \theta_2 = 0 \tag{A.29}$$

To derive the Hamiltonian system of equations from these conditions we take the partial differential operator through the integral sign in (A.27) and then apply the Chain Rule.

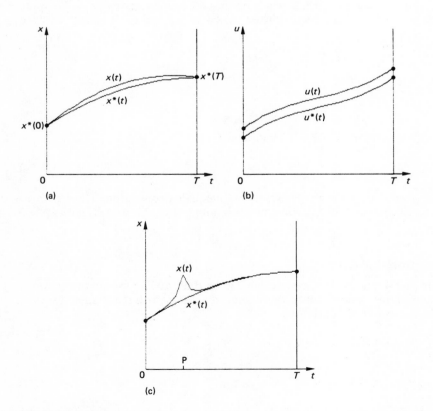

Fig. A.15

Precisely, we have

$$\frac{\partial L}{\partial \theta_k} = \int_0^T \left(\frac{\partial H}{\partial x} \frac{\partial x}{\partial \theta_k} + \frac{\partial H}{\partial u} \frac{\partial u}{\partial \theta_k} + \lambda \frac{\partial \dot{x}}{\partial \theta_k} \right) dt = 0 \qquad \text{for } k = 1, 2, \tag{A.30}$$

since λ and t do not depend on θ and x_T and x_0 are fixed. With $k = 1$ the integral reduces to:

$$\frac{\partial L}{\partial \theta_1} = \int_0^T \left(\frac{\partial H}{\partial x} + \dot{\lambda} \right) y(t) \, dt = 0 \tag{A.31}$$

Since $y(t)$ can be chosen essentially as we please, let us take it to be zero except close to an arbitrary point P in the interval $0 \leqslant t \leqslant T$ (Fig. A.15c). From (A.31) in this case it must be that the continuous function $\partial H/\partial x + \dot{\lambda}$ is zero at P, i.e.

$$\frac{\partial H}{\partial x} + \dot{\lambda} = 0 \tag{A.32}$$

otherwise we could always choose the interval in which y is non-zero small enough that (A.31) is contradicted. (See, for example, J. C. Clegg, *Calculus of Variations*, Oliver and Boyd (1968) for a more detailed argument.)

With $k = 2$ in (A.30) we have

$$\frac{\partial L}{\partial \theta_k} = \int_0^T \frac{\partial H}{\partial u} v(t) \, dt = 0$$

and hence we conclude

$$\frac{\partial H}{\partial u} = 0 \tag{A.33}$$

The multiplier function $\lambda(t)$ on the interval $0 \leqslant t \leqslant T$ is determined by applying the dynamic condition $\dot{x} = \partial H/\partial \lambda$ which, with (A.32) and (A.33), defines the Hamiltonian system of equations for a basic continuous optimal control problem. These equations give necessary conditions for a local maximum. To derive sufficient conditions for a local maximum we would proceed by examining the convexity properties of the Lagrangean $L(\theta_1, \theta_2)$ at the stationary point identified by these equations (in the manner for example of Question 5 of Problem Set 4). Sufficient conditions, however, are difficult to obtain because of the degeneracies that can arise. One set of sufficient conditions imposes a negative second derivative $\partial^2 H/\partial u^2$ (to ensure a local maximum of H with respect to u) and what are called Normality and Jacobi conditions whose function is to eliminate degeneracies. Details can be found in A. E. Bryson and Y. C. Ho, *Applied Optimal Control*, Wiley (1975). For the Calculus of Variations Problem with $\dot{x} = u$ the Euler-Lagrange equations can be derived from the Hamiltonian equations by eliminating the multiplier function λ or directly from the Lagrangean $L = \int_0^T f(x, \dot{x}, t) \, dt$ on elimination of the control variable u. In fact

$$\frac{\partial L}{\partial \theta} = \int_0^T \left(\frac{\partial f}{\partial x} \frac{\partial x}{\partial \theta} + \frac{\partial f}{\partial \dot{x}} \frac{\partial \dot{x}}{\partial \theta} \right) dt$$

$$= \int_0^T \left(\frac{\partial f}{\partial x} - \frac{d}{dt}\left(\frac{\partial f}{\partial \dot{x}} \right) \right) y(t) \, dt + \frac{\partial x}{\partial \theta} \cdot \frac{\partial f}{\partial \dot{x}} \bigg|_0^T$$

on integrating the second term by parts. The integrated term is zero because x_0 and x_T are fixed and the integrand is zero since $y(t)$ is arbitrary.

For a full discussion of the Calculus of Variations see I. M. Gelfand and S. V. Fomin, *Calculus of Variations*, Prentice-Hall (1963). These authors in particular take up two important issues that we have not had time to discuss; the difference between strong and weak variations and the degeneracies that can arise from the existence of conjugate points.

(d) Terminal Conditions

The terminal condition

$$\lambda_T^* = \frac{\partial \phi}{\partial x_T} \tag{A.34}$$

can be obtained from the optimal Lagrangean for the problem with asset function ϕ and freed end-point (T, x_T):

$$L^*(x_T, T) = \int_0^T (H^* + \dot{\lambda}^* x^*) \, dt + \phi(x_T, T) - (\lambda_T^* x_T - \lambda_0^* x_0)$$

by differentiating with respect to x_T. With x^*, λ^* functions of the parameter x_T we have

$$\frac{\partial L^*}{\partial x_T} = \int_0^T \left(\frac{\partial H^*}{\partial x_T} + \dot{\lambda}^* \frac{\partial x^*}{\partial x_T}\right) dt + \left(\frac{\partial \phi}{\partial x_T} - \lambda_T^*\right) + \int_0^T \frac{\partial \dot{\lambda}^*}{\partial x_T} x^* \, dt - \frac{\partial \lambda_T^*}{\partial x_T} x_T + \frac{\partial \lambda_0^*}{\partial x_T} x_0$$

$$= \int_0^T \left(\frac{\partial H^*}{\partial x_T} + \dot{\lambda}^* \frac{\partial x^*}{\partial x_T} - \dot{x}^* \frac{\partial \lambda^*}{\partial x_T}\right) dt + \left(\frac{\partial \phi}{\partial x_T} - \lambda_T^*\right)$$

on integrating by parts. Using the Chain Rule to evaluate the Hamiltonian derivative we obtain:

$$\frac{\partial H^*}{\partial x_T} = \frac{\partial H^*}{\partial x^*} \frac{\partial x^*}{\partial x_T} + \frac{\partial H^*}{\partial u^*} \frac{\partial u^*}{\partial x_T} + \frac{\partial H^*}{\partial \lambda^*} \frac{\partial \lambda^*}{\partial x_T}.$$

Substitution in the integral then yields a zero integrand and the required result. The condition

$$H_T + \frac{\partial \phi}{\partial T} = 0$$

can be obtained in a similar fashion by differentiating L^* with respect to T. The H_T term comes from differentiating the upper limit on the integral.

(e) Corner Conditions

An optimal path can possess corners at those points where an inequality bound on the state variable commences or ceases to be binding. The conditions to be satisfied at such points can be established by a simple extension of the arguments that led to the terminal conditions (7.8), (7.9). We optimise the expression obtained by adding to the Lagrangean the exit and entry constraints, i.e.

$$L' = \int_0^{t_1} (H - \lambda \dot{x}) \, dt + \int_{t_1}^{t_2} (H - \lambda \dot{x}) \, dt + \int_{t_1}^T (H - \lambda \dot{x}) \, dt$$
$$+ v_1 S(x(t_1), t_1) + v_2 S(x(t_2), t_2) + \text{final asset term}$$

where $S \leq 0$ is the state inequality constraint, t_1, t_2 the entry and exit times and v_1, v_2 the associated multipliers. Optimisation of L' with respect to $t_1, t_2, x_1 = x(t_1)$ and $x_2 = x(t_2)$ leads to the conditions:

$$\Delta_i H = v_i \frac{\partial S}{\partial t_i}; \quad H = f + \lambda g + \mu S \text{ with } \mu = 0 \text{ if } S < 0$$

$$\Delta_i \lambda = -v_i \frac{\partial S}{\partial x_i} \quad i = 1, 2 \tag{A.35}$$

where Δ_i indicates the change at time t_i. From these *jump conditions* we can regain the terminal conditions by taking $t_1 = T$ and supposing that H and λ are both zero for $t > T$.

If there are several state variables then the inclusion of the state constraint $S(x, t) = 0$ on L' may not be sufficient to guarantee that the corners are approached correctly. For example consider the two state variable problem with dynamic equations:

$$\dot{x}_1 = x_2; \quad \dot{x}_2 = u \tag{A.36}$$

An optimal path will not stay on the boundary $S = x_i = 0$ in this case unless $x_2 = 0$ also. (If x_1 denotes the position of a particle moving under Newton's laws of Motion then the velocity, x_2, must also be zero for the particle to be stationary at $x_1 = 0$.) This extra state variable condition must also be added to the Lagrangean, with its own multiplier.

In general a sufficient number of constraints are obtained by repeatedly differentiating the constraint function $S(x, t)$ with respect to t using the Chain Rule and replacing each time derivative of the state variables by the right hand side of the appropriate dynamic equation until a function that explicitly contains the control variable is obtained. The corresponding constraint can then be satisfied by a suitable choice for that control. For our example (A.36) we have:

$$S(x_1, x_2, t) = x_1$$

$$\frac{dS}{dt} = \frac{\partial S}{\partial t} + \frac{\partial S}{\partial x_1}\dot{x}_1 + \frac{\partial S}{\partial x_2}\dot{x}_2 = 0 + x_2 + 0 \equiv S^{(1)}(x, t)$$

$$\frac{dS^{(1)}}{dt} = \frac{\partial S^{(1)}}{\partial t} + \frac{\partial S^{(1)}}{\partial x_1}\dot{x}_1 + \frac{\partial S^{(1)}}{\partial x_2}\dot{x}_2 = 0 + 0 + u \equiv S^{(2)}(x, t, u).$$

If the control variable emerges in the pth differentiation then the previous $p - 1$ derivatives of S are to be added to the Lagrangean. The jump conditions then become:

$$\Delta_i H = \sum_{j=0}^{p-1} v_{ij} \frac{\partial S^{(j)}}{\partial t_i}$$

$$\Delta_i \lambda_k = -\sum_{j=0}^{p-1} v_{ij} \frac{\partial S^{(j)}}{\partial x_{ki}}, \quad x_{ki} \equiv x_k(t_i) \tag{A.37}$$

In the case of a control variable constraint the parameter p equals zero since the control variable already appears in the constraint and hence the jump conditions (A.35), (A.37) reduce to continuity conditions on H and λ, i.e.

$$\Delta H = 0, \quad \Delta \lambda = 0 \tag{A.38}$$

Together with continuity in the derivative of H with respect to u one can obtain from (A.38) the Weierstrass-Erdmann conditions for a problem in the Calculus of Variations:

$$\text{maximise} \int_0^T f(x, \dot{x}, t) \, dt.$$

These conditions state that at a corner the functions:

$$\frac{\partial f}{\partial \dot{x}_k} \quad \text{and} \quad f - \Sigma \, \dot{x}_k \frac{\partial f}{\partial \dot{x}_k} \quad (k = 1, \ldots, n)$$

must be continuous.

Solutions

Note that because of space constraints we have been forced to omit some of the simpler graphs and contour maps in the following solution sets.

Chapter 1

Exercise Set 1
(i)   Minimum $(15/8)$ at $x = 1/2$, maximum at $x = \pm\infty$ (see section 1.3).
(iii) Minimum $(\frac{1}{2}(\ln 2 + 1))$ at $x = \frac{1}{2}\ln 2$, maximum at $-\infty$.
(v) Minimum (0) at $x = 0$, maxima $(1/e)$ at $x = \pm\sqrt{e-1}$.

Exercise Set 2
1(i) $x_1^2 + 5(x_2 - 2/5)^2 + 1/5$: minimum $(1/5)$ at $(0, 2/5)$, maximum (∞) at ∞.
(iii) Maximum (1) at $(0, 0)$, minimum (0) at ∞.
(v) Maximum (0) at $(\pm 1, 0)$, minimum $(-\infty)$ at ∞. (The function takes the value -1 at the other stationary point $(0, 0)$.)

2 (i) Stationary points at $(0, 1)$, $(0, -1)$.
As $x_1 \to \infty$ $f \to \infty$ (max) if $|x_2| < 1$
                $f \to -\infty$ (min) if $|x_2| > 1$
(iii) Stationary points at $(0, 0)$, $(0, \pm 1)$.
As $x_1 \to \infty$ $f \to -\infty$ (min) for finite x_2;
as $x_2 \to \infty$ $f \to \infty$ (max) for finite x_1.

Exercise Set 3
Contour equations
(i) $x_2 = 2x_1 + 3 - c$ (parallel lines with gradient 2) (Fig. S1.1).
(iii) $x_1 = c - 1 - x_2^2$ (parabolae) (Fig. S1.2).

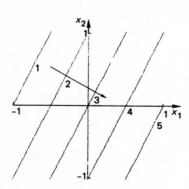

Fig. S1.1

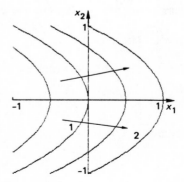

Fig. S1.2

(v) $(x_1 + 1)^2 + x_2^2 = c - 2$ (circles, centred at $(-1, 0)$) (Fig. S1.3).
(vii) $x_1 x_2 = \ln c$ (hyperbolae, centred at $(0,0)$) (Fig. S1.4).
(ix) $(x_1 + 1)^2 - x_2^2 = c - 1$ (hyperbolae centred at $(-1, 0)$) (Fig. S1.5).

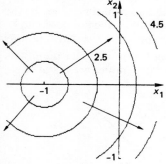

Fig. S1.3

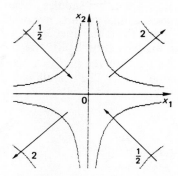

Fig. S1.4

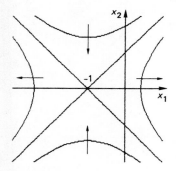

Fig. S1.5

Exercise Set 4
(i) This problem was analysed in Exercise Set 2, question 2(i). The stationary points at $(0, \pm 1)$ are found to be saddles by determining the signs of f in the regions defined by the zero contour $x_1 = 0$, $x_2 = \pm 1$ through the stationary points (Fig. S1.6).

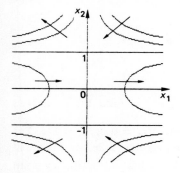

Fig. S1.6

(iii) There are stationary points when either $\cos x_1$ and $\cos x_2$ are both zero or $\sin x_1$ and $\sin x_2$ are both zero; i.e. at the points $(k\pi, l\pi), (\frac{1}{2}\pi + m\pi, \frac{1}{2}\pi + n\pi)$, where k, l, m, n are integers. At the first set $f = 0$ and at the second $f = (-1)^{m+n}$. Hence the maximum value of 1 is taken at the second set when $m+n$ is even and the minimum value of -1 when $m+n$ is odd. The first set of points are saddles as can be deduced by a sign analysis (Fig. S1.7).

(v) There are nine stationary points, five of them lying on the zero contour defined by the circle $x_1^2 + x_2^2 = 1$ and the radial lines $x_2 = \pm x_1$. These five are all saddle points as can be deduced by sign analysis (Fig. S1.8). The other four are trapped by the zero contour at $(0, \pm 1/\sqrt{2}), (\pm 1/\sqrt{2}, 0)$, the first two being local maxima (value 1/4) and the others local minima (value $-1/4$). The overall maximum and minimum values are taken at ∞.

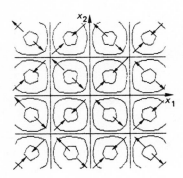

Fig. S1.7

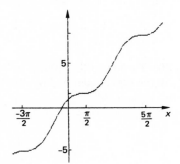

Fig. S1.8

Problem Set 1

1(a)(i) Minimum $(1/\sqrt{5})$ at $x = 2/5$. Maximum (∞) as $x \to \pm \infty$.

(iii) Inflections at the solutions of $\sin x = 1$: $x = \pi/2 + 2n\pi$, where n is any integer. Maximum (∞) as $x \to \infty$, minimum $(-\infty)$ as $x \to -\infty$ (Fig. S1.9).

Fig. S1.9

(v) Maximum (2.13) at $2 - \sqrt{3}$, local minimum (1.44) at $2 + \sqrt{3}$, minimum $(-\pi/2)$ as $x \to -\infty$.

(b)(i) $\partial f/\partial x_1 = 0$ gives $x_2 = 0$ or $x_1^2 = 1/2$: $\partial f/\partial x_2 = 0$ gives $x_1 = 0$ or $x_2^2 = 1/2$. The stationary points at $(1/\sqrt{2}, 1/\sqrt{2}), (-1/\sqrt{2}, -1/\sqrt{2})$ are maxima $(1/2e)$, and those at $(1/\sqrt{2}, -1/\sqrt{2}), (-1/\sqrt{2}, 1/\sqrt{2})$ are minima $(-1/2e)$. The additional stationary point $(0, 0)$ will later be shown to be a saddle.

(iii) $\partial f/\partial x_1 = 0$ gives $x_2 = 0$ or $3x_1^2 + x_2^2 = 1$: $\partial f/\partial x_2 = 0$ gives $x_1 = 0$ or $3x_2^2 + x_1^2 = 1$. The stationary points at $(1/2, -1/2)$, $(-1/2, 1/2)$ are maxima $(1/8)$, and those at $(1/2, 1/2)$, $(-1/2, -1/2)$ are minima $(-1/8)$. The other stationary points at $(0, 0)$, $(0, \pm 1)$, $(\pm 1, 0)$ are saddles as can be shown by sign analysis.

(v) $\partial f/\partial x_1 = 0$ gives $x_1 = 0$ or $x_1^2 + x_2^2 = 13$: $\partial f/\partial x_2 = 0$ gives $x_2 = 0$ or $x_1^2 + x_2^2 = 5$. The stationary point $(0, 0)$ is a local maximum with value 25, whereas the points $(\sqrt{13}, 0)$, $(-\sqrt{13}, 0)$ are minima with value -144. The overall maximum (∞) is at infinity. The remaining stationary points $(0, \pm\sqrt{5})$ can be shown to be saddles.

(c)(i) $x_1^2 + (x_2 - 2)^2 = c$; circles centred at $(0, 2)$.

(iii) $2(x_1 - 1)^2 - x_2 = c + 2$; parabolae with common axis $x_1 = 1$.

(v) $(x_1 + 5/2)^2 - 2x_2^2 = c + 29/4$; hyperbolae centred at $(-5/2, 0)$.

2(i) See Fig. S1.10.
(iii) See Fig. S1.11.
(v) See Fig. S1.12.

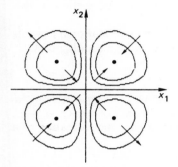

Fig. S1.10

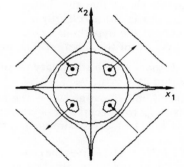

Fig. S1.11

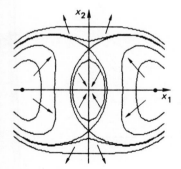

Fig. S1.12

3(i) The map is shown in Fig. S1.13. This example shows that there can be more than two 'valleys' and two 'mountains' meeting at a saddle.

(iii) See Fig. S1.14. This is a counter example to the belief that a function has a local minimum at a point if it has a local minimum along every straight line through that point.

4(i) $\partial f/\partial x_1 = 0$ and $\partial f/\partial x_2 = 0$ both give $x_1 + x_2 = 0$; i.e. all points on this line are stationary points. They are in fact a line of minima, with the maximum taken at ∞ to either side of this line.

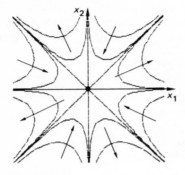
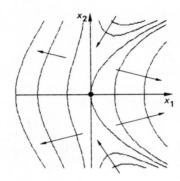

Fig. S1.13   Fig. S1.14

(iii) The zero contour lines $x_1 = 0$, $x_2 = 0$ are lines of minima with the maximum taken at ∞. This example shows that the stationary point defined by the intersection of two sections of a contour is not necessarily a saddle.
(v) The branches of the hyperbola $x_1^2 - x_2^2 = 1$ define two curves of stationary points defining the minimum of the function. The origin can be seen to be a saddle point.

5(a)(i) $a = -1$: there are no stationary points.
$a = 0$: an inflection at $x = 0$.
$a = 1$: a local maximum $(2a\sqrt{a})$ at $x = -\sqrt{a}$, and a local minimum $(-2a\sqrt{a})$ at $x = \sqrt{a}$. The overall maximum (∞) occurs as $x \to \infty$, and the overall minimum $(-\infty)$ as $x \to -\infty$.
(ii) See Fig. S1.15. The local maximum and minimum emerge after the formation of a point of inflection.
(b) See Fig. S1.16. Fig. S1.15 gives the cross-sections of this figure at $x_2 = -1, 0, 1$. This surface defines the cusp catastrophy in Catastrophy Theory (see *Catastrophy Theory*, E. C. Zeeman, Addison Wesley 1977). It has been used to model such diverse behaviour as OPEC pricing policy and stock market behaviour.

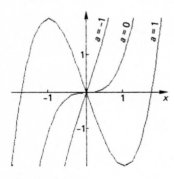
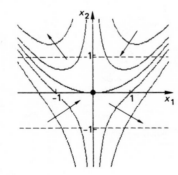

Fig. S1.15   Fig. S1.16

7(a) $(a_3 - a_3) = b_1(a_1 - a_1) + b_2(a_2 - a_2)$ for all b_1, b_2.
(b) Vertical cross-section parallel to the x_1 axis has $x_2 = a_2$. Therefore the tangent line must be: $z - a_3 = b_1(x_1 - a_1)$, which has gradient b_1. Hence $\partial f/\partial x_1 = b_1$. Similarly $b_2 = \partial f/\partial x_2$.
(c) Referring to Fig. 1.12(a) in main text: if h represents the distance along the $x_1 x_2$ plane in direction θ, then $x_1 - a_1 = h\cos\theta$, $x_2 - a_2 = h\sin\theta$. Substituting these into the equation

of the tangent plane gives: $\cos\theta\,\partial f/\partial x_1 + \sin\theta\,\partial f/\partial x_2 = (z-a_3)/h$. But this last term is what we mean by $D_\theta f$, hence the result.

Chapter 2

Exercise Set 1

(i) Maximum ($\sqrt{2}$) at $(1/\sqrt{2}, 1/\sqrt{2})$, minimum ($-\sqrt{2}$) at $(-1/\sqrt{2}, -1/\sqrt{2})$, the exact positions of the points being obtained by symmetry (Fig. S2.1).

(iii) There is one constrained stationary point which is a maximum. The point is in fact $(2, 0)$ as can be determined by the analysis of the next section (Fig. S2.2).

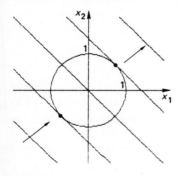

Fig. S2.1   **Fig. S2.2**

Exercise Set 2

(i) $L = x_1 + x_2 - \lambda(x_1^2 + x_2^2 - 1)$. $\partial L/\partial x_1 = 1 - 2\lambda x_1$; $\partial L/\partial x_2 = 1 - 2\lambda x_2$, giving $\lambda(x_1 - x_2) = 0$. One solution is $x_1 = x_2 = \pm 1/\sqrt{2}$, as can be seen from substitution into the constraint equation. The other solution ($\lambda = 0$) gives no solution as there are no points at infinity. Hence the maximum value is $\sqrt{2}$ at $(1/\sqrt{2}, 1/\sqrt{2})$, and the minimum $(-\sqrt{2})$ at $(-1/\sqrt{2}, -1/\sqrt{2})$ (Fig. S2.1). This checks with solution (i) of the previous Exercise Set.

(iii) $\partial L/\partial x_1 = -1/2 - \lambda = 0$; $\partial L/\partial x_2 = -x_2 - 1 - \lambda = 0$. So $\lambda = -1/2$, $x_2 = -1/2$, and $x_1 = 3/2$ using the constraint equation. The maximum $(-3/8)$ is taken at $(3/2, -1/2)$, and the minimum at infinity (Fig. S2.3).

(v) From $\partial L/\partial x_1 = 2x_1 - 2\lambda x_1 = 0$ we obtain $x_1 = 0$ or $\lambda = 1$. In the first case $x_2 = 2$ or 0 from the constraint. In the second case $x_2 = 1/2$ from $\partial L/\partial x_2 = -2x_2 - 2\lambda(x_2 - 1) = 0$ and $x_1 = \pm\sqrt{3/2}$ from the constraint. There are no points at infinity. From the values of

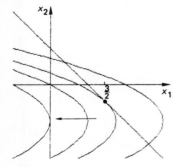

Fig. S2.3

the objective function we deduce that there is a local minimum (0) at (0, 0), an overall minimum (-4) at (0, 2), and an overall maximum (1/2) at ($\pm\sqrt{3/2}$, 1/2) (Fig. S2.4).

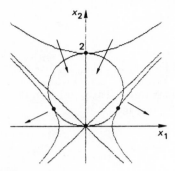

Fig. S2.4

Exercise Set 3

1(i) The Lagrange equations give $\lambda(x_1 + x_2) = \lambda(x_1 - x_2) = 0$. One solution gives $x_1 = -x_2 = x_3 = \pm 1/\sqrt{3}$ using the constraint equation. The other solution ($\lambda = 0$) does not correspond to a stationary point since there are no points at infinity (the constraint equation defines a sphere in three dimensional space). The maximum value ($\sqrt{3}$) is taken at $(1/\sqrt{3}, -1/\sqrt{3}, 1/\sqrt{3})$, and the minimum ($-\sqrt{3}$) at $(-1/\sqrt{3}, 1/\sqrt{3}, -1/\sqrt{3})$.

(ii) $\lambda_1 = 200$, $\lambda_2 = 2400$, $x_1 = 880$, $x_2 = 120$, $x_3 = 2600$, corresponding to a maximum. The problem is discussed in the next section, and the solution given following equation (2.27).

2(i) $\partial L/\partial x_i = -2a_i x_i - \lambda = 0$ ($i = 1 \ldots n$), giving $x_i = -\lambda/2a_i$. From the constraint, $\lambda = -2/(\sum_i 1/a_i)$, hence $x_i = a_i^{-1}(\sum_i 1/a_i)^{-1}$. The maximum value of f is $-(\sum_i 1/a_i)^{-1}$. The function takes the value $-\infty$ at all points at infinity.

Problem Set 2

1(a)(i) $\partial L/\partial x_1 = x_2 - 2\lambda x_1 = 0$; $\partial L/\partial x_2 = x_1 - 4\lambda x_2 = 0$, giving $\lambda = x_2/2x_1 = x_1/4x_2$, hence $2x_2^2 = x_1^2$ and so $x_1 = \pm 1/\sqrt{2}$ and $x_2 = \pm 1/2$ from the constraint equation ($x_1 = 0$, $x_2 = 0$ is inconsistent with the constraint). There are no points at infinity: the maximum ($1/2\sqrt{2}$) is taken at $(1/\sqrt{2}, 1/2)$, $(-1/\sqrt{2}, -1/2)$, and the minimum ($-1/2\sqrt{2}$) at $(1/\sqrt{2}, -1/2)$, $(-1/\sqrt{2}, 1/2)$ (Fig. S2.5).

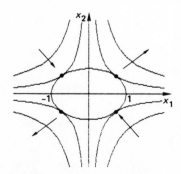

Fig. S2.5

(iii) $\partial L/\partial x_1 = 2x_1 + 2x_2 - 2\lambda x_1 = 0$: $\partial L/\partial x_2 = 2x_1 + 4x_2 - 2\lambda x_2 = 0$. $\lambda = (x_1 + x_2)/x_1$
$= (x_1 + 2x_2)/x_2$ ($x_1 = x_2 = 0$ does not satisfy the constraint). Equating gives a quadratic in x_1/x_2 whose roots are $-1/2 \pm \sqrt{5}/2$. Using the constraint gives the stationary points: the maximum (2.63) is at (0.53, 0.85), (−0.53, −0.85) and the minimum (0.38) at (−0.85, 0.53), (0.85, −0.53).

(b)(i) $L = -3x_1^2 - 2x_2^2 - x_3^2 - \lambda_1(x_1 + x_2 - 1) - \lambda_2(2x_2 + x_3 - 1)$. $\partial L/\partial x_1 = -6x_1 - \lambda_1 = 0$: $\partial L/\partial x_2 = -4x_2 - \lambda_1 - 2\lambda_2 = 0$: $\partial L/\partial x_3 = -2x_3 - \lambda_2 = 0$. These, together with the constraint equations, give: $\lambda_1 = -8/3$, $\lambda_2 = 2/9$. The minimum ($-\infty$) is at infinity, and the maximum ($-11/9$) at $(4/9, 5/9, -1/9)$.

(iii) $\partial L/\partial x_1 = 2x_1 + 2x_2 - 2\lambda x_1 = 0$: $\partial L/\partial x_2 = 2x_1 + 4x_2 - 2\lambda x_2 = 0$: $\partial L/\partial x_3 = 6x_3 - 2\lambda x_3 = 0$. $\lambda = 3$ gives no solution, $x_3 = 0$ gives $x_1 + x_2 = 1$, and the problem becomes that of part (iii) of (a).

3(a)(i) $\partial L/\partial x_1 = -x_1 + 2\lambda x_1 = 0$: $\partial L/\partial x_2 = -x_2 - 2\lambda x_2 = 0$. The minimum ($-\infty$) is at infinity, and maximum ($-1/2$) at $(\pm 1, 0)$ with $\lambda = 1/2$.

(ii),(iii) The circular contours change first to ellipses (a single maximum at (0, 0)), these then elongate further to parallel straight lines (a line of maxima along the x_1 axis) when $\lambda = \lambda^*$. These, in turn, transform into hyperbolae (with a saddle at the origin).

4(a) $\partial L/\partial x_1 = -2(x_1 - a) - \lambda(2x_1 - x_2^2)$: $\partial L/\partial x_2 = -2(x_2 - b) + 2\lambda x_1 x_2$.
$a = -1$, $b = 0$: there is no solution unless $\lambda = \infty$, which corresponds to the origin. The infinity occurs because the origin is also a stationary point of the constraint function (Fig. S2.6a).
$a = 0$, $b = 0$: the only solution is $(0, 0)$ (Fig. S2.6b). Here λ is undetermined due to the coincidence of the stationary points of f and g.

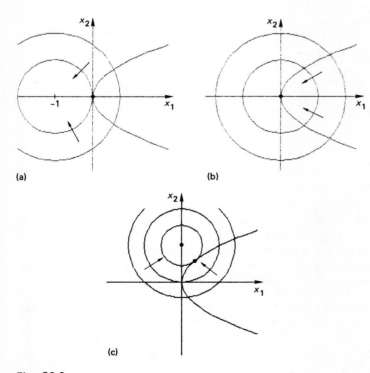

(a)   (b)

(c)

Fig. S2.6

$a = 0, b = 1$: gives (0, 1) as one stationary point, obviously the maximum, as it coincides with the maximum of the unconstrained function. $\lambda = -1$ results in $2x_2^3 - x_2 + 1 = 0$, a cubic which has only one real root near 0.68 (plot the function!), and so (0.46, 0.68) is a local maximum (Fig. S2.6c).

5(a) Compare 1a(iii). Both values have the same sign, so the contours must be ellipses with axes through the origin and the points (0.53, 0.85) and (0.85, -0.53) (Fig. S2.7).

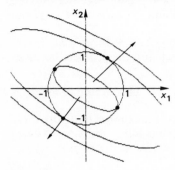

Fig. S2.7

(b)(i) The Lagrange equations give: $x_1(a-\lambda) + bx_2 = 0$ and $bx_1 + x_2(c-\lambda) = 0$. Hence we can write $x_1/x_2 = (\lambda - a)/b = b/(\lambda - c)$, and so $\lambda^2 - \lambda(a+c) + (ac - b^2) = 0$. Making the substitution $\tan\theta = x_2/x_1 = (\lambda - a)/b$ gives $\cos(2\theta - \phi) = 0$, where $\tan\phi = (c-a)/2b$. Hence $\theta = \phi/2 + \pi/4$ or $\theta = \phi/2 + 3\pi/4$. After some algebra we find $f = (a+c)/2 + \sin(2\theta - \phi)((c-a)^2/4 + b^2)^{1/2}$, which for each of the two choices for θ must be a root of the quadratic in λ, i.e. $f = \lambda_1$ or λ_2. If $ac > b^2$ then from this quadratic we see that $\lambda_1\lambda_2 > 0$, i.e. the values f_1, f_2 of f at the two stationary points defined by λ_1, λ_2 are of the same sign.

6(i) $\partial L/\partial x_1 = -2x_1 - \lambda = 0$: $\partial L/\partial x_2 = -2x_2 - \lambda = 0$.
$\lambda = -2x_1 = -2x_2$. $x_1 = x_2 = b/2$. $\lambda^* = -b$.
Hence $f^* = -b^2/2$, and so $df^*/db = -b = \lambda^*$.
(iii) $\partial L/\partial x_1 = 1 - 2\lambda x_1 = 0$: $\partial L/\partial x_2 = 1 - 2\lambda x_2 = 0$.
$\lambda = 1/2x_1 = 1/2x_2$. $x_1 = x_2 = \sqrt{b/2}$, $\lambda^* = \sqrt{(1/2b)}$
$f^* = \sqrt{2b}$, and hence $df^*/db = \sqrt{(1/2b)} = \lambda^*$.

Chapter 3

Exercise Set 1
(a) $df/dx = 2(1-x) = 0$ gives $x = 1$, $f(1) = 1$.
(i) $x = 1$ not in interval. $f(-1) = -3$ (min), $f(0) = 0$ (max).
(ii) $f(1) = 1$ (max), $f(0) = f(2) = 0$ (min).
(iii) $x = 1$ not in interval. $f(2) = 0$ (max), $f(\infty) = -\infty$ (min).
(c) $df/dx = 2(c-x) = 0$ gives $x = c, f(c) = c$.
(i) $x = c$ not in interval. $f(-1) = -4$ (min), $f(1) = 2$ (max).
(ii) $f(-1) = -3$ (min), $f(1) = 1$ (max).
(iii) $f(-1) = 0, f(-1/2) = 1/4$ (max), $f(1) = -2$ (min).
(iv) $x = c$ not in interval. $f(-1) = 2$ (max), $f(1) = -4$ (min).

Exercise Set 2
(i) $L = x_1 - 2x_2 - \lambda(x_1^2 + x_2^2 - 1)$.
$\partial L/\partial x_1 = 1 - 2\lambda x_1 = 0$; $\partial L/\partial x_2 = -2(1 + \lambda x_2) = 0$.

Interior ($\lambda = 0$): no solution.
Boundary: $x_2 = -2x_1 = -1/\lambda$, so $x_2 = -2x_1$ and since $x_1^2 + x_2^2 = 1$ on the boundary $x_1 = \pm 1/\sqrt{5}$. Hence maximum ($\sqrt{5}$) at ($1/\sqrt{5}, -2/\sqrt{5}$), minimum ($-\sqrt{5}$) at ($-1/\sqrt{5}, 2/\sqrt{5}$) (Fig. S3.1).
(iii) $\quad L = -(x_1 - c)^2 - x_2^2 - \lambda(-x_1)$.
$\partial L/\partial x_1 = -2(x_1 - c) + \lambda = 0; \partial L/\partial x_2 = -2x_2 = 0$.
Interior ($\lambda = 0$): $x_1 = c$, $x_2 = 0$, $f(c, 0) = 0$.
Boundary: $x_2 = 0$, $x_1 = 0$, $f(0, 0) = -c^2$.
$c = -1$: ($-1, 0$) not in feasible region, (0, 0) gives maximum, minimum ($-\infty$) at infinity (Fig. S3.2).
$c = 1$: maximum (0) at (1, 0), minimum ($-\infty$) at infinity.
(v) $\quad L = -(x_1 + x_2/2 + x_2^2/2) - \lambda(2 - x_1 - x_2)$.
$\partial L/\partial x_1 = -1 + \lambda = 0: \partial L/\partial x_2 = -1/2 - x_2 + \lambda = 0$.
Interior ($\lambda = 0$): no solution.
Boundary: $\lambda = 1 = 1/2 + x_2$ gives: (3/2, 1/2), $f = -15/8$ (max), minimum at infinity (Fig. S3.3).

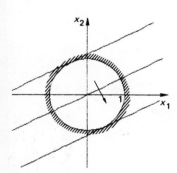

Fig. S3.1

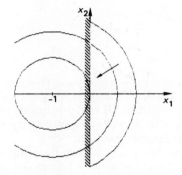

Fig. S3.2

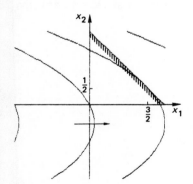

Fig. S3.3

Exercise Set 3
(i) $\quad L = x_1^2 + x_2^2 - \lambda_1(x_2 - 1) - \lambda_2(x_1^2 - 1 - x_2)$.
$\partial L/\partial x_1 = 2x_1 - 2\lambda_2 x_1 = 0: \partial L/\partial x_2 = 2x_2 - \lambda_1 + \lambda_2 = 0$.
Interior ($\lambda_1 = \lambda_2 = 0$): (0, 0), $f = 0$.

Boundary 1 ($\lambda_2 = 0$): $x_2 = 1$, $x_1 = 0$, giving $f = 1$, $\lambda_1 = 2$.
Boundary 2 ($\lambda_1 = 0$): $x_1 = 0$ or $\lambda_2 = 1$.
(a) $x_1 = 0$: $x_2 = x_1^2 - 1 = -1$, $\lambda_2 = 2$, $f = 1$.
(b) $\lambda_2 = 1$: $x_2 = -1/2$, $x_1^2 = 1 + x_2 = 1/2$ giving $(\pm 1/\sqrt{2}, -1/2)$, $f = 3/4$.
Corner: $x_2 = 1$, $x_1 = \pm\sqrt{2}$, $\lambda_1 = 3$, $\lambda_2 = 1$, $f = 3$.
There are no points at infinity. The minimum (0) is taken at (0, 0), the maximum (3) at $(\pm\sqrt{2}, 1)$ (Fig. S3.4).

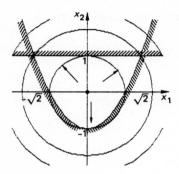

Fig. S3.4

(iii) $L = x_1 x_2 - \lambda_1 (x_1^2 + x_2^2 - 2) - \lambda_2 (1 - x_1)$.
$\partial L/\partial x_1 = x_2 - 2\lambda_1 x_1 + \lambda_2 = 0$: $\partial L/\partial x_2 = x_1 - 2\lambda_1 x_2 = 0$.
Interior ($\lambda_1 = \lambda_2 = 0$): (0, 0) which is not in the feasible region.
Boundary 1 ($\lambda_2 = 0$): $x_2 = 2\lambda_1(2\lambda_1 x_2)$, so $x_2 = 0$ or $\lambda_1 = \pm 1/2$.
(a) $x_1 = 0$: no feasible solution.
(b) $\lambda_1 = 1/2$: $x_1 = x_2 = \pm 1$ from the constraint, $f = 1$,
    $\lambda_1 = -1/2$: $x_1 = -x_2 = \pm 1$ from the constraint, $f = -1$.
Boundary 2 ($\lambda_1 = 0$): $x_1 = 0$, no solution.
Corner: $x_1 = 1$, $x_2 = \pm 1$, $\lambda_1 = \pm 1/2$, $\lambda_2 = 0$, $f = \pm 1$.
There are no points at infinity. The minimum value (-1) is taken at $(1, -1)$, and the maximum value (1) at (1, 1) (Fig. S3.5).

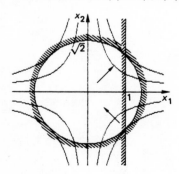

Fig. S3.5

(v) $L = (x_1 - 1/2)^2 + (x_2 - 1/2)^2 - \lambda_1(2 - x_1 - x_2) - \lambda_2(x_1 - 2) - \lambda_3(x_2 - 2)$.
$\partial L/\partial x_1 = 2(x_1 - 1/2) + \lambda_1 - \lambda_2 = 0$: $\partial L/\partial x_2 = 2(x_2 - 1/2) + \lambda_1 - \lambda_3 = 0$.
Interior ($\lambda_1 = \lambda_2 = \lambda_3 = 0$): (1/2, 1/2), not in the feasible region.
Boundary 1 ($\lambda_2 = \lambda_3 = 0$): $\lambda_1 = -2(x_1 - 1/2) = -2(x_2 - 1/2)$, giving $x_1 = x_2 = 1$, $\lambda_1 = -1$ (using boundary equation), $f = 1/2$.

Boundary 2 ($\lambda_1 = \lambda_3 = 0$): (2, 1/2), $f = 9/4$, $\lambda_2 = 3$.
Boundary 3 ($\lambda_1 = \lambda_2 = 0$): (1/2, 2), $f = 9/4$, $\lambda_3 = 3$.
Corner 1 ($\lambda_1 = 0$): (2, 2), $f = 9/2$, $\lambda_2 = \lambda_3 = 3$.
Corner 2 ($\lambda_2 = 0$): (0, 2), $f = 5/2$, $\lambda_1 = 1$, $\lambda_3 = 4$.
Corner 3 ($\lambda_3 = 0$): (2, 0), $f = 5/2$, $\lambda_1 = 1$, $\lambda_2 = 4$.
There are no points at infinity. The maximum (9/2) is taken at (2, 2) and the minimum (1/2) is taken at (1, 1) (Fig. S3.6).

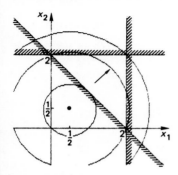

Fig. S3.6

(vii) $L = 2x_1 + x_2 - \lambda_1(1 - x_1) - \lambda_2(x_1 - 1 - x_2) - \lambda_3(x_2 - 1)$.
$\partial L/\partial x_1 = 2 + \lambda_1 - \lambda_2 = 0$: $\partial L/\partial x_2 = 1 + \lambda_2 - \lambda_3 = 0$.
Interior ($\lambda_1 = \lambda_2 = \lambda_3 = 0$): no solution.
Boundary 1 ($\lambda_2 = \lambda_3 = 0$): no solution.
Boundary 2 ($\lambda_1 = \lambda_3 = 0$): no solution.
Boundary 3 ($\lambda_1 = \lambda_2 = 0$): no solution.
Corner 1 ($\lambda_1 = 0$): $\lambda_2 = 2$, $\lambda_3 = 3$, $x_2 = 1$, $x_1 = 2$, $f = 5$.
Corner 2 ($\lambda_2 = 0$): $\lambda_1 = -2$, $\lambda_3 = 1$, $x_1 = 1$, $x_2 = 1$, $f = 3$.
Corner 3 ($\lambda_3 = 0$): $\lambda_2 = -1$, $\lambda_1 = -3$, $x_1 = 1$, $x_2 = 0$, $f = 2$.
There are no points at infinity. The maximum (5) is taken at (2, 1), the minimum (2) at (1, 0).

Problem Set 3
(a) $f = x^3 - 3x$: $df/dx = 3x^2 - 3 = 3(x-1)(x+1) = 0$ when $x = \pm 1$.
(i) $(-2, -1/2)$: $f(-2) = -2$ (min), $f(-1) = 2$ (max), $f(-1/2) = 11/8$.
(ii) $(-1/2, -1/2)$: $f(-1/2) = 11/8$ (max), $f(1/2) = -11/8$ (min).
(iii) $(1/2, 2)$: $f(1/2) = -11/8$, $f(1) = -2$ (min), $f(2) = 2$ (max).
(c) $f = x^3 - 3cx$: $df/dx = 3(x^2 - c) = 0$ when $x = \pm\sqrt{c}$ for $c \geqslant 0$.
(i) $c = -1$: $f(-1) = -4$ (min), $f(1) = 4$ (max).
(ii) $c = 1/4$: $f(-1) = -1/4$ (min), $f(-1/2) = 1/4$ (max), $f(1/2) = -1/4$ (min), $f(1) = 1/4$ (max).
(iii) $c = 9/4$: $f(-1) = 23/4$ (max), $f(1) = -23/4$ (min).

2(a) $L = x_1 + x_2 - \lambda_1((x_1 - 1)^2 + x_2^2 - 4) - \lambda_2((x_1 + 1)^2 + x_2^2 - 4)$.
$\partial L/\partial x_1 = 1 - 2\lambda_1(x_1 - 1) - 2\lambda_2(x_1 + 1)$: $\partial L/\partial x_2 = 1 - 2\lambda_1 x_2 + 2\lambda_2 x_2$.
Interior ($\lambda_1 = \lambda_2 = 0$): no solution.
Boundary 1 ($\lambda_2 = 0$): $x_1 - 1 = x_2 = 1/2\lambda_1$. From the constraint $x_2 = \pm\sqrt{2}$ and so $x_1 = \pm\sqrt{2} + 1$. Only $(1 - \sqrt{2}, -\sqrt{2})$ is feasible, with $f = 1 - 2\sqrt{2}$, and $\lambda_1 = -1/2\sqrt{2}$.
Boundary 2 ($\lambda_1 = 0$): $x_2 = x_1 + 1 = 1/2\lambda_2$, and from the constraint $x_2 = \pm\sqrt{2}$, giving $x_1 = \pm\sqrt{2} - 1$. Only $(\sqrt{2} - 1, \sqrt{2})$ is feasible, with $f = 2\sqrt{2} - 1$ and $\lambda_2 = 1/2\sqrt{2}$.
Corners: $x_1 = 0$, $x_2 = \pm\sqrt{3}$, giving $f = \pm\sqrt{3}$ and $\lambda_1 = (\pm 1 - \sqrt{3})/4\sqrt{3}$, $\lambda_2 = (\sqrt{3} \pm 1)/4\sqrt{3}$.

There are no points at infinity. The maximum $(2\sqrt{2}-1)$ is taken at $(\sqrt{2}-1, \sqrt{2})$ and the minimum $(-2\sqrt{2}+1)$ at $(1-\sqrt{2}, -\sqrt{2})$.

(c) $L = x_1 + x_2 - \lambda_1(x_1^2 - x_2^2) - \lambda_2(x_1^2 + x_2^2 - 1)$.
$\partial L/\partial x_1 = 1 - 2\lambda_1 x_1 - 2\lambda_2 x_1$: $\partial L/\partial x_2 = 1 + 2\lambda_1 x_2 - 2\lambda_2 x_2$.
Interior $(\lambda_1 = \lambda_2 = 0)$: no solution.
Boundary 1 $(\lambda_2 = 0)$: $x_1 = -x_2 = 1/2\lambda_1$ subject to $x_1^2 + x_2^2 \leqslant 1$, i.e. $x_1^2 \leqslant 1/2$ or $-1/\sqrt{2} \leqslant x_1 \leqslant 1/\sqrt{2}$. Hence there is an infinity of solutions along $x_1 + x_2 = 0$ for $|x_1| \leqslant 1/\sqrt{2}$ with $f = 0$.
Boundary 2 $(\lambda_1 = 0)$: $x_1 = x_2 = 1/2\lambda_2$ and $2x_1^2 = 1$ from the constraint, giving $x_2 = \pm 1/\sqrt{2}$, and $f = \pm\sqrt{2}$.
Corners: $x_1 = \pm x_2$ and $x_1 = \pm 1/\sqrt{2}$.
Hence the maximum $(\sqrt{2})$ is taken at $(1/\sqrt{2}, 1/\sqrt{2})$, and the minimum $(-\sqrt{2})$ at $(-1/\sqrt{2}, -1/\sqrt{2})$.

3(a) $L = -(x_1-a)^2/2 - (x_2-a)^2/2 - \lambda_1(x_1-2) - \lambda_2(x_2-2) + \lambda_3(x_1+x_2-2)$.
$\partial L/\partial x_1 = -2(x_1-a) - \lambda_1 + \lambda_3$: $\partial L/\partial x_2 = -2(x_2-a) - \lambda_2 + \lambda_3$.

		$a = 0$	$a = 1$	$a = 3$
Interior	$(\lambda_1 = \lambda_2 = \lambda_3 = 0)$	$(0, 0)$	$(1, 1)$	$(3, 3)$
	(a, a)	not feasible	$f = 0$	not feasible
Boundary 1	$(\lambda_2 = \lambda_3 = 0)$	$(2, 0)$	$(2, 1)$	$(2, 3)$
	$(2, a)$	$f = -2$	$f = -1/2$	not feasible
Boundary 2	$(\lambda_1 = \lambda_3 = 0)$	$(0, 2)$	$(1, 2)$	$(3, 2)$
	$(a, 2)$	$f = -2$	$f = -1/2$	not feasible
Boundary 3	$(\lambda_1 = \lambda_2 = 0)$	$(1, 1)$	$(1, 1)$	$(1, 1)$
	$(1, 1)$	$f = -1$	$f = 0$	$f = -4$
Corner 1	$(\lambda_1 = 0)$	$(0, 2)$	$(0, 2)$	$(0, 2)$
	$(0, 2)$	$f = -2$	$f = -1$	$f = -5$
Corner 2	$(\lambda_2 = 0)$	$(2, 0)$	$(2, 0)$	$(2, 0)$
	$(2, 1)$	$f = -2$	$f = -1$	$f = -5$
Corner 3	$(\lambda_3 = 0)$	$(2, 2)$	$(2, 2)$	$(2, 2)$
	$(2, 2)$	$f = -4$	$f = -1$	$f = -1$
Maximum:		(-1) at $(1, 1)$	(0) at $(1, 1)$	(-1) at $(2, 2)$

Generally, $\max f = f(1, 1) = -(a-1)^2$ for $a \leqslant 1$
$ = f(a, a) = 0$ for $1 \leqslant a \leqslant 2$
$ = f(2, 2) = -(a-2)^2$ for $a \geqslant 2$.

4(i) $L = -x_1 - \lambda_1(-x_2) - \lambda_2(x_2 - x_1^3/3)$.
$\partial L/\partial x_1 = -1 + \lambda_2 x_1^2 = 0$: $\partial L/\partial x_2 = \lambda_1 - \lambda_2 = 0$.
Since the maximum has $x_1 = x_2 = 0$, we must have $\lambda_1 = \lambda_2 = \infty$ such that as x_1 tends to 0 the limit of $\lambda_2 x_1$ is 1. In terms of vectors, ∇f is perpendicular to ∇g_1 and ∇g_2, and hence cannot be written as a linear combination of these vectors with finite coefficients (Fig. S3.7).

(iii) $L = -x_1^2 - (x_2+1)^2 - \lambda_1(-x_1) - \lambda_2(-x_2) - \lambda_3(x_1 - x_2)$.
$\partial L/\partial x_1 = -2x_1 + \lambda_1 - \lambda_3 = 0$: $\partial L/\partial x_2 = -2(x_2+1) + \lambda_2 + \lambda_3 = 0$.
The maximum is seen to be at the corner $(0, 0)$. $\lambda_1 = \lambda_3$, and $\lambda_2 = 2 - \lambda_3$ for any value of λ_3. This ambiguity in the values of the multipliers arises because one of the constraints is redundant.

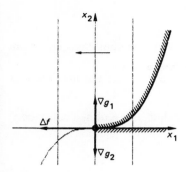

Fig. S3.7

(v) The feasible region is just the single point $(0, 0)$.
$L = x_1 + x_2 - \lambda_1(x_1^2 + x_2^2) - \lambda_2(-x_1)$.
$\partial L/\partial x_1 = 1 - 2\lambda_1 x_1 + \lambda_2 = 0$: $\partial L/\partial x_2 = 1 - 2\lambda_1 x_2 = 0$.
$\lambda_1 \to \infty$ such that $2\lambda_1 x_2 \to 1$ as $x_2 \to 0$.

5(a)(i)   Maximum (2) at $(1, 1)$ (Fig. S3.8).
(iii)   Maximum $(5/2)$ at $(3/2, -1/2)$ (Fig. S3.9).
(b)(i)   Maximum (1) at $(1, 0)$, a point at which 3 constraints meet (Fig. S3.10).
(iii)   Maximum (1) at $(0, 1)$. The feasible region is a line segment (Fig. S3.11).

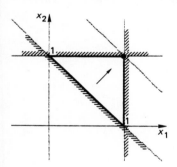

Fig. S3.8

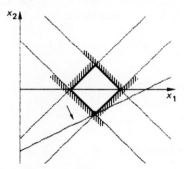

Fig. S3.9

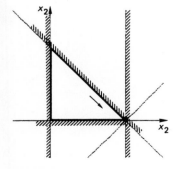

Fig. S3.10

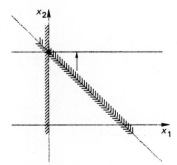

Fig. S3.11

7(a)(i) $L = x_1 + x_2 - \lambda_1(x_1^2 + x_2^2 - 1) - \lambda_2(-x_1)$.
$\partial L/\partial x_1 = 1 - 2\lambda_1 x_1 + \lambda_2$: $\partial L/\partial x_2 = 1 - 2\lambda_1 x_2$.
Boundary ($\lambda_2 = 0$): $x_1 = x_2 = 1/2\lambda_1$ with $\lambda_1 = \pm 1/\sqrt{2}$ from the constraint equation, but $x_1 \geq 0$ so $x_1 = x_2 = 1/\sqrt{2}$ and $f = \sqrt{2}$.
Corner: $x_1 = 0$, $x_2 = \pm 1$, $\lambda_2 = -1$, $\lambda_1 = \pm 1/2$ and $f = \pm 1$. Hence the maximum is taken at $(1/\sqrt{2}, 1/\sqrt{2})$.

(b) If $L = f(\mathbf{x}) - \sum_{j=1}^{m} \lambda_j g_j(\mathbf{x})$ (the reduced Lagrangean) then the Kuhn–Tucker conditions corresponding to equations 3.24a and 3.25a are:

$x_i \geq 0$, $\partial L/\partial x_i \leq 0$, $x_i \partial L/\partial x_i = 0$ for $i = 1 \ldots n$
$\partial L/\partial \lambda_j = 0$ for $j = 1 \ldots m$.

Note we no longer require the λs to be non-negative.

Chapter 4

Abbreviation: s.t. = subject to.

Exercise Set 1

1(i) $h(\lambda) = \max(-x^2 - \lambda(1-x)) = \lambda^2/4 - \lambda$ with $x = \tfrac{1}{2}\lambda$. Dual solution: $h(2) = -1$. Primal solution: $f(1) = -1$. L contours: $(x-1)(\lambda - (x+1)) = c+1$. Saddle at $(1, 2)$ (Fig. S4.1).

(iii) $h(\lambda) = \max(x^2(1-\lambda) + \lambda)$; $h(\lambda) = \lambda$ if $\lambda \geq 1$ (with $x = 0$) or ∞ if $0 \leq \lambda < 1$ (with $x = \infty$). Dual solution: $h(1) = 1$ (Fig. S4.2a). Primal solution: $f(\pm 1) = 1$. L contours: $(x^2 - 1)(1 - \lambda) = c - 1$. Saddles at $(\pm 1, 1)$ (Fig. S4.2b).

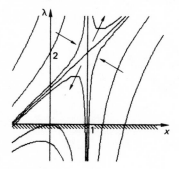

Fig. S4.1

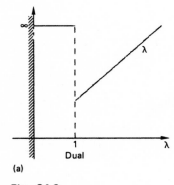
(a) Dual

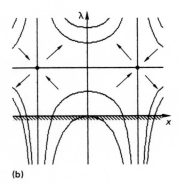
(b)

Fig. S4.2

2(i) $h(\lambda) = \max(x + 3 - \lambda(x-1))$ with $x \geq 0$; $h(\lambda) = \infty$ if $0 \leq \lambda < 1$ (with $x = \infty$) or $(3 + \lambda)$ if $\lambda \geq 1$ (with $x = 0$). Dual solution: $h(1) = 4$. Primal solution: $f(1) = 4$. L contours: $(x-1)(1-\lambda) = c - 4$. Saddle at $(1, 1)$ (Fig. S4.3).
(iii) $h(\lambda) = \max(\ln(x+1) - \lambda(x-1))$ with $x \geq 0$; $h(\lambda) = -\ln\lambda - 1 + 2\lambda$ if $\lambda \leq 1$ (with $x = \lambda^{-1} - 1$) or λ if $\lambda \geq 1$ (with $x = 0$). Dual solution: $h(1/2) = \ln 2$. Primal solution: $f(1) = \ln 2$. L contours: $\lambda = (\ln(x+1) - c)/(x-1)$. Saddle at $(1, 1/2)$ (Fig. S4.4).

3(i) $h(\lambda) = \max(x + 1 - \lambda(x^2 - \varepsilon^2))$; $h(\lambda) = (4\lambda)^{-1} + 1 + \lambda\varepsilon^2$ if $\lambda > 0$ (with $x = \frac{1}{2}\lambda^{-1}$) or ∞ if $\lambda = 0$ (with $x = \infty$). Dual solution: $h(\frac{1}{2}\varepsilon^{-1}) = \varepsilon + 1$. Primal solution: $f(\varepsilon) = \varepsilon + 1$. L contours: $(x \pm \varepsilon)(1 - \lambda(x \mp \varepsilon)) = c - (1 \mp \varepsilon)$. Saddles at $(\pm\varepsilon, \pm\frac{1}{2}\varepsilon^{-1})$ (Fig. S4.5).
Note: as ε tends to zero, i.e. as the primal feasible region collapses onto a point, the saddles move off to infinity.

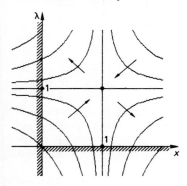

Fig. S4.3

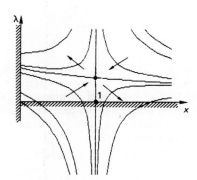

Fig. S4.4

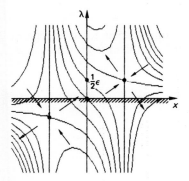

Fig. S4.5

Exercise Set 2

1(a)(i) Primal solution: $f(0, 1) = -1/2$ (Fig. S4.6); $h(\lambda) = \max(-\frac{1}{2}(x_1 + 1)^2 - \frac{1}{2}(x_2 - 1)^2 - \lambda(x_2 - 2)) = \frac{1}{2}\lambda^2 + \lambda - \frac{1}{2}$ with $x_1 = 0$, $x_2 = 1 - \lambda$. Dual solution: $h(0) = -\frac{1}{2}$.
(b)(i) Primal solution: $f(0, 0) = 0$ (Fig. S4.7); $h(\lambda) = \max(-(x_2 + x_1^2) - \lambda(x_1 - x_2)) = \infty$ if $\lambda \neq 1$ (with $x = \pm\infty$) or 0 if $\lambda = 1$ (with $x_1 = x_2 = 0$). Dual solution: $h(1) = 0$.

2(a)(i) Primal solution: $f(0, 2) = -1$. Dual objective function as before but λ no longer constrained to be non-negative. Dual solution: $h(-1) = -1$.
(b)(i) As in 1(b)(i) since the second constraint is binding there.

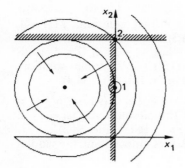

Fig. S4.6

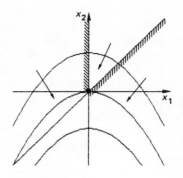

Fig. S4.7

Exercise Set 3

(i) Primal: $\max -(3x_1/2 + 2x_2)$ s.t. $-(x_1 + 2x_2) \leq -1$, $-(x_1 + x_2) \leq 1$, $x_1 \geq 0$, $x_2 \geq 0$.
Dual: $\min -\lambda_1 + \lambda_2$ s.t. $-(\lambda_1 + \lambda_2) \geq -3/2$, $-2\lambda_1 - \lambda_2 \geq -2$, $\lambda_1 \geq 0$, $\lambda_2 \geq 0$. Primal solution: $f(0, 1/2) = -1$. Dual solution: $h(1, 0) = -1$.
Note: this example illustrates the fact that the dual of the dual of an L.P. is just the L.P. itself—our primal here being the dual of the L.P. discussed in the text (after relabelling and mapping to a max problem).

(iii) Primal: $\max x_1$ s.t. $x_1 + x_2 \leq 1$, $x_1 + \frac{1}{2}x_2 \leq 1$, $x_1 \geq 0$, $x_2 \geq 0$. Primal solution: $f(1, 0) = 1$.
Dual: $\min \lambda_1 + \lambda_2$ s.t. $\lambda_1 + \lambda_2 \geq 1$, $\lambda_1 + \lambda_2/2 \geq 0$, $\lambda_1 \geq 0$, $\lambda_2 \geq 0$. Dual solution: $h(\lambda_1, 1 - \lambda_1) = 1$ for $0 \leq \lambda_1 \leq 1$ (Fig. S4.8).
Note. A primal redundant constraint generates an infinity of dual solutions.

(v) Primal: $\max (x_1 + x_2)$ s.t. $x_1 - x_2 \leq 1$, $-x_1 + x_2 \leq 1$, $x_1 \geq 0$, $x_2 \geq 0$.
Dual: $\min (\lambda_1 + \lambda_2)$ s.t. $\lambda_1 - \lambda_2 \geq 1$, $-\lambda_1 + \lambda_2 \geq 1$, $\lambda_1 \geq 0$, $\lambda_2 \geq 0$.
Note: the primal optimal value is infinite because of its unbounded feasible region (Fig. S4.9); the dual L.P. has no feasible point hence the dual problem from which it is derived has an everywhere infinite objective function.

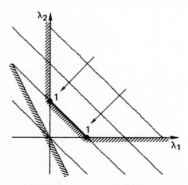

Fig. S4.8

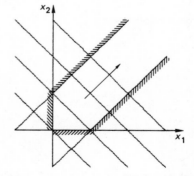

Fig. S4.9

Exercise Set 4

1(i) $f''(x) = -\frac{1}{4}x^{-3/2} < 0$ for $x > 0$; concave.
(iii) $f''(x) = -x^{-2} - \frac{1}{4}x^{-3/2} < 0$ for $x > 0$; concave.
(v) $f''(x) = (9x^4 + 6x)\exp(x^3)$; $f''(1) > 0$, $f''(-1/2) < 0$; neither concave nor convex.

2(i) $F = -(a_1 - a_2(s_1 - s_2)u)^2$; $F'' = -2(s_1 - s_2)^2 \leq 0$; concave.
(iii) $f = 2x_1 + 6x_2 - x_2^2$; $F = (2a_1 + 6a_2) + u(2s_1 + 6s_2) - (a_1 + s_1 u)^2$; $F'' = -2s_2^2$; concave; sum of two concave functions is concave.
(v) $F = \exp\frac{1}{2}(\ln(a_1 + us_1) + \ln(a_2 + us_2))$; $F''/F = -\frac{1}{4}(s_1/(a_1 + us_1) - s_2/(a_2 + us_2))^2 \leq 0$; concave.

Exercise Set 5
1(i) Dual: max $V = \lambda_1^{-\lambda_1}(\frac{1}{2}\lambda_2)^{-\lambda_2}\lambda_3^{-\lambda_3}$ s.t. $\lambda_1 - \frac{1}{2}\lambda_3 = 0$, $-\lambda_1 + \lambda_2 = 0$, $\lambda_1 + \lambda_2 + \lambda_3 = 1$. Dual solution: $h^* = h(1/4, 1/4, 1/2) = 2^{7/4}$. Primal from (4.16): $x_1 x_2^{-1} = \lambda_1 h^* = 2^{-1/4}$; $2x_2 = \lambda_2 h^*$; $x_1^{-1/2} = \lambda_3 h^*$. Primal solution: $f(2^{-3/2}, 2^{-5/4}) = 2^{7/4}$.

2(i) Dual: max $\lambda_1^{-\lambda_1}(\frac{1}{2}\lambda_2)^{-\lambda_2}\lambda_3^{-\lambda_3}\lambda_4^{-\lambda_4}$ s.t. $\lambda_1 - \frac{1}{2}\lambda_3 + \lambda_4 = 0$, $-\lambda_1 + \lambda_2 = 0$, $\lambda_1 + \lambda_2 + \lambda_3 + \lambda_4 = 1$. Use dual constraints to eliminate $\lambda_2, \lambda_3, \lambda_4$ to obtain a one-variable problem: max $2^{\lambda_1}\lambda_1^{-2\lambda_1}(2(1-\lambda_1)/3)^{-2(1-\lambda_1)/3}(1-4\lambda/3)^{-(1-4\lambda_1)/3}$ s.t. $0 \leq \lambda_1 \leq 1/4$ to preserve non-negativity of all dual variables. Plot the graph of the objective function and read off maximum 3.66 at $\lambda_1 = 0.195$. From (4.16) $x_1 = 0.36$, $x_2 = 0.25$.

3(i) Dual: max $\lambda_1^{-\lambda_1}\lambda_2^{-\lambda_2}\lambda_3^{-\lambda_3}$ s.t. $\lambda_1 + \lambda_3 = 0$, $\lambda_2 + \lambda_3 = 0$, $\lambda_1 + \lambda_2 + \lambda_3 = 1$. Dual solution: $\lambda_1 = \lambda_2 = 1$, $\lambda_3 = -1$. A negative solution is not acceptable as it corresponds to a primal solution on the boundary, in fact at $(0, 0)$. To see this, draw the contour map $(x_1 + 1)(x_2 + 1) = c + 1$ (Fig. S4.10) (otherwise obvious from form of primal objective function).

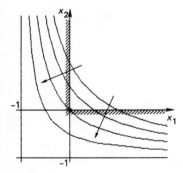

Fig. S4.10

Problem Set 4:
1(i) $h(\lambda) = \max(-x - \lambda(1 - x^2))$ with $x \geq 0$; $h(\lambda) = \infty$ with $x = \infty$ if $\lambda > 0$ or 0 with $x = 0$ if $\lambda = 0$. Dual solution: $h(0) = 0$. Primal solution: $f(1) = -1$. Primal and dual optimal values are not equal; the constraint function is not convex. Note: L has a col (saddle) at $(1, 1/2)$ but its valleys and mountains are so oriented that the min in λ and max in x takes us away from that col.
(iii) $h(\lambda_1, \lambda_2) = \max(x_2 - \lambda_1(x_1 + x_2^2 - 1) + \lambda_2(x_1 + 1))$; $h = \infty$ if $\lambda_1 \neq \lambda_2$ because of linearity in x_1; $h = \frac{1}{4}\lambda_1^{-1} + 2\lambda_1$ with $x_2 = \frac{1}{2}\lambda_1^{-1}$ if $\lambda_1 = \lambda_2$. Dual solution: $h(2^{-3/2}, 2^{-3/2}) = 2^{1/2}$, primal solution $f(-1, 2^{1/2}) = 2^{1/2}$. All conditions satisfied.
(v) Dual: min $2\lambda_1 + \lambda_2$ s.t. $2\lambda_1 + \lambda_2 \geq 1$, $\lambda_1 \geq 1$, $\lambda_1 \geq 0$, $\lambda_2 \geq 0$. Dual solution: $h(1, 0) = 2$. Primal: $f(0, 2) = 2$.
Note: redundancy in dual constraints but not an infinity of solutions in primal.

2(i) Dual: max $\lambda_1^{-\lambda_1}(\frac{1}{2}\lambda_2)^{-\lambda_2}(\frac{1}{3}\lambda_3)^{-\lambda_3}$ s.t. $\frac{1}{2}\lambda_1 - \lambda_2 - \frac{1}{3}\lambda_3 = 0$, $-\lambda_2 + 2\lambda_3 = 0$, $\lambda_1 + \lambda_2 + \lambda_3 = 1$. Dual solution: $h(14/23, 6/23, 3/23) = 23(14^{-14/23})(3^{-6/23})$. Primal solution with $x_1 = 14^{18/23}3^{-12/23}$, $x_2 = 14^{-4/23}3^{-5/23}$.

(iii) Dual: max $(\lambda_1^{-\lambda_1}\lambda_2^{-\lambda_2}\lambda_3^{-\lambda_3})\lambda_3^{\lambda_3}$ s.t. $\lambda_1 - \lambda_3 = 0$, $\lambda_2 - \lambda_3 = 0$, $\lambda_1 + \lambda_2 = 1$. Dual solution: $h(1/2, 1/2, 1/2) = 2$. Primal solution: $f(1, 1) = 2$.
(v) Dual: max $(\lambda_1^{-\lambda_1}\lambda_2^{-\lambda_2}\lambda_3^{-\lambda_3})\lambda_2^{\lambda_2}\lambda_3^{\lambda_3}$ s.t. $\frac{1}{2}\lambda_1 - \lambda_2 = 0$, $\frac{2}{3}\lambda_1 - \lambda_3 = 0$, $\lambda_1 = 1$. Dual solution: $h(1, 1/3, 2/3) = 1$. Primal solution: $f(1, 1) = 1$ from (4.16).
(vii) Dual: max $\lambda_1^{-\lambda_1}\lambda_2^{-\lambda_2}\lambda_3^{-\lambda_3}$ s.t. $\frac{1}{2}\lambda_1 + \lambda_2 - \lambda_3 = 0$, $\frac{1}{2}\lambda_1 + \lambda_2 - \lambda_3 = 0$, $\lambda_1 + \lambda_2 + \lambda_3 = 1$. The second constraint is redundant, reflecting the fact that the primal problem can be written in terms of the single variable $y = x_1 x_2$ i.e. min $y^{1/2} + y + y^{-1}$ with $y \geq 0$. Solve this revised problem (e.g. graphically to obtain $y = 0.8$) or solve the dual problem after removing the redundancy—a one-variable problem s.t. $0 \leq \lambda_1 \leq 2/3$.

3(c) The first order stationary conditions for

$$L = -\ln\left(\sum_i \exp(Y_i)\right) + \sum_i \lambda_i\left(Y_i - \sum_j a_{ij} X_j - \ln C_i\right)$$

are: $\partial L/\partial Y_i = -\exp Y_i / \left(\sum_k \exp Y_k\right) + \lambda_i = 0$,

and $\partial L/\partial X_j = -\sum_i \lambda_i a_{ij} = 0$ (otherwise $h = \infty$).

Add the first set of conditions to obtain $\sum_i \lambda_i = 1$. Also: $Y_i = \ln \lambda_i + \ln\left(\sum_k \exp Y_k\right)$.

Substitute in L to obtain the dual objective function: $h(\lambda) = \sum_i^k \lambda_i \ln \lambda_i - \sum_i \lambda_i \ln C_i = -\ln \prod_i (C_i/\lambda_i)^{\lambda_i}$.

Take the exponential to arrive at (4.15).

5(a) (i) $f'' = 1$: local (in fact overall) min.
(iii) $f'' = 4 \exp(-2x) > 0$; local min.
(v) $f'' = 2((1 - 5x^2) - (1 - 3x^2)\ln(1 + x^2))(1 + x^2)^{-3}$; $\quad f''(0) = 2$: local min.
$f''(\pm(e - 1)^{1/2}) < 0$: local (in fact overall) max.
(b)(iii) Let $f_{11} = \partial^2 f/\partial x_1^2$, $f_{22} = \partial^2 f/\partial x_2^2$, $f_{12} = \partial^2 f/\partial x_1 \partial x_2$.
(i) $f_{11} = 2 > 0$; conditions not met.
(iii) $f_{11} = f_{22} = -2$, $f_{12} = 0$ at $\mathbf{x} = (0, 0)$; conditions met: local (in fact overall) max.
(v) $f_{11} = 4 - 12x_1^2$, $f_{22} = -2$, $f_{12} = 0$; at $(\pm 1, 0)$ $f_{11} = -8$; conditions met: local (overall) max; at $(0, 0)$ $f_{11} = 4$; conditions not met.

6(b)(i) $Q = \begin{bmatrix} -1 & 0 \\ 0 & -1 \end{bmatrix}$, $q = \begin{bmatrix} 1 \\ -1 \end{bmatrix}$, $A = \begin{bmatrix} 1 & 1 \\ 0 & 1 \end{bmatrix}$, $b = \begin{bmatrix} 1 \\ 1 \end{bmatrix}$;

after some algebra

$$h(\lambda) = \tfrac{1}{4}(2\lambda_1^2 + 2\lambda_1\lambda_2 + \lambda_2^2 + 4\lambda_1 + 6\lambda_2 + 2).$$

Chapter 5

Exercise Set 1
1(b) See Fig. S5.1.
A(0, 2), B(1, 1), C(3/5, 4/5), D(0, 1)
$x_1 - 2x_2 + x_3 \qquad\qquad = -1$
$x_1 + x_2 \quad + x_4 \qquad = 2$
$x_1 + 3x_2 \qquad - x_5 = 3$
$\qquad x_1, x_2, x_3, x_4, x_5 \geq 0$
A(x_1, x_4), B(x_3, x_4), C(x_3, x_5), D(x_1, x_5).

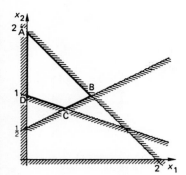

Fig. S5.1

2

	A	B	C	D	Optimum Point
(a)	4	3	11/5	2	D
(c)	8	5	19/5	4	C
(e)	−2	−2	−7/5	−1	All points on AB

Exercise Set 2

1(a)(i) See Fig. S5.2.

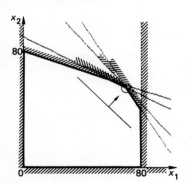

Fig. S5.2

(ii)

Basis		P	11	10	0	0	0	0	0	
μ	x		x_1	x_2	x_3	x_4	x_5	x_6	RHS	Ratio
0	x_3		[1]	0	1	0	0	0	80	⑧⓪
0	x_4		3	2	0	1	0	0	320	106.7
0	x_5		1	3	0	0	1	0	240	240
0	x_6		1	2	0	0	0	1	180	180
	L		⑪	10	0	0	0	0	0	

Table cont.

Basis μ	x	P	11 x_1	10 x_2	0 x_3	0 x_4	0 x_5	0 x_6	0 RHS	Ratio
11	x_1		1	0	1	0	0	0	80	∞
0	x_4		0	☐2	−3	1	0	0	80	㊵
0	x_5		0	3	−1	0	1	0	160	53.3
0	x_6		0	2	−1	0	0	1	100	50
	L		0	⑩	−11	0	0	0	−880	
11	x_1		1	0	1	0	0	0	80	80
10	x_2		0	1	−3/2	1/2	0	0	40	<0
0	x_5		0	0	7/2	−3/2	1	0	40	11.4
0	x_6		0	0	☐2	−1	0	1	20	⑩
	L		0	0	④	−5	0	0	−1280	
11	x_1		1	0	0	1/2	0	−1/2	70	
10	x_2		0	1	0	−1/4	0	3/4	55	
0	x_5		0	0	0	1/4	1	−7/4	5	
0	x_3		0	0	1	−1/2	0	1/2	10	
	L		0	0	0	−3	0	−2	−1320	

The solution (1320) is at the point (70, 55; 10, 0, 5, 0).

2(ii)

Basis μ	x	P	3 x_1	2 x_2	4 x_3	0 x_4	0 x_5	0 x_6	0 RHS	Ratio
0	x_4		3	1	4	1	0	0	60	15
0	x_5		1	2	☐3	0	1	0	30	⑩
0	x_6		2	2	3	0	0	1	60	20
	L		3	2	④	0	0	0	0	
0	x_4		☐5/3	−5/3	0	1	−4/3	0	20	⑫
4	x_3		1/3	2/3	1	0	1/3	0	10	30
0	x_6		1	0	0	0	−1	1	30	30
	L		⑤/3	−2/3	0	0	−4/3	0	−40	
3	x_1		1	−1	0	3/5	−4/5	0	12	<0
4	x_3		0	☐1	1	−1/5	3/5	0	6	⑥
0	x_6		0	1	0	−3/5	−1/5	1	18	18
	L		0	①	0	−1	0	0	−60	
3	x_1		1	0	1	2/5	−1/5	0	18	
2	x_2		0	1	1	−1/5	3/5	0	6	
0	x_6		0	0	−1	−2/5	−4/5	1	12	
	L		0	0	−1	−4/5	−3/5	0	−66	

The solution (66) is at the point (18, 6, 0; 0, 0, 12)

Exercise Set 3

(i)

Basis μ	x	P	1 x_1	1 x_2	1 x_3	0 x_4	0 x_5	0 x_6	0 RHS	Ratio
0	x_4		⬛2	−3	2	1	0	0	2	①
0	x_5		−3	2	2	0	1	0	2	<0
0	x_6		2	2	−3	0	0	1	2	1
	L		①	1	1	0	0	0	0	
1	x_1		1	−3/2	1	1/2	0	0	1	<0
0	x_5		0	−5/2	5	3/2	1	0	5	<0
0	x_6		0	⬛5	−5	−1	0	1	ε	ⓔ/5
	L		0	ⓢ5/2	0	−1/2	0	0	−1	
1	x_1		1	0	−1/2	1/5	0	3/10	$1+3\varepsilon/10$	<0
0	x_5		0	0	⬛5/2	1	1	1/2	$5+\varepsilon/2$	②
1	x_2		0	1	−1	−1/5	0	1/5	$\varepsilon/5$	<0
	L		0	0	ⓢ5/2	0	0	−1/2	$-1-\varepsilon/2$	
1	x_1		1	0	0	2/5	1/5	2/5	2	
1	x_3		0	0	1	2/5	2/5	1/5	2	
1	x_2		0	1	0	1/5	2/5	2/5	2	
	L		0	0	0	−1	−1	−1	−6	

(iii)

Basis μ	x	P −60 x_1	−30 x_2	−60 x_3	0 x_4	0 x_5	0 x_6	−M x_7	−M x_8	−M x_9	0 RHS	Ratio
−M	x_7	3	1	2	−1	0	0	1	0	0	3	1
−M	x_8	1	2	2	0	−1	0	0	1	0	2	2
−M	x_9	⬛4	3	3	0	0	−1	0	0	1	4	①
	L	ⓢ$-60+8M$	$-30+6M$	$-60+7M$	−M	−M	−M	0	0	0	9M	
−M	x_7	0	−5/4	−1/4	−1	0	3/4	1	0		ε	<0
−M	x_8	0	5/4	⬛5/4	0	−1	1/4	0	1		1	④/5
−60	x_1	1	3/4	3/4	0	0	−1/4	0	0		1	4/3
	L	0	15	ⓢ$-15+M$	−M	−M	$-15+M$	0	0		$M+M\varepsilon+60$	
−M	x_7	0	−1	0	−1	−1/5	⬛4/5	1			1/5	①/4
−60	x_3	0	1	1	0	−4/5	1/5	0			4/5	4
−60	x_1	1	0	0	0	3/5	−2/5	0			2/5	<0
	L	0	$30-M$	0	−M	−12	ⓢ$4M/5-12$ $-M/5$	0			$M/5+72$	

278 Solutions

Table cont.

Basis μ	x	P -60 x_1	-30 x_2	-60 x_3	0 x_4	0 x_5	0 x_6	-M x_7	-M x_8	-M x_9	0 RHS	Ratio
0	x_6	0	-5/4	0	-5/4	-1/4	1				1/4	<0
-60	x_3	0	[5/4]	1	1/4	-3/4	0				3/4	③/5
-60	x_1	1	-1/2	0	-1/2	1/2	0				1/2	<0
	L	0	⑮	0	-15	-15	0				75	
0	x_6	0	0	1	-1	-1	1				1	
-30	x_2	0	1	4/5	1/5	-3/5	0				3/5	
-60	x_1	1	0	2/5	-2/5	1/5	0				4/5	
	L	0	0	-12	-18	-6	0				66	

Note that we need not calculate the entries corresponding to the artificial variables after they leave the basis, as they will never return.

(v)

Basis μ	x	P -3 x_1	3 x_2	4 x_3	-1 x_4	0 x_5	0 x_6	0 x_7	0 RHS	Ratio
0	x_5	0	0	1	1	1	0	0	4	4
0	x_6	2	-2	[3]	-1	0	1	0	8	⑧/3
0	x_7	-1	1	2	4	0	0	1	11	11/2
	L	-3	3	④	-1	0	0	0	0	
0	x_5	-2/3	[2/3]	0	4/3	1	-1/3	0	4/3	②
4	x_3	2/3	-2/3	1	-1/3	0	1/3	0	8/3	<0
0	x_7	-7/3	7/3	0	14/3	0	-2/3	1	17/3	17/7
	L	-17/3	⑰/3	0	1/3	0	-4/3	0	-32/3	
3	x_2	-1	1	0	2	3/2	-1/2	0	2	<0
4	x_3	0	0	1	1	1	0	0	4	∞
0	x_7	0	0	0	0	-7/2	[1/2]	1	1	②
	L	0	0	0	-11	-17/2	③/2	0	-22	
3	x_2	-1	1	0	2	-2	0	1	3	<0
4	x_3	0	0	1	1	[1]	0	0	4	④
0	x_6	0	0	0	0	-7	1	2	2	<0
	L	0	0	0	-11	②	0	-3	-25	
3	x_2	-1	1	2	4	0	0	1	-11	
0	x_5	0	0	1	1	1	0	0	4	
0	x_6	0	0	7	7	0	1	2	30	
	L	0	0	-2	-13	0	0	-3	-33	

Problem Set 5

1(i)

Basis μ	x	P	-9 x_1	-1 x_2	1 x_3	-3 x_4	0 x_5	0 x_6	0 x_7	$-M$ x_8	0 RHS	Ratio
0	x_5		-3	0	4	-1	1	0	0	0	3	∞
$-M$	x_8		2	⑤	1	1	0	-1	0	1	1	①/⑤
0	x_7		1	1	0	-1	0	0	1	0	2	2
	L		-9 $+2M$	$\widehat{-1}$ $+5M$	1 $+M$	-3 $+M$	0	$-M$	0	0	M	
0	x_5		-3	0	$\boxed{4}$	-1	1	0	0		3	③/④
-1	x_2		2/5	1	1/5	1/5	0	$-1/5$	0		1/5	1
0	x_7		3/5	0	$-1/5$	$-6/5$	0	1/5	1		9/5	<0
	L		$-43/5$	0	⑥/⑤	$-14/5$	0	$-1/5$	0		1/5	
1	x_3		$-3/4$	0	1	$-1/4$	1/4	0	0		3/4	
-1	x_2		11/20	1	0	1/4	$-1/20$	$-1/5$	0		1/20	
0	x_7		9/20	0	0	$-5/4$	1/20	1/5	1		39/20	
	L		$-77/10$	0	0	$-5/2$	$-3/10$	$-1/5$	0		$-7/10$	

The multipliers are $(3/10, 1/5, 0)$.

3 Since we know a feasible vertex there is no need to use artificial variables. We can put the equations in standard form with respect to the known basis formed by x_1, x_2, x_3. This can be achieved by pivoting first on x_2 in the second constraint and then on x_3 in the third. The problem now becomes:

$$\begin{aligned} \text{minimise} \quad & x_1 + 2x_2 + x_3 \\ \text{subject to} \quad & x_1 + 4x_4 = 8 \\ & x_2 - x_4 = 6 \\ & x_3 + 2x_4 = 3 \\ & x_1, \ldots, x_4 \geq 0. \end{aligned}$$

The tabular solution can now be found as usual.

Basis μ	x	P	-1 x_1	-2 x_2	-1 x_3	0 x_4	0 RHS	Ratio
-1	x_1		1	0	0	4	8	2
-2	x_2		0	1	0	-1	6	<0
-1	x_3		0	0	1	$\boxed{2}$	3	③/②
	L		0	0	0	④	23	
-1	x_1		1	0	-2	0	2	
-2	x_2		0	1	1/2	0	15/2	
0	x_4		0	0	1/2	1	3/2	
	L		0	0	-2	0	17	

4 There are 5 sensible knife settings for sheets 2.3 m wide.

Widths	Knife Settings				
	1	2	3	4	5
1.20	1	1	0	0	0
0.80	1	0	2	1	0
0.45	0	2	1	3	5
Loss	0.3	0.2	0.25	0.15	0.05

If x_1 rolls are produced with knife setting 1, etc., then we can write the problem as:

$$\text{minimise} \quad 0.3x_1 + 0.2x_2 + 0.25x_3 + 0.15x_4 + 0.05x_5$$
$$\text{subject to} \quad x_1 + x_2 \geq 30$$
$$x_1 + 2x_3 + x_4 \geq 40$$
$$2x_2 + x_3 + 3x_4 + 5x_5 \geq 50$$
$$x_1, \ldots, x_5 \geq 0.$$

The solution can be obtained quite straightforwardly by the use of artificial variables as (26, 4, 0, 0, 14, 0). It is very convenient that this is an integer solution. Usually this would not be the case and so either approximations would have to be made or else an integer algorithm used. (See, for example, E. M. L. Beale, *Mathematical Programming in Practice*, Pitman 1968.)

7(i) Rewriting the last constraint as $x_1 - x_2 - x_3 \leq 0$, and adding one artificial variable, a tabular solution is as follows:

Basis μ	x	P	3 x_1	2 x_2	1 x_3	0 x_4	0 x_5	0 x_6	$-M$ x_7	0 RHS	Ratio
0	x_4		1	1	1	1	0	0	0	3	3
$-M$	x_7		2	0	⊡1⊡	0	-1	0	1	2	②
0	x_6		1	-1	-1	0	0	1	0	$\phi\varepsilon$	<0
	L		$3+2M$	2	①$1+M$①	0	$-M$	0	0	$2M$	
0	x_4		-1	⊡1⊡	0	1	1	0		1	①
1	x_3		2	0	1	0	-1	0		2	∞
0	x_6		3	-1	0	0	-1	1		$2+\varepsilon$	<0
	L		1	②	0	0	1	0		-2	
2	x_2		-1	1	0	1	1	0		1	<0
1	x_3		⊡2⊡	0	1	0	-1	0		2	①
0	x_6		2	0	0	1	0	1		3	$3/2$
	L		③	0	0	-2	-1	0		-4	
2	x_2		0	1	$1/2$	1	$1/2$	0		2	2
3	x_1		1	0	$1/2$	0	$-1/2$	0		1	<0
0	x_6		0	0	-1	1	⊡1⊡	1		1	①
	L		0	0	$-3/2$	-2	①$1/2$①	0		-5	

Table cont.

Basis		P	3	2	1	0	0	0	$-M$	0	
μ	x		x_1	x_2	x_3	x_4	x_5	x_6	x_7	RHS	Ratio
2	x_2		0	1	1	1/2	0	$-1/2$		3/2	
3	x_1		1	0	0	1/2	0	1/2		3/2	
0	x_5		0	0	-1	1	1	0		1	
	L		0	0	-1	$-5/2$	0	$-1/2$		$-15/2$	

(iii) Remembering that we rewrote the third constraint, the column of new coefficients is:
$\begin{pmatrix} 1 \\ -1 \\ -1 \end{pmatrix}$ which can be written as $\begin{pmatrix} 1 \\ 0 \\ 0 \end{pmatrix} + \begin{pmatrix} 0 \\ -1 \\ 0 \end{pmatrix} - \begin{pmatrix} 0 \\ 0 \\ 1 \end{pmatrix}$,

the columns of x_4, x_5 and x_6 in the first tableau. In the final tableau this combination of columns becomes:

$\begin{pmatrix} 1/2 \\ 1/2 \\ 1 \end{pmatrix} + \begin{pmatrix} 0 \\ 0 \\ 1 \end{pmatrix} - \begin{pmatrix} -1/2 \\ 1/2 \\ 0 \end{pmatrix} = \begin{pmatrix} 1 \\ 0 \\ 2 \end{pmatrix}$,

which must be the corresponding y-column in the final tableau. The corresponding value for the coefficient in L is $3 - (2 - 0 - 0) = 1$, so that the solution is no longer optimal. Continuing the tableaux with the y-column included gives the following solution.

Basis		P	3	2	1	0	0	0	3	0	
μ	x		x_1	x_2	x_3	x_4	x_5	x_6	y	RHS	Ratio
2	x_2		0	1	1	1/2	0	$-1/2$	1	3/2	3/2
3	x_1		1	0	0	1/2	0	1/2	0	3/2	∞
0	x_5		0	0	-1	1	1	0	②	1	①/2
	L		0	0	-1	$-5/2$	0	$-1/2$	①	$-15/2$	
2	x_2		0	1	3/2	0	$-1/2$	$-1/2$	0	1	
3	x_1		1	0	0	1/2	0	1/2	0	3/2	
3	y		0	0	$-1/2$	1/2	1/2	0	1	1/2	
	L		0	0	$-1/2$	-3	$-1/2$	$-1/2$	0	-8	

Chapter 6

Abbreviations: MP = Maximum Principle; HEs = Hamilton Equations; ECs = End Conditions; ISs = Isoclines; EP = equilibrium point.

Exercise Set 1

(i) $H_k = -(x_k + \tfrac{1}{2}u_k^2) + \lambda_k u_k$. MP: $u_k = \lambda_k$. HEs: $\lambda_{k-1} - \lambda_k = \partial H_k/\partial x_k = -1$, $x_{k+1} - x_k = \partial H_k/\partial \lambda_k = u_k$. Solve uncoupled λ equation: $\lambda_1 = \lambda_0 + 1$, $\lambda_2 = \lambda_0 + 2$. Substitute in x equation and solve: $x_1 = x_0 + \lambda_0$, $x_2 = x_0 + 2\lambda_0 + 1$, $x_3 = x_0 + 3\lambda_0 + 3$. ECs: $x_0 = 4$, $x_3 = 1$ give $\lambda_0 = -2$, i.e. $\lambda = \mathbf{u} = (-2, -1, 0)$, $\mathbf{x} = (4, 2, 1, 1)$.

(iii) MP: $u_k = \frac{1}{2}\lambda_k$. HEs: $\lambda_k = \lambda_{k-1}$, $x_{k+1} = x_k + \frac{1}{2}\lambda_k$. Solve λ equation: $\lambda_k = \lambda_0$ all k. Substitute in x equation: $x_1 = x_0 + \frac{1}{2}\lambda_0$, $x_2 = x_0 + \lambda_0$, $x_3 = x_0 + \frac{3}{2}\lambda_0$, $x_4 = x_0 + 2\lambda_0$. ECs: $x_0 = 0$, $x_4 = 4$ give $\lambda_0 = 2$, i.e. $\mathbf{u} = \frac{1}{2}\lambda = (1, 1, 1, 1)$, $\mathbf{x} = (0, 1, 2, 3, 4)$.

(v) Pioneer problem with $c = 2$, $\alpha = \frac{1}{2}$. MP: $u_k = \frac{1}{16}\lambda_k^{-2}$. HEs: $\lambda_k = \frac{1}{2}\lambda_{k-1}$, $x_{k+1} = 2x_k - 2u_k$. Solve λ equation: $\lambda_k = (\frac{1}{2})^k\lambda_0$. Solve x equation: $x_1 = 2 - A$, $x_2 = 4 - 6A$, $x_3 = 8 - 28A$ with $A = \frac{1}{8}\lambda_0^{-2}$, $x_0 = 1$. But $x_3 = 4$, hence $A = \frac{1}{4}$ and $\lambda = (\frac{7}{8})^{1/2}(1, \frac{1}{2}, \frac{1}{4})$, $\mathbf{u} = \frac{1}{14}(1, 4, 16)$, $\mathbf{x} = \frac{1}{6}(7, 13, 22, 28)$.

Exercise Set 2

(i) $H = -(x + \frac{1}{2}u^2) + \lambda u$. MP: $u = \lambda$. HEs: $-\dot\lambda = \partial H/\partial x = -1$, $\dot x = \partial H/\partial \lambda = u$. Solve λ equation: $\lambda = t + A$. Solve x equation: $x = \frac{1}{2}t^2 + At + B$. ECs give $A = -\frac{3}{2}$, $B = 2$.

(iii) $H = -(1 + u^2)^{1/2} + \lambda u$. MP: $u = \lambda/(1 - \lambda^2)^{1/2}$. HEs: $\dot\lambda = 0$, $\dot x = u$. Solve λ equation: $\lambda = A$. Solve x equation: $x = A(1 - A^2)^{-1/2}t + B$. ECs give $A = 2^{-1/2}$, $B = 0$. Note: this can be interpreted as the problem of finding the shortest curve between $(0, 0)$, $(1, 1)$ in x, t space. This solution of course is the straight line $x = t$ (see Section 1 of Appendix for details and Problem (iv) of Set 1 for discrete analogue).

(v) $H = -\frac{1}{2}(x^2 + u^2) + \lambda(x + u)$. MP: $u = \lambda$. HEs: $\dot\lambda = x - \lambda$, $\dot x = x + u$. Not uncoupled; use the fact that H is constant to give $\lambda = -x \pm 2^{1/2}(x^2 + C)^{1/2}$. Substitute in x equation: $\dot x = \pm 2^{1/2}(x^2 + C)^{1/2}$, a separable equation with solution $\int (x^2 + C)^{-1/2}dx = \pm 2^{1/2}t + A$. Substitute $x = C^{1/2}\sinh(\pm 2^{1/2}t + A)$, integrate, and apply ECs: $x = \sinh 2^{1/2}t/\sinh 2^{1/2}$ with $u = \lambda = \dot x - x$. Note: an alternative method is to eliminate λ to obtain a 2nd order differential equation for x. In fact from the dynamic equation $\lambda = \dot x - x$ we have $\dot\lambda = \ddot x - \dot x = x - \lambda = x - (\dot x - x)$, i.e. $\ddot x - 2x = 0$. Solution: $x = A_1\exp(2^{1/2}t) + B_1\exp(-2^{1/2}t)$ with A_1, B_1 found from ECs.

Exercise Set 3

(i) HEs: $\dot\lambda = \frac{3}{2}x - \frac{3}{2}\lambda$, $\dot x = \frac{3}{2}x + \frac{1}{2}\lambda$. ISs: $\lambda = x$, $\lambda = -3x$ (Fig. S6.1).

(iii) HEs: $\dot\lambda = -3\lambda$, $\dot x = 3x - \exp(-2t)\lambda^{-2}$. Let $\lambda_1 = \lambda\exp t$ to obtain $\dot\lambda_1 = -2\lambda_1$, $\dot x = 3x - \lambda_1^{-2}$. ISs: $\lambda_1 = 0$, $x = \frac{1}{3}\lambda_1^{-2}$ (Fig. S6.2). The λ_1-isocline is itself a trajectory. Note: this is an example of an optimal continuous harvesting model of, say, a forest with x the stock of timber in situ and u the rate at which the timber is cut. If the amount of timber at the end is equal to that at the beginning then the optimal trajectory has the shape shown in Fig. S6.3a. The cutting rate is progressively increased from a value low enough for the forest to initially grow, to a value high enough to return to its original size (Fig. S6.3b).

(v) HEs: $\dot\lambda = 4\lambda(x - \frac{1}{2})$, $\dot x = 2x(1 - x) - \lambda^{-2}$. λIS: $\lambda = 0$ or $x = \frac{1}{2}$. xIS: $\lambda = \pm(2x(1 - x))^{-1/2}$ (Fig. S6.4). Note: this is the continuous harvesting model with saturation, modelling the fact that timber cannot grow without limit. The Phase Plot now possesses

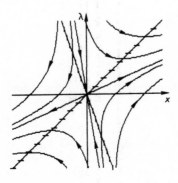

Fig. S6.1

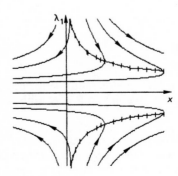

Fig. S6.2

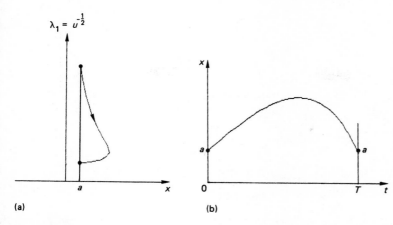

Fig. S6.3

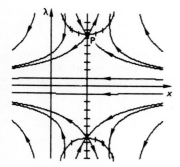

Fig. S6.4

an EP, P, where the cutting rate equals the maximum growth rate of the forest with stock x constant. The optimal path in any given scenario is to remain as close to P for as long as possible.

Problem Set 6

1(i) HEs: $\lambda_{k-1} = 0$ for $k = 1, 2, 3$, i.e. $\lambda = (0, 0, 0)$; $x_{k+1} = \lambda_k$ for $k = 0, 1, 2, 3$, i.e. $x_1 = x_2 = x_3 = 0$, $x_4 = \lambda_3 = 1$. MP: $u_k = \lambda_k - x_k$ for $k = 0, 1, 2, 3$, i.e. $\mathbf{u} = (-1, 0, 0, 1)$.
(iii) HEs: $\lambda_k = \lambda_{k-1}$, $x_{k+1} = x_k + u_k$. MP: $u_k = \lambda_k(1 - \lambda_k^2)^{-1/2}$. Solve λ equation: $\lambda_k = \lambda_0$ (all k). Solve x equation: $x_k = x_0 + ku_k$. ECs: $u_0 = 1$, $\lambda_0 = 2^{-1/2}$; $x_k = k$.

2(i) HEs: $\dot{\lambda} = 1$, $\dot{x} = u$. MP: $u = \ln \lambda$. Solve: $\lambda = t + A$, $x = (t + A)\ln(t + A) - t + B$. ECs give: $1 = A \ln A + B$, $2\ln 2 = (1 + A)\ln(1 + A) - 1 + B$, satisfied by $A = B = 1$.
(iii) $H = -u^2 + \lambda$. HEs: $\dot{\lambda} = 0$, $\dot{x} = 1$. MP: $u = 0$. Solution: $x = t$, $\lambda = A$, $u = 0$ for $0 \leq t \leq 1$. Note: u is not in dynamic equation, i.e. x is not controllable; it is coincidental that ECs on x can be satisfied; they would not be if, e.g. $x(1) = 2$.

3(i) HEs: $\dot{\lambda} = 1 - \lambda$, $\dot{x} = x + \lambda$. ISs: $\lambda = 1$, $x = -\lambda$ (Fig. S6.5). Note: the λ-isocline coincides with two trajectories of the system. EP at $(-1, 1)$.
(iii) From 2(i): $\dot{\lambda} = 1$, $\dot{x} = \ln \lambda$. No λ-isocline; x-isocline: $\lambda = 1$; lower half of phase space not accessible (Fig. S6.6). Note: the trajectories meet the x-axis at finite points, a fact not immediately obvious from the HEs but deducible from the analytic solution in 2(i).

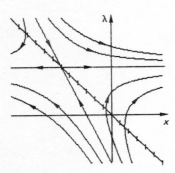

Fig. S6.5

Fig. S6.6

4(a)(i) Consider $x > 0$. MP: $u = \lambda(x^2 - \lambda^2)^{-1/2}$. Substitute in H: $H = -(x^2 - \lambda^2)^{1/2} =$ constant (say $-\hat{C}$). Solve for λ: $\lambda = \pm(x^2 - \hat{C}^2)^{1/2}$, i.e. $u = \pm\hat{C}^{-1}(x^2 - \hat{C}^2)^{1/2}$, $\dot{x} = C^{-1}(x^2 - C^2)^{1/2}$ where $C = \pm\hat{C}$.

(ii) Separable equation: $\int (x^2 - C^2)^{-1/2} dx = C^{-1}t + A = \cosh^{-1}(x/C)$ using suggested substitution. Hence $x = C\cosh(C^{-1}t + A)$. ECs give $a = C\cosh A = C\cosh(C^{-1}T + A)$, i.e. $A = -\tfrac{1}{2}T/C$, $a/C = \cosh\tfrac{1}{2}T/C$.

(iii) See Fig. S6.7. With $a/T = \tfrac{1}{2}$ there are two points of intersection, the lower one optimal (evaluate the objective function to check); with $a/T \approx \tfrac{3}{4}$ these points coincide; with $a/T = \tfrac{1}{2}$ there is no intersection and hence no solution of the given form (see Chapter 7 for further discussion).

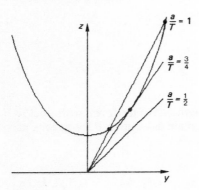

Fig. S6.7

5(a) Form the Lagrangean: $L = \sum_{k=0}^{n}\{-(x_k + \tfrac{1}{2}u_k^2) - \lambda_k(x_{k+1} - x_k - u_k)\} = \sum_{k=1}^{n} x_k$ $(-1 + \lambda_k - \lambda_{k-1}) + \sum_{k=0}^{n}(\lambda_k u_k - \tfrac{1}{2}u_k^2)$. Maximise with respect to u_k: $\partial L/\partial u_k = -u_k + \lambda_k = 0$. Maximise with respect to x_k: the objective function is infinite unless $-1 + \lambda_k - \lambda_{k-1} = 0$. Dual $h(\lambda) = \tfrac{1}{2}\sum_{k=0}^{n}\lambda_k^2$ s.t. $\lambda_k = \lambda_{k-1} + 1$. Solve the dual constraints: $\lambda_k = \lambda_0 + k$ and substitute in $h(\lambda)$ to obtain the one-variable problem: minimise $\tfrac{1}{2}\sum_{k=0}^{n}(\lambda_0 + k)^2$. Solution $\lambda_0 = -\tfrac{1}{2}n$ since $\sum_{k=0}^{n} k = \tfrac{1}{2}n(n+1)$.

6(b)(i) The dual is obtained from the master formula of Chapter 4 or more simply from first principles: $L = \sum_{k=0}^{n} (\rho^k u_k - \lambda_k(x_{k+1} - cx_k + cu_k)) = \sum_{k=0}^{n} u_k(\rho^k - c\lambda_k) + \sum_{k=1}^{n} x_k(c\lambda_k - \lambda_{k-1}) + c\lambda_0 x_0 - \lambda_n x_{n+1}$. Finite optimum with respect to u_k, x_k is achieved when these variables are zero and coefficients non positive; hence the required result.

(ii) With x_0, x_{n+1} positive the objective is to choose multipliers so that λ_0 is minimised and λ_n maximised. For a given λ_0 the latter is achieved by making each of the first set of constraints binding, i.e. $\lambda_k = \lambda_0 c^{-k}$. The second set of inequalities then becomes: $(\rho c)^k \leqslant c\lambda_0$. The minimum value of λ_0 consistent with these constraints with $\rho c < 1$ is $1/c$ with the $k=0$ constraint binding. The other constraints are not binding and hence their multipliers u_k ($k = 1, \ldots, n$) are zero, i.e. there is consumption only in the first year. The discounting is so high that consumption in later years becomes unimportant.

8(a) $H = f + \lambda u$. MP: $-\lambda = \partial f/\partial u$, (1). HEs: $-\dot\lambda = \partial f/\partial x$, $\dot x = u$. Differentiate (1) to obtain: $-\dot\lambda = d/dt(\partial f/\partial u) = \partial f/\partial x$ as required.
(b) $f = -x(1+\dot x^2)^{1/2}$: $\partial f/\partial u = \partial f/\partial \dot x = -x\dot x(1+\dot x^2)^{1/2}$: $\partial f/\partial x = -(1+\dot x^2)^{1/2}$: $d/dt(\partial f/\partial \dot x) = -\dot x^2(1+\dot x^2)^{-1/2} - x\ddot x(1+\dot x^2)^{-3/2}$. Substitute in the Euler–Lagrange equation to obtain: $x\ddot x = 1 + \dot x^2$.

Chapter 7

Further abbreviation: AF = asset function.

Exercise Set 1

(i) HEs: $\dot\lambda = 1 - \lambda$, $\dot x = x + u$. (a) If $0 \leqslant \lambda \leqslant 2$ then by MP $u = \lambda$; trajectories in this strip are those constructed in Fig. S6.5. (b) If $\lambda \geqslant 2$ then $u = 2$ at the boundary. IS: $x = -2$. (c) If $\lambda \leqslant 0$ then $u = 0$ at the boundary. ISs: $x = 0$ (Fig. S7.1).
(iii) HEs: $\dot\lambda = 2x$, $\dot x = u$. (a) If $\lambda > 0$ then $u = 1$; λ-isocline: $x = 0$. (b) If $\lambda < 0$ then $u = -1$; λ-isocline: $x = 0$ (Fig. S7.2). Note: EP at $(0, 0)$ corresponding to knife-edge case with $H(u)$ horizontal. If $x(0) = x(T) = 1$ (say) then the trajectory for low T is shown in Fig. S7.3a with the $x(t)$ graph formed from two line segments with gradients $1, -1$ (Fig. S7.3b). For larger values of T the optimal trajectory passes through the EP (Fig. S7.4a), the $x(t)$ graph now containing a third section with $x(t) = 0$ (Fig. S7.4b).
(v) HEs: $\dot\lambda = 4\lambda(x - \tfrac{1}{2})$, $\dot x = 2x(1-x) - u$. (a) If $\lambda^2 \leqslant 4$ then $u = \tfrac{1}{4}$. ISs: $x = \tfrac{1}{2} \pm 2^{-3/2} = x^*_\pm$, $\lambda = 0$, $x = \tfrac{1}{2}$. (b) If $\lambda^2 \geqslant 4$ then $u = \lambda^{-2}$. ISs: $x = \tfrac{1}{2}$, $\lambda = \pm(2x(1-x))^{-1/2}$ (Fig. S7.5). Note: lower bound $u = 0$ is not achieved. Comparison with Fig. S6.4 shows the effect of a strong upper bound on cutting resources. The forest can no longer be maintained at a level where growth is maximised. There are two sustainable levels, at $x^*_\pm$, with growth less than maximum.

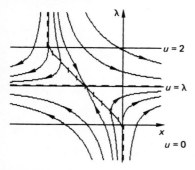

Fig. S7.1

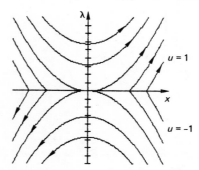

Fig. S7.2

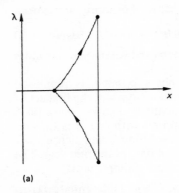

(a) (b)

Fig. S7.3

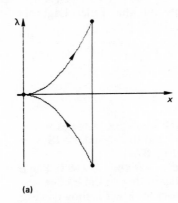

(a)

Fig. S7.4

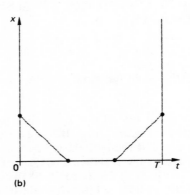

(b)

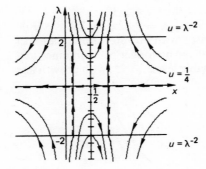

Fig. S7.5

Exercise Set 2

1(i) Solution: $x = \tfrac{1}{2}t^2 + At + B$, $\lambda = u = t + A$ (see Exercise Set 2, Chapter 6). ECs: $x(0) = 2$, $\lambda(1) = 0$, i.e. $B = 2$, $A = -1$.

(iii) Solution: $x = C \sinh(2^{1/2}t)$, $\lambda = C(2^{1/2} \cosh(2^{1/2}t) - \sinh(2^{1/2}t))$ when $x(0) = 0$ (see

Exercise Set 2, Chapter 6). EC: $\lambda(1) = 0$ gives $C = 0$ since $\cosh y > \sinh y$ (all y). Hence $u = \lambda = x = 0$ for $0 \leqslant t \leqslant 1$.

2(a) MP: $\exp(-rt)(D + uD') = \lambda$. Upper EC gives $H_T = u_T(\exp(-rT)D(u_T) - \lambda_T) = 0$, i.e. $u_T = 0$ or $\lambda_T = \exp(-rT)D(u_T)$. In the former case MP gives $\exp(-rT)D(0) = \lambda_T$ at $t = T$. In the latter case MP gives $u_T D'(u_T) = 0$ at $t = T$, i.e. $u_T = 0$ (unless exceptionally $D'(u_T) = 0$). Note: at exhaustion u with x is reduced to zero; λ equals maximum (discounted) price that consumers will pay.
(b)(i) Solution: $x = (\ln A)t + \frac{1}{2}t^2 + B$, $u = -(\ln A + t)$. ECs: $x(0) = 1$, $u(T) = x(T) = 0$, i.e. $T = 2^{1/2}$, $A = \exp(-2^{1/2})$, $B = 1$.

3(i) From equation (7.14) $x = A(1 - A^2)^{-1/2}T + 1$, $\lambda_T = A$, $H_T = -(1 - A^2)^{1/2}$, ECs with $\phi = \mu(T^2 - x_T)$ give $\lambda_T = \partial\phi/\partial x_T = -\mu$, (1); $H_T = -\partial\phi/\partial T = -2\mu T$, (2); $x_T = T^2$, (3). Solve: (1) gives $\mu = -A$; substitution in (2) gives $-(1 - A^2)^{1/2} = 2AT$; substitution in (3) gives $T = 2^{-1/2}$ (with $T \geqslant 0$). Hence $x = -2^{1/2}t + 1$, $\lambda = -3^{-1/2}$, $u = -2^{-1/2}$. Note: the solution tells us that the shortest distance from $(0, 1)$ to the parabola $x = t^2$ is along the line to point $(2^{-1/2}, 2^{-1})$ on the parabola. There is a singular non-optimal solution with $T = 0$, $A = 1$.
(iii) ECs with $\phi = \mu((T - 2)^2 + x_T^2 - 1)$ give $\lambda_T = A = \partial\phi/\partial x_T = 2\mu x_T$, (1); $H_T = -(1 - A^2)^{1/2} = -\partial\phi/\partial T = -2\mu(T - 2)$, (2); $(T - 2)^2 + x_T^2 = 1$, (3). Solve: divide (1) by (2) to give $A = 5^{-1/2}$; substitute in (3) to give $T = 2 \pm 2(5^{-1/2})$. Hence $x = -\frac{1}{2}t + 1$, $\lambda = -5^{-1/2}$, $u = -\frac{1}{2}$. The positive root for T is optimal (substitute in objective function to check). Note: the solution tells us that the shortest path from $(0, 1)$ to the circle centred at $(2, 0)$ radius 1 is along the line from $(0, 1)$ to $(2, 0)$.

Exercise Set 3

1(i) $H = \lambda u - v(1 + u^2)^{1/2}$. MP: $u = \lambda(v^2 - \lambda^2)^{-1/2}$. HEs: $\dot{\lambda} = 0$, $\dot{x} = u$. Solution: $\lambda = A$, $x = A(v^2 - A^2)^{-1/2}t + B$. AF: x_T. ECs: $x(0) = 0$, $\lambda_T = 1$, i.e. $A = 1$, $B = 0$. The multiplier v is found from the integral constraint: $\int_0^1 (1 + u^2)^{1/2} dt = v(v^2 - 1)^{-1/2} = 2^{1/2}$, i.e. $v = 2^{1/2}$. Note: this is the problem of finding how far up the line $t = 1$ in the x, t plane the end of a piece of string of length $2^{1/2}$ can reach when the other end is fixed at the origin.
(iii) Solution of HEs: $x = A(v^2 - A^2)^{-1/2}t + B$, $\lambda = A$. AF: $\mu(x_T - T) + T$. ECs: $x(0) = 1$, $\lambda_T = \mu$, $x_T = T$, $H_T = -(v^2 - \lambda^2)^{1/2} = -1 + \mu$. The first two conditions give $B = 1$, $\mu = A$. The last two conditions with integral constraint give: $A(v^2 - A^2)^{-1/2}T = T - 1$, (1); $(v^2 - A^2)^{1/2} = 1 - A$, (2); $v(v^2 - A^2)^{-1/2} = 1$, (3). Solve: divide (1) by (3) to give $A = (T - 1)v$; substitute in (3) to give $T = 1$, hence $A = \mu = 0$, $v = 1$. Note: this is the problem of finding how far to the right one end of a piece of string of length 1 can be moved on the line $x = t$ when the other end is fixed at $(0, 1)$.

2(i) $H = -u^2 + \lambda_1 x_2 + \lambda_2 u$. MP: $u = \frac{1}{2}\lambda_2$. HEs: $\dot{\lambda}_1 = 0$, $\dot{\lambda}_2 = -\lambda_1$, $\dot{x}_2 = \frac{1}{2}\lambda_2$, $\dot{x}_1 = x_2$. Solve in sequence to give: $\lambda_1 = A$, $\lambda_2 = -At + B$, $x_2 = \frac{1}{2}Bt - \frac{1}{4}At^2 + C$, $x_1 = \frac{1}{4}Bt^2 - \frac{1}{12}At^3 + Ct + D$. ECs: $A = 12$, $B = 8$, $C = D = 0$.

Problem Set 7

1(i) HEs: $\dot{\lambda} = 2x + u - \lambda$, $\dot{x} = x - u$. MP: (a) if $-2 \leqslant x + \lambda \leqslant 0$ then $u = -\frac{1}{2}(x + \lambda)$; (b) if $x + \lambda \geqslant 0$ then $u = 0$; (c) if $x + \lambda \leqslant -2$ then $u = 1$ (Fig. S7.6).
(iii) HEs: $\dot{\lambda} = 0$, $\dot{x} = u$. MP: H linear in u; (a) if $\lambda > -2$ then $u = 1$; (b) if $\lambda < -2$ then $u = 0$ (Fig. S7.7). Note: all points below the boundary are EPs. At any point on the boundary u can take any value in the range $0 \leqslant u \leqslant 1$. There is, therefore, an infinite number of optimal solutions for any choice of end-point values corresponding to these 'boundary trajectories'. This is also clear from the objective function. If we substitute for u we find its value as $2(x(T) - x(0))$ depending only on the end-point values.

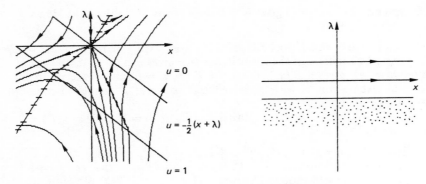

Fig. S7.6

Fig. S7.7

2(a) MP: $u = 1 + \frac{1}{2}\lambda x^{-2}$. HEs: $\dot{\lambda} = 2x(u-1)^2$, $\dot{x} = u$. Standard procedure gives $x^2 - (t - \frac{3}{8})^2 = -\frac{9}{64}$ as a solution satisfying ECs. This solution is not acceptable since it defines a hyperbola with initial and end-points on different branches. The optimal solution is given by $x = 0$ for $0 \leqslant t \leqslant \frac{1}{2}$, $x = t - \frac{1}{2}$ for $\frac{1}{2} \leqslant t \leqslant 1$ and corresponds to trajectory AB in Fig. S7.8. Point A is generated by the multiple solution of the MP when $x = 0$. That this is the solution is obvious from the form of the objective function which achieves its minimum value of zero if at any point of the trajectory either $x = 0$ or $u = 1$.

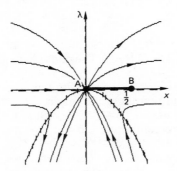

Fig. S7.8

(b)(i) MP: (a) $u = 1$ if $\lambda > 0$; (b) $u = -1$ if $\lambda < 0$; (c) $u = \pm 1$ if $\lambda = 0$. The argument used for problem (7.36) in the main text now holds—to satisfy the ECs $\lambda = 0$ to generate a solution formed by line segments with gradients ± 1. Note: this is the problem of finding the longest curve between points $(0, 0)$, $(1, \frac{1}{2})$ where the gradient (defined almost everywhere) is nowhere greater than 1 in magnitude.
(iii) MP: (a) $u = 1$ if $\lambda < 0$; (b) $u = -1$ for $\lambda > 0$; (c) $u = \pm 1$ for $\lambda = 0$. Argument now as in (i).
(c) (i) MP: $u = \infty$ if $\lambda \geqslant 0$, $u = 0$ if $\lambda \leqslant 0$. The optimal solution has $x = 0$ for $0 \leqslant t < 1$ and $x(1) = 1$ with a discontinuity at $t = 1$.
(d) (i) HEs: $\dot{\lambda} = x(x^2 - \lambda^2)^{-1/2}$, $\dot{x} = \lambda(x^2 - \lambda^2)^{-1/2}$, i.e. only the region with $x^2 \geqslant \lambda^2$ can be accessed (Fig. S7.9a). The discontinuous solution corresponds to boundary ABC. Section AB gives the jump in x from a to 0 ($u = -\infty$); origin B (see comment in Sections 7.5(b) and 7.7 on non-negative constraint on x) gives $x = 0$ in $0 < t < T$; BC gives the jump in x from 0 to a at $x = 1$ ($u = +\infty$) (Fig. S7.9b).

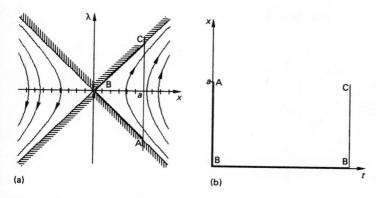

Fig. S7.9

3(a) (i) $H = \lambda_1 x_2 + \lambda_2 u$. HEs: $-\dot{\lambda}_1 = \partial H/\partial x_1 = 0$, $-\dot{\lambda}_2 = \partial H/\partial x_2 = \lambda_1$, (1); $\dot{x}_1 = \partial H/\partial \lambda_1 = x_2$, $\dot{x}_2 = \partial H/\partial \lambda_2 = u$, (2). Solve (1) by integration: $\lambda_1 = A$, $\lambda_2 = -At + B$. By MP: $u = 1$ if $\lambda_2 > 0$, $u = -1$ if $\lambda_2 < 0$ (H linear in u). Equations (2) give: $x_2 = ut + \hat{C}$, $x_1 = \frac{1}{2}u(t+C)^2 + D$ with $C = \hat{C}/u$, (3). To find if there is switching in u between ± 1 note first from the linearity of λ_2 in t that there can be at most one switch. To obtain a visualisation of the process plot curves (3) in x_1, x_2 (state) space, eliminate t in (3) to obtain two families of parabolae: $x_1 = \frac{1}{2}ux_2^2 + D$ with $u = \pm 1$ (Fig. S7.10). The objective is to get from point P: $(-1, 1)$ to $(0, 0)$ in shortest time, moving along the parabolae, with at most one jump between parabolae. Desired solution is PQO. Point Q: $(-\frac{3}{4}, \sqrt{\frac{3}{2}})$ is at the intersection of parabolae: $x_1 = \frac{1}{2}x_2^2 - \frac{3}{2}$ through P and $x_1 = -\frac{1}{2}x_2^2$ through O. The time of switching and arrival at O is found from (3). For the first parabola $C = 1$ since $x_2 = 1$ when $t = 0$, hence $t = \sqrt{\frac{3}{2}} - 1$ at Q. On the second parabola $C = 1 - 2\sqrt{\frac{3}{2}}$. The optimal time at O is $t = 2\sqrt{\frac{3}{2}} - 1$. Note: this models the problem of stopping a vehicle in minimum time when the driving force is limited in magnitude. Position is x_1, velocity x_2 and the dynamic equation is just Newton's equation of motion. The vehicle is initially moving slowly; the speed is increased with the accelerator full on; the vehicle is then brought to a halt with the brakes slammed on.

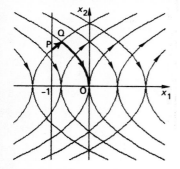

Fig. S7.10

(b)(i) $H = \lambda_1 u + \lambda_2(1+u^2)^{1/2}$. HEs: $-\dot{\lambda}_1 = \partial H/\partial x_1 = 0$, $-\dot{\lambda}_2 = \partial H/\partial x_2 = 0$, $\dot{x}_1 = u$, $\dot{x}_2 = (1+u^2)^{1/2}$. MP: $u = -\lambda(1-\lambda^2)^{-1/2}$ with $\lambda = -\lambda_1/\lambda_2$. Solution obtained by integration: $\lambda_1 = A$, $\lambda_2 = B$, $x_1 = -\lambda(1-\lambda^2)^{-1/2}t + C$, $x_2 = (1-\lambda^2)^{-1/2}t + D$ with $\lambda = -A/B$. ECs: $x_1(0) = x_2(0) = 0$, $\lambda_1(1) = 1$, $x_2(1) = 2^{1/2}$ give $C = D = 0$, $A = 1$, $B = \pm 2^{1/2}$ with the negative root giving the maximum (see Exercise set 3 question 1(i)).

5 (a) See Fig. S7.11. If $x < \frac{1}{4}$ then there is no way to save the fish.

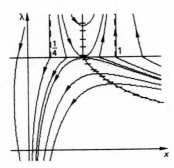

Fig. S7.11

8(a) The Hamiltonian systems of equations for the two problems are:

$$H = f + \lambda g - vh$$
$$\partial H/\partial u = \partial f/\partial u + \lambda \partial g/\partial u - v\, \partial h/\partial u = 0$$
$$-\dot\lambda = -\partial f/\partial x - \lambda \partial g/\partial x + v\partial h/\partial x$$
$$\dot x = g$$

$$H' = -h + \lambda' g + v'f$$
$$\partial H'/\partial u = -\partial h/\partial u + \lambda' \partial g/\partial u + v'\partial f/\partial u = 0$$
$$-\dot\lambda' = \partial h/\partial x - \lambda' \partial g/\partial x - v'\partial f/\partial x$$
$$\dot x = g$$

The equation sets are identical if λ' replaced by $\lambda v'$ and v' by v^{-1}. Hence the optimal solutions come from a common family of solutions. If B is the optimal value for first problem and A for the second then the solutions are identical.

Index

Allocation problem, 219
applications and worked problems
   assignment problem, 114
   blending problem, 114
   brachistochrone problem, 248
   capital investment problem, 219
   chemical tank design, 30, 54, 103
   coal production, 10
   coal transportation, 72, 75
   Cobb–Douglas cost function, 109
   commodity dealing, 49
   computer manufacturer, 114
   cross country race, 187
   diet problem, 114
   electricity generation and allocation, 1, 10, 30, 48, 55, 66, 79
   energy conservation, 199
   exhaustible resources, 160
   geodesic problem, 175, 247
   hydroelectricity problem, 196
   linear regression, 17, 135
   minimum time problem, 201
   pioneer problem, 142, 150–53, 175, 221, 224
   procurement/inventory, 58
   Ramsey's growth model, 171
   renewable resources, 182
   road design, 19
   soap film problem, 174, 205, 248
   university admissions, 47
artificial variable, 128
asset function, 186
assignment problem, 114
autonomous control problem, 159

Balanced growth path, 172
bang-bang, 181, 183
bang-sit-bang, 185
basic variable, 117
basis, 117
binding constraint, 60
blending problem, 114
boundary saddle point, 84
bounds
   upper, lower, 55–9
   on control variables, 177–82
   on state variables, 193–6

brachistochrone problem, 248

Calculus of variations, 202–7
capital investment problem, 219
catastrophe theory, 260
chain rule, 249
chemical tank design, 30, 54, 103
coal production problem, 10
coal transportation problem, 72, 75
Cobb–Douglas function, 109, 171
column vector, 232
commodity dealing, 49
complementary slackness, 69–72
computer manufacturer problem, 114
concavity, 94–9
conic section, 240
constrained Lagrangean, 88
constrained stationary point, 31–3, 38
constraint qualification, 101
constraint sensitivity, 40
contour map, 20–26
control variable, 143
convex set, 112
convexity, 94–9
corner conditions, 203, 254
cross-country race, 187

Decreasing returns to scale, 172
degree of difficulty, 106
derivative
   directional, 13
   partial, 13
diet problem, 114
difference equations, 146
   uncoupled 148
differential equations, 155, 242–4
   linear, 243
   second order, 158
   separable, 243
   uncoupled, 157
directional derivative, 13, 28
discontinuities, 204–7
discount factor, 150, 160
dual problem, 80, 81, 100
   economic interpretation, 79–80
duality, 79–81
dynamic equation, 144

Index

dynamic programming, 214, 220

Electricity generation and allocation problem, 1, 10, 30, 48, 55, 66, 79
ellipse, 240
energy conservation, 199
equilibrium point, 164
Euler–Lagrange equation, 176
exhaustible resources, 160
exponential function, 234

Feasible region, 60
   boundary, 59
   interior, 60
feasible vertex, 119

Geometric programming, 103–8
   degree of difficulty, 106
   normality conditions, 105
   orthogonality conditions, 105
geodesic problem, 175, 247
geometric vector, 237
gradient vector, 67, 239

Hamilton equation, 144
Hamilton–Jacobi equation, 223
Hamiltonian, 145
Hotelling's demand function, 162
hyperbola, 241
hyperbolic function, 244–6

Inflexion, 6, 32
integral constraint, 192
integrating factor, 243
iso-elastic demand function, 161
isocline, 165, 166

Jacobi conditions, 253
jump conditions, 195, 255

Kuhn–Tucker conditions, 74–6, 118

Lagrange equations, 34, 41, 59
Lagrange multiplier, 33, 34
   economic interpretation, 48, 186
Lagrange surface, 38
Lagrangean, 34, 80
linear dependence, 231
linear equation, 229
linear programming, 70–74, 90–93, 102, 111, 114–41
   alternative solution, 130
   cycling, 129
   degeneracy, 129
   dual, 90–94
   infinite solution, 129
   unconstrained variable, 129

limit at infinity, 4, 15, 234–7
local maximum/minimum, 3, 4, 6, 15
local multiplier, 122
logarithmic function, 235

Marginal cost, 49
master Lagrangean, 63
mathematical program, 225
matrix, 232
maximum principle, 146
maximum value
   local, 6, 15
   overall, 3, 15
minimum value
   local, 6, 15
   overall, 4, 15
multiplier function, 155

Normality condition, 105, 253

Objective function, 6
objective functional, 155
optimal control
   continuous, 153–9
   discrete, 144–9
   vector control, 196–9
orthogonality condition, 105

Parabola, 241
partial derivative, 13
peak, 3, 6, 15
penalty function, 128
phase space, 162
pivot equation, 122
pivot variable, 122
pivoting, 122
polynomial function, 234
posynomial, 104
present value, 160
primal problem, 80, 81, 100
primal/dual structure, 81, 100
principle of optimality, 214, 220
procedures
   1.1 one variable unconstrained maximisation, 4
   1.2 two variable unconstrained maximisation, 15
   2.1 maximisation with a single equality constraint, 35
   2.2 maximisation with equality constraints, 42
   3.1 maximisation with bounded variables, 56
   3.2 maximisation with inequality constraints, 70
   4.1 primal/dual analysis, 86
   4.2 geometric program, 107

procedures (*Contd*)
   5.1 simplex tabular algorithm, 125
   6.1 discrete optimal control, 149
   6.2 continuous optimal control, 159
   6.3 phase space method, 170
   7.1 terminal conditions, 190
   7.2 generalised optimal control, 198

Quadratic programming, 112, 136–9

Ramsey's growth model, 171
reduced Lagrangean, 74
renewable resource problem, 182
road design problem, 19
row vector, 232

Saddle point, 15, 84, 101
saddle-point problem, 79, 81
saddle-point properties, 100–102
scalar product, 232
Schwarz inequality, 45
second order conditions, 111
sensitivity analysis, 133, 251
separable objective function, 144
shadow price, 49
sign of Lagrange multiplier, 66–8
simplex algorithm, 115–32
singularity, 37
slack variable, 116
soap-film problem, 174, 205, 248
standard form, 121
state variable, 143

stationary points, 3–9, 12–16, 31–3, 38
stochastic programming, 224

Tangent plane, 13
terminal conditions, 186–90, 254
trajectory, 163
trim-loss problem, 140
trough, 4, 6, 15

University admission problem, 47
utility, 143

Variables
   basic, 117
   control, 143
   pivot, 122
   slack, 116
   state, 143
   unconstrained, 129
variational method, 251
variational principle, 249
variations: strong and weak, 253
vectors
   column, 232
   geometric, 237
   gradient, 67, 239
   row, 232
vertex
   feasible, 119
   neighbouring, 71

Weierstrass–Erdmann conditions, 255